Coarse Wavelength Division Multiplexing

Technologies and Applications

OPTICAL SCIENCE AND ENGINEERING

Founding Editor
Brian J. Thompson
University of Rochester
Rochester, New York

1. Electron and Ion Microscopy and Microanalysis: Principles and Applications, *Lawrence E. Murr*
2. Acousto-Optic Signal Processing: Theory and Implementation, *edited by Norman J. Berg and John N. Lee*
3. Electro-Optic and Acousto-Optic Scanning and Deflection, *Milton Gottlieb, Clive L. M. Ireland, and John Martin Ley*
4. Single-Mode Fiber Optics: Principles and Applications, *Luc B. Jeunhomme*
5. Pulse Code Formats for Fiber Optical Data Communication: Basic Principles and Applications, *David J. Morris*
6. Optical Materials: An Introduction to Selection and Application, *Solomon Musikant*
7. Infrared Methods for Gaseous Measurements: Theory and Practice, *edited by Joda Wormhoudt*
8. Laser Beam Scanning: Opto-Mechanical Devices, Systems, and Data Storage Optics, *edited by Gerald F. Marshall*
9. Opto-Mechanical Systems Design, *Paul R. Yoder, Jr.*
10. Optical Fiber Splices and Connectors: Theory and Methods, *Calvin M. Miller with Stephen C. Mettler and Ian A. White*
11. Laser Spectroscopy and Its Applications, *edited by Leon J. Radziemski, Richard W. Solarz, and Jeffrey A. Paisner*
12. Infrared Optoelectronics: Devices and Applications, *William Nunley and J. Scott Bechtel*
13. Integrated Optical Circuits and Components: Design and Applications, *edited by Lynn D. Hutcheson*
14. Handbook of Molecular Lasers, *edited by Peter K. Cheo*
15. Handbook of Optical Fibers and Cables, *Hiroshi Murata*
16. Acousto-Optics, *Adrian Korpel*
17. Procedures in Applied Optics, *John Strong*
18. Handbook of Solid-State Lasers, *edited by Peter K. Cheo*
19. Optical Computing: Digital and Symbolic, *edited by Raymond Arrathoon*
20. Laser Applications in Physical Chemistry, *edited by D. K. Evans*

21. Laser-Induced Plasmas and Applications, *edited by Leon J. Radziemski and David A. Cremers*
22. Infrared Technology Fundamentals, *Irving J. Spiro and Monroe Schlessinger*
23. Single-Mode Fiber Optics: Principles and Applications, Second Edition, Revised and Expanded, *Luc B. Jeunhomme*
24. Image Analysis Applications, *edited by Rangachar Kasturi and Mohan M. Trivedi*
25. Photoconductivity: Art, Science, and Technology, *N. V. Joshi*
26. Principles of Optical Circuit Engineering, *Mark A. Mentzer*
27. Lens Design, *Milton Laikin*
28. Optical Components, Systems, and Measurement Techniques, *Rajpal S. Sirohi and M. P. Kothiyal*
29. Electron and Ion Microscopy and Microanalysis: Principles and Applications, Second Edition, Revised and Expanded, *Lawrence E. Murr*
30. Handbook of Infrared Optical Materials, *edited by Paul Klocek*
31. Optical Scanning, *edited by Gerald F. Marshall*
32. Polymers for Lightwave and Integrated Optics: Technology and Applications, *edited by Lawrence A. Hornak*
33. Electro-Optical Displays, *edited by Mohammad A. Karim*
34. Mathematical Morphology in Image Processing, *edited by Edward R. Dougherty*
35. Opto-Mechanical Systems Design: Second Edition, Revised and Expanded, *Paul R. Yoder, Jr.*
36. Polarized Light: Fundamentals and Applications, *Edward Collett*
37. Rare Earth Doped Fiber Lasers and Amplifiers, *edited by Michel J. F. Digonnet*
38. Speckle Metrology, *edited by Rajpal S. Sirohi*
39. Organic Photoreceptors for Imaging Systems, *Paul M. Borsenberger and David S. Weiss*
40. Photonic Switching and Interconnects, *edited by Abdellatif Marrakchi*
41. Design and Fabrication of Acousto-Optic Devices, *edited by Akis P. Goutzoulis and Dennis R. Pape*
42. Digital Image Processing Methods, *edited by Edward R. Dougherty*
43. Visual Science and Engineering: Models and Applications, *edited by D. H. Kelly*
44. Handbook of Lens Design, *Daniel Malacara and Zacarias Malacara*
45. Photonic Devices and Systems, *edited by Robert G. Hunsberger*
46. Infrared Technology Fundamentals: Second Edition, Revised and Expanded, *edited by Monroe Schlessinger*
47. Spatial Light Modulator Technology: Materials, Devices, and Applications, *edited by Uzi Efron*
48. Lens Design: Second Edition, Revised and Expanded, *Milton Laikin*

49. Thin Films for Optical Systems, *edited by Francoise R. Flory*
50. Tunable Laser Applications, *edited by F. J. Duarte*
51. Acousto-Optic Signal Processing: Theory and Implementation, Second Edition, *edited by Norman J. Berg and John M. Pellegrino*
52. Handbook of Nonlinear Optics, *Richard L. Sutherland*
53. Handbook of Optical Fibers and Cables: Second Edition, *Hiroshi Murata*
54. Optical Storage and Retrieval: Memory, Neural Networks, and Fractals, *edited by Francis T. S. Yu and Suganda Jutamulia*
55. Devices for Optoelectronics, *Wallace B. Leigh*
56. Practical Design and Production of Optical Thin Films, *Ronald R. Willey*
57. Acousto-Optics: Second Edition, *Adrian Korpel*
58. Diffraction Gratings and Applications, *Erwin G. Loewen and Evgeny Popov*
59. Organic Photoreceptors for Xerography, *Paul M. Borsenberger and David S. Weiss*
60. Characterization Techniques and Tabulations for Organic Nonlinear Optical Materials, *edited by Mark G. Kuzyk and Carl W. Dirk*
61. Interferogram Analysis for Optical Testing, *Daniel Malacara, Manuel Servin, and Zacarias Malacara*
62. Computational Modeling of Vision: The Role of Combination, *William R. Uttal, Ramakrishna Kakarala, Spiram Dayanand, Thomas Shepherd, Jagadeesh Kalki, Charles F. Lunskis, Jr., and Ning Liu*
63. Microoptics Technology: Fabrication and Applications of Lens Arrays and Devices, *Nicholas Borrelli*
64. Visual Information Representation, Communication, and Image Processing, *edited by Chang Wen Chen and Ya-Qin Zhang*
65. Optical Methods of Measurement, *Rajpal S. Sirohi and F. S. Chau*
66. Integrated Optical Circuits and Components: Design and Applications, *edited by Edmond J. Murphy*
67. Adaptive Optics Engineering Handbook, *edited by Robert K. Tyson*
68. Entropy and Information Optics, *Francis T. S. Yu*
69. Computational Methods for Electromagnetic and Optical Systems, *John M. Jarem and Partha P. Banerjee*
70. Laser Beam Shaping, *Fred M. Dickey and Scott C. Holswade*
71. Rare-Earth-Doped Fiber Lasers and Amplifiers: Second Edition, Revised and Expanded, *edited by Michel J. F. Digonnet*
72. Lens Design: Third Edition, Revised and Expanded, *Milton Laikin*
73. Handbook of Optical Engineering, *edited by Daniel Malacara and Brian J. Thompson*

74. Handbook of Imaging Materials: Second Edition, Revised and Expanded, *edited by Arthur S. Diamond and David S. Weiss*
75. Handbook of Image Quality: Characterization and Prediction, *Brian W. Keelan*
76. Fiber Optic Sensors, *edited by Francis T. S. Yu and Shizhuo Yin*
77. Optical Switching/Networking and Computing for Multimedia Systems, *edited by Mohsen Guizani and Abdella Battou*
78. Image Recognition and Classification: Algorithms, Systems, and Applications, *edited by Bahram Javidi*
79. Practical Design and Production of Optical Thin Films: Second Edition, Revised and Expanded, *Ronald R. Willey*
80. Ultrafast Lasers: Technology and Applications, *edited by Martin E. Fermann, Almantas Galvanauskas, and Gregg Sucha*
81. Light Propagation in Periodic Media: Differential Theory and Design, *Michel Nevière and Evgeny Popov*
82. Handbook of Nonlinear Optics, Second Edition, Revised and Expanded, *Richard L. Sutherland*
83. Polarized Light: Second Edition, Revised and Expanded, *Dennis Goldstein*
84. Optical Remote Sensing: Science and Technology, *Walter Egan*
85. Handbook of Optical Design: Second Edition, *Daniel Malacara and Zacarias Malacara*
86. Nonlinear Optics: Theory, Numerical Modeling, and Applications, *Partha P. Banerjee*
87. Semiconductor and Metal Nanocrystals: Synthesis and Electronic and Optical Properties, edited by *Victor I. Klimov*
88. High-Performance Backbone Network Technology, *edited by Naoaki Yamanaka*
89. Semiconductor Laser Fundamentals, *Toshiaki Suhara*
90. Handbook of Optical and Laser Scanning, *edited by Gerald F. Marshall*
91. Organic Light-Emitting Diodes: Principles, Characteristics, and Processes, *Jan Kalinowski*
92. Micro-Optomechatronics, *Hiroshi Hosaka, Yoshitada Katagiri, Terunao Hirota, and Kiyoshi Itao*
93. Microoptics Technology: Second Edition, *Nicholas F. Borrelli*
94. Organic Electroluminescence, *edited by Zakya Kafafi*
95. Engineering Thin Films and Nanostructures with Ion Beams, *Emile Knystautas*
96. Interferogram Analysis for Optical Testing, Second Edition, *Daniel Malacara, Manuel Sercin, and Zacarias Malacara*
97. Laser Remote Sensing, *edited by Takashi Fujii and Tetsuo Fukuchi*
98. Passive Micro-Optical Alignment Methods, *edited by Robert A. Boudreau and Sharon M. Boudreau*
99. Organic Photovoltaics: Mechanism, Materials, and Devices, *edited by Sam-Shajing Sun and Niyazi Serdar Saracftci*

100. Handbook of Optical Interconnects, *edited by Shigeru Kawai*
101. GMPLS Technologies: Broadband Backbone Networks and Systems, *Naoaki Yamanaka, Kohei Shiomoto, and Eiji Oki*
102. Laser Beam Shaping Applications, *edited by Fred M. Dickey, Scott C. Holswade and David L. Shealy*
103. Electromagnetic Theory and Applications for Photonic Crystals, *Kiyotoshi Yasumoto*
104. Physics of Optoelectronics, *Michael A. Parker*
105. Opto-Mechanical Systems Design: Third Edition, *Paul R. Yoder, Jr.*
106. Color Desktop Printer Technology, *edited by Mitchell Rosen and Noboru Ohta*
107. Laser Safety Management, *Ken Barat*
108. Optics in Magnetic Multilayers and Nanostructures, *Štefan Višňovský*
109. Optical Inspection of Microsystems, *edited by Wolfgang Osten*
110. Applied Microphotonics, *edited by Wes R. Jamroz, Roman Kruzelecky, and Emile I. Haddad*
111. Organic Light-Emitting Materials and Devices, *edited by Zhigang Li and Hong Meng*
112. Silicon Nanoelectronics, *edited by Shunri Oda and David Ferry*
113. Image Sensors and Signal Processor for Digital Still Cameras, *Junichi Nakamura*
114. Encyclopedic Handbook of Integrated Circuits, *edited by Kenichi Iga and Yasuo Kokubun*
115. Quantum Communications and Cryptography, *edited by Alexander V. Sergienko*
116. Optical Code Division Multiple Access: Fundamentals and Applications, *edited by Paul R. Prucnal*
117. Polymer Fiber Optics: Materials, Physics, and Applications, *Mark G. Kuzyk*
118. Smart Biosensor Technology, *edited by George K. Knopf and Amarjeet S. Bassi*
119. Solid-State Lasers and Applications, *edited by Alphan Sennaroglu*
120. Optical Waveguides: From Theory to Applied Technologies, *edited by Maria L. Calvo and Vasudevan Lakshiminarayanan*
121. Gas Lasers, *edited by Masamori Endo and Robert F. Walker*
122. Lens Design, Fourth Edition, *Milton Laikin*
123. Photonics: Principles and Practices, *Abdul Al-Azzawi*
124. Microwave Photonics, *edited by Chi H. Lee*
125. Physical Properties and Data of Optical Materials, *Moriaki Wakaki, Keiei Kudo, and Takehisa Shibuya*
126. Microlithography: Science and Technology, Second Edition, *edited by Kazuaki Suzuki and Bruce W. Smith*
127. Coarse Wavelength Division Multiplexing: Technologies and Applications, *edited by Hans-Jörg Thiele and Marcus Nebeling*

Coarse Wavelength Division Multiplexing

Technologies and Applications

Edited by

Hans-Jörg Thiele
Marcus Nebeling

CRC Press
Taylor & Francis Group
Boca Raton London New York

CRC Press is an imprint of the
Taylor & Francis Group, an **informa** business

CRC Press
Taylor & Francis Group
6000 Broken Sound Parkway NW, Suite 300
Boca Raton, FL 33487-2742

© 2007 by Taylor & Francis Group, LLC
CRC Press is an imprint of Taylor & Francis Group, an Informa business

No claim to original U.S. Government works
Printed in the United States of America on acid-free paper
10 9 8 7 6 5 4 3 2 1

International Standard Book Number-10: 0-8493-3533-7 (Hardcover)
International Standard Book Number-13: 978-0-8493-3533-4 (Hardcover)

This book contains information obtained from authentic and highly regarded sources. Reprinted material is quoted with permission, and sources are indicated. A wide variety of references are listed. Reasonable efforts have been made to publish reliable data and information, but the author and the publisher cannot assume responsibility for the validity of all materials or for the consequences of their use.

No part of this book may be reprinted, reproduced, transmitted, or utilized in any form by any electronic, mechanical, or other means, now known or hereafter invented, including photocopying, microfilming, and recording, or in any information storage or retrieval system, without written permission from the publishers.

For permission to photocopy or use material electronically from this work, please access www.copyright.com (http://www.copyright.com/) or contact the Copyright Clearance Center, Inc. (CCC) 222 Rosewood Drive, Danvers, MA 01923, 978-750-8400. CCC is a not-for-profit organization that provides licenses and registration for a variety of users. For organizations that have been granted a photocopy license by the CCC, a separate system of payment has been arranged.

Trademark Notice: Product or corporate names may be trademarks or registered trademarks, and are used only for identification and explanation without intent to infringe.

Library of Congress Cataloging-in-Publication Data

Coarse wavelength division multiplexing : technologies and applications / editors: Hans-Jörg Thiele, Marcus Nebeling.
   p. cm.
Includes bibliographical references and index.
ISBN-13: 978-0-8493-3533-4 (alk. paper)
ISBN-10: 0-8493-3533-7 (alk. paper)
1. Wavelength division multiplexing. I. Thiele, Hans-Jörg. II. Nebeling, Marcus.

TK5103.592.W38C67 2007
621.382'7--dc22   2006101008

Visit the Taylor & Francis Web site at
http://www.taylorandfrancis.com

and the CRC Press Web site at
http://www.crcpress.com

Dedication

To our families

Contents

Chapter 1 CWDM Standards 1
Mike Hudson

Chapter 2 Optical Fibers to Support CWDM 19
Kai H. Chang, Lars Grüner-Nielsen, and David W. Peckham

Chapter 3 CWDM Transceivers 57
Marcus Nebeling

Chapter 4 WDM Filters for CWDM 91
Ralf Lohrmann

Chapter 5 Optimizing CWDM for Nonamplified Networks 125
Charles Ufongene

Chapter 6 Amplifiers for CWDM 171
Leo H. Spiekman

Chapter 7 CWDM — Upgrade Paths and Toward 10 Gb/s 199
Hans-Jörg Thiele and Peter J. Winzer

Chapter 8 CWDM in Metropolitan Networks 251
Jim Aldridge

Chapter 9 CWDM in CATV/HFC Networks 269
Jim D. Farina

Chapter 10 CWDM for Fiber Access Solutions 285
Carlos Bock and Josep Prat

List of Abbreviations 313

Index 319

Introduction

In recent years, the field of optical communications has shown tremendous growth where the progress in optical transmission is reflected by long-haul transmission systems operating with up to 40 Gb/s line rate and with more than 80 densely spaced channels, bridging terrestrial and submarine distances of thousands of kilometers. Despite the increasing bit-rate distance product of such dense wavelength division multiplexing (DWDM) systems and their high complexity, lower cost wavelength division multiplexing (WDM) transmission has become another important factor, particularly in the shorter reaching metro networks. WDM in the metro space in many ways is a simplified, less complex form of DWDM. Closer to the end user, a more cost-effective optical transport layer is of greater importance, while at the same time flexibility and upgradeability is often desired. The term "Metro DWDM" was thus born to differentiate multi-protocol products based on relaxed DWDM technologies from the long-haul products based on more precise DWDM technologies. For metro applications requiring lower network capacity and cost, solutions to DWDM technology problems required new WDM technologies with simpler, wider tolerance laser manufacturing practices, less accurate laser temperature control, and reduced design complexity and cost of optical filters. The concepts of "wavelength banding" or "coarsely spaced wavelengths" emerged as technology solutions; however, these were still fundamentally based on DWDM, with 200 GHz and higher wavelength spacing in the C-band. More coarsely spaced DWDM wavelengths, such as 200- and 400-GHz spacing, reduced the metro product cost by requiring less stringent temperature and current control for the lasers and a more relaxed filter design. Expensive amplification was still required, however, to compensate for node losses in the all-optical metro networks.

The challenge of these new system design requirements was met by a technique known as coarse wavelength division multiplexing (CWDM). The basic idea of CWDM is the use of low cost optical components for amplifier-less transmission. In its currently adopted form, this technology defines up to 18 wavelengths channels on the precise grid specified by ITU-T G.692 across the entire low-loss spectral region of the optical fiber, a bandwidth of more than 300 nm. Due to the restricted performance and less complex design of the components used, there are some physical limitations of CWDM systems, which set them apart from DWDM systems. In terms of performance, CWDM is not to be compared with DWDM technology based on long-haul optical transport, but rather with some specific applications where a combination of capacity, flexibility, and low cost is demanded.

The original concept of CWDM prior to its ITU standardization is closely linked to the history of WDM in general, because originally there was no distinction between coarse and dense WDM. In the early 1980s, the continuous progress in the fabrication of both optical fiber and fiber-based components boosted the adoption of optical communication systems. The concept of WDM for transmitting multiple wavelengths over a single fiber had already been introduced at an early stage of this development. These systems contained multiple channels with typically 25-nm spacing in the 850-nm window using only multimode fiber. At that time, these systems were termed WDM systems, although this configuration is closer to what we understand as coarse WDM or CWDM today. The original operating wavelengths of CWDM transmission were located in the first telecommunications window, which was defined by the availability of components. First applications included multi-channel video distribution and bidirectional point-to-point data transmission over a single fiber. When single mode fiber became the medium of choice over the years, the typical operating wavelength of WDM systems also moved toward longer wavelength in the 1300-nm region and finally around 1550 nm. Some of the first single mode WDM systems that were deployed used fused biconic taper (FBT) WDMs in the second and third window. In particular, the third window around 1550 nm, also known as the C-band (1530 to 1565 nm), provided low fiber loss and since the early 1990s also the availability of erbium-doped fiber amplifiers (EDFAs), which played a major role in the definition and following success of dense WDM systems.

Despite the move toward longer wavelengths and the emerging DWDM systems, CWDM continued to be targeted at 850 nm multimode LAN applications where new vertical cavity surface-emitting laser (VCSEL) and thin-film filter technologies helped to reduce cost and increase packaging density. However, by the late 1990s, CWDM was gaining interest within the IEEE 802.3 High Speed Study Group for solving dispersion and loss problems for 10 Gb Ethernet LANs and some $10 \times$ GbE WAN applications. For the 10 GbE LAN applications, four wavelengths in the 850 or 1310 nm windows were proposed to extend the life of the installed base of multimode fiber in building and campus environments. To differentiate between the two LAN windows, the 802.3 High Speed Study Group referred to the 850 nm wavelengths as CWDM and the 1310 nm wavelengths as wide WDM (WWDM). Throughout the mid-late 1990s, there were also references to CWDM in carrier access network applications, such as passive optical networks (PONs); however, by today's standards, these were really just band splitters for multiplexing upstream and downstream traffic into the 1310 and 1550 nm windows. Such 2-wavelength PONs were referred to as coarse WDM or WWDM since DWDM in the 1550 nm window was the technology of choice for carrier applications requiring more than two wavelengths. With the further evolution of narrower, 4-wavelength CWDM and WWDM LAN technologies, the 1310/1550 nm band splitters were eventually categorized as simply WDM technology to avoid confusion and to reflect the much wider spectral passband. A summary of early WDM technologies is shown in Figure 1.

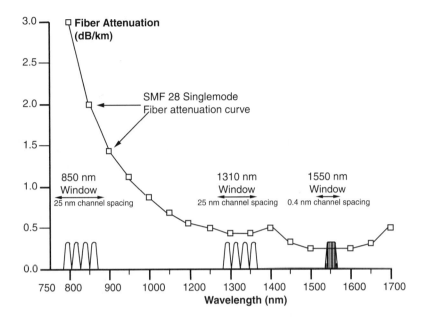

FIGURE 1 Overview of spectrum used for different CWDM-based applications.

For metro applications requiring less network capacity and lower cost than found in existing DWDM systems, solutions based on CWDM required new technologies with simpler, wider tolerance laser manufacturing practices, no laser temperature control, reduced design complexity, and lower cost of optical filters. The CWDM technologies developed for the LAN market were an obvious choice; however, these needed to be re-engineered to provide a range of wavelengths more suited to the transmission distance requirements of metro applications. From the year 2000, papers started to appear promoting CWDM technologies and standards for metro applications with 20 to 25 nm filter spacing and "uncooled" distributed feedback (DFB) lasers. Recognition by carriers and the ITU of the need for low-cost CWDM and the technology differences between CWDM and DWDM resulted in two separate WDM standards: **ITU-T G.694.1** "Spectral Grids for WDM applications: DWDM Frequency Grid" and **ITU-T G.694.2** "Spectral Grids for WDM Applications: CWDM Wavelength Grid," which are discussed in this book. These standards paved the way to "Metro CWDM" and helped to bring DWDM to the carrier market. Metro CWDM technologies now comprise optical filters and uncooled lasers with 20-nm spacing. There are 18 wavelengths currently specified with nominal wavelengths ranging from 1270 to 1610 nm inclusive. While DWDM systems typically use the C-band and if more capacity is required to extend the bandwidth into the L-band at longer wavelengths, CWDM requires more spectral bandwidth due to the coarse wavelength spacing. Figure 2 provides an overview of the ITU-based CWDM wavelength grid spanning the O- (original), E- (extended), S- (short), C- (conventional), and L- (long) bands.

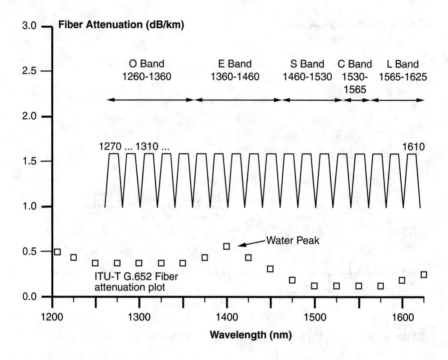

FIGURE 2 Metro CWDM wavelength grid defined by ITU-T G.694.

As it is evident from Figure 2, five of the CWDM wavelengths fall within the E-band. This band is normally not used for standard G.652 type fiber due to the higher optical attenuation at the water peak. These losses are typically 0.5 dB/km, which is not large; however, the maximum loss can be 2 dB/km or greater and no amplification is available to compensate for this attenuation. Carriers are not willing to take the risk that purchased equipment may not operate on some or all of their metro G.652 type fiber; consequently, the first CWDM products to be implemented and deployed in metro applications focused on the O-, S-, C-, and L-bands only. For all-optical metro applications where the O-band has not yet been provisioned for a 1310-nm transport service, there is a preference not to utilize these short wavelengths at all and add new capacity via eight CWDM wavelengths in the S-, C-, and L-bands alone. For this reason, the first CWDM lasers and filters developed to meet the emerging ITU-T G.694.2 standard were designed for eight CWDM wavelengths between 1470 and 1610 nm inclusive. The fiber attenuation in this region is at or near its minimum, which is needed to maximize the optical link budget, thereby enabling an increased number of optical add/drop multiplexer (OADM) nodes and/or maximized transmission distance between OEO regeneration sites since cost-effective amplification for all channels is not available. CWDM transmitters in this region can also use higher power lasers than the O-band while meeting Class 1 eye safety requirements.

To date, CWDM has been commercialized and many vendors offer CWDM as modular "pay as you grow" systems, sometimes also in combination with DWDM. Significant manufacturing cost reductions in CWDM components are expected over the next years due to automated manufacturing processes and increased component integration. Long wavelength VCSEL transmitter and APD/TIA receiver arrays with integral thin-film multiplexing filters are likely to be developed for multi-channel point–point Metro CWDM applications. This will further reduce the cost per channel. This book is intended to provide a comprehensive collection of various aspects of CWDM in a single place and puts these aspects into perspective. It is structured to cover component issues, system architectures, and concepts for extensions and upgrades as well as practical applications. Although the book is intended to be self-contained, starting with standards, components for CWDM and then moving to applications, each chapter by itself provides sufficient information in order to be used independently and is followed by references to relevant papers for further study. The focus of the book is directed on practical information related to CWDM, a topic that is considered to be a subset of optical communication systems.

We start this book with a chapter by Mike Hudson giving a summary of the ITU-T standards defining CWDM. Here, we find a detailed listing of reference link topologies, power budget, distances, penalties, channels plans, and other important parameters that define CWDM systems. After this overview, we address in the following three chapters essential component classes, the optical fibers (Lars Grüner-Nielsen, Kai Chang, and David Peckham), transceivers (Marcus Nebeling), and WDM filters (Ralf Lohrmann) that, combined, form the basis for the CWDM transmission links. The relevant fiber parameters are discussed with emphasis on commonly used fibers for CWDM applications, particularly low water peak fibers that allow operation across the entire 1300- to 1600-nm wavelength band. The availability of simple transceivers across this wavelength range was another driving factor for the success of CWDM. Here, we present an overview of suitable laser sources and diodes used in transceivers and also focus on the various transceiver standards. Finally, wavelength filters are discussed in Chapter 4, which enable CWDM and are also building blocks for more complex network structures in fiber rings or add–drop network elements. After reviewing different technical options for these filters, the thin-film filter approach is discussed in detail. We follow the component discussion with CWDM system issues and start with a presentation of work on optimizing CWDM for nonamplified networks by Charles Ufongene in Chapter 5. In this chapter, different architectures are considered, such as hubbed rings and meshed networks. Chapter 6 and Chapter 7 point out upgrade paths to overcome limitations of current CWDM systems. In Chapter 6, Leo Spiekman outlines the feasibility of optically amplified CWDM systems and investigates the challenges that are present with integrating CWDM and optical amplification. The application of three different amplification technologies is also reviewed. Chapter 7 addresses the question of transmission capacity beyond the 18-channel and typical OC-48 limit of conventional CWDM systems. Here, Hans-Jörg Thiele and Peter Winzer investigate two different approaches involving DWDM sub-bands and increased channel bit-rates by either optimized driving conditions or electronic equalization techniques.

Chapter 8, Chapter 9, and Chapter 10 deal with the wide range of CWDM applications. First, the role of the CWDM concept is discussed for metro networks by Jim Aldridge, where architecture and scalability are addressed. More specialized applications using CWDM are also covered here. Jim Farina looks into the specific use of CWDM in the CATV regime in Chapter 9. Josep Prat and Carlos Bock explain the role of CWDM with the current FTTH initiatives in Chapter 10, where CWDM acts both as a network and service overlay.

Marcus Nebeling and Hans-Jörg Thiele

Editors

Marcus Nebeling earned his BA in math and physics from Lewis and Clark College, Portland, Oregon and his MA in physics from the University of Utah. He started working on CWDM transmission systems in 1985 at Quante Corporation as a systems engineer designing optical networks for CATV applications. His work in fiber optics continued at Raynet Corporation, where he worked as a design engineer doing pioneering work on the first  passive optical network systems. In 1995, he moved to Nynashamn, Sweden, where he worked for Ericsson Telecom on HFC and FTTH systems. For the past 8 years, he has been president/CEO of Fiber Network Engineering, a company that designs and develops optical transmission products for Datacom, Telecom, and CATV applications. He has written papers on FTTH and CWDM systems for various trade journals. He lives with his wife Sara and their three children in Half Moon Bay, California.

Hans-Jörg Thiele earned his MSc in physics from Dortmund University, Germany, and his Ph.D. in electrical engineering from University College, London (UK), working on fiber nonlinearities in high-speed transmission systems. He started as a member of the technical staff at Bell Laboratories, Crawford Hill, New Jersey, in 2000, working on optical transmission issues such as Raman for DWDM long-haul systems and different aspects for metro/CWDM. After holding a guest scientist position at Heinrich Hertz Institute, Berlin, Germany, he is now with Siemens AG, Munich, where he is in charge of link  and network optimization in WDM transport systems. He has been author and co-author of some 50 journal and conference papers, including invited contributions and one book chapter. Hans is married and lives in Munich.

Contributors

Jim Aldridge
Optics Central
Palo Alto, California, U.S.A.
jima@opticscentral.com

Carlos Bock
University Polytechnica
  Catalonia (UPC)
Barcelona, Spain
Bock@tsc.upc.edu

Kai H. Chang
Heraeus Tenevo, Inc.
Buford, Georgia, U.S.A.
kai.chang@heraeus.com

Jim D. Farina
Farina Consulting
Howell, New Jersey, U.S.A.
jim@farinaweb.com

Mike Hudson
RBN Inc.
Sydney, Australia
mhudson@rbni.com

Ralf Lohrmann
Cube Optics AG
Mainz, Germany
Lohrmann@cubeoptics.com

Marcus Nebeling
Fiber Network Engineering Inc.
Half Moon Bay, California, U.S.A.
Marcus@fn-eng.com

Lars Grüner-Nielsen
OFS
Brøndby, Denmark
Lgruner@ofsoptics.com

David W. Peckham
OFS
Norcross, Georgia, U.S.A.
dpeckham@ofsoptics.com

Josep Prat
University Polytechnica
  Catalonia (UPC)
Barcelona, Spain
Jprat@tsc.upc.edu

Leo H. Spiekman
Alphion Corp.
Princeton Junction,
  New Jersey, U.S.A.
Spiekman@ieee.org

Hans-Jörg Thiele
Siemens AG
Munich, Germany
Hj.thiele@siemens.com

Charles Ufongene
United Technologies Corp.
Hartford, Connecticut, U.S.A.
ufongene@mit.edu

Peter J. Winzer
Lucent Technologies
Bell Laboratories
Holmdel, New Jersey, U.S.A.
Peter.winzer@ieee.org

List of Tables

Chapter 1

Table 1.1 G.694.2 CWDM Wavelength Grid (Defined by the Center Filter Wavelengths)
Table 1.2 G.695 Channel Plans
Table 1.3 G.695 Application Code Summary

Chapter 2

Table 2.1 Transmission Distances on SSMF of NRZ Signal for 1-dB Eye-Closure Penalty Due to Dispersion Using Transmitter without Chirp
Table 2.2 ITU-T G.652 Fiber Geometry Requirements
Table 2.3 Fiber Properties of Typical Matched Cladding and Depressed Cladding Fiber Meeting G.652B Recommendation
Table 2.4 Cabled Fiber Loss Requirements of the ITU-T G.652 Recommendation
Table 2.5 Macrobend Loss Performance (in dB) of SSMF (60, 50, and 15 mm Diameters are for 100 turns, 30 and 32 mm for 1 turn)

Chapter 3

Table 3.1 Defining the Permissible CWDM Wavelength Range under Operating Conditions
Table 3.2 Overview of Light Sources
Table 3.3 DFB Laser Device Data
Table 3.4 Worst-Case Reach (in km) for Different Laser Types (10 Gb/s) and Transmission Fibers
Table 3.5 Comparing the Two Principal Types of Detectors
Table 3.6 Overview of 10-Gb/s Transceiver Standards

Chapter 4

Table 4.1 Comparison of Different Filter Type Fused Coupler, FBG, AWG, and TFF
Table 4.2 Performance Data of TF Filters for CWDM

Chapter 5

Table 5.1 ITU-Based CWDM Channel Plan for 12 CWDM Channels
Table 5.2 A 16-CWDM Channel Grid Divided into Four Wavelength Bands
Table 5.3 Wavelength Band Assignment in 4-Node Unidirectional Hubbed Ring
Table 5.4 Filter Loss Calculation for 4-Node Unidirectional Hubbed Ring, Incrementing Node Assignment
Table 5.5 Losses and Ring Perimeter for 4-Node Unidirectional Hubbed Ring Case, Incrementing Node Assignment
Table 5.6 Filter Loss Calculation for 4-Node Unidirectional Hubbed Ring, Attenuation Slope Compensating Wavelength Assignment
Table 5.7 Losses and Ring Perimeter for 4-Node Unidirectional Hubbed Ring Case, Attenuation Slope Compensating Wavelength Assignment
Table 5.8 Wavelength Band Assignment in 8-Node Unidirectional Hubbed Ring
Table 5.9 An 8-Node Ring Channel to Node Assignment for Figure 5.12a
Table 5.10 An 8-Node Ring Channel to Node Assignment for Figure 5.12b
Table 5.11 Nonattenuation Slope Compensating Wavelength Assignment in 6-Node Meshed Network
Table 5.12 Wavelength Assignment in 6-Node Meshed Network: (Top) Number of Traversed Ring Spans and (Bottom) Attenuation Slope Compensation Assignment

Chapter 6

Table 6.1 Summary of Amplifier Properties

Chapter 7

Table 7.1 Transmission Penalties for 1550-nm DWDM Sub-Band
Table 7.2 Overview of 10-Gb/s Transmission Results

Chapter 8

Table 8.1 Overview of Common Network Properties and Usage of CWDM

Chapter 10

Table 10.1 Distances and Available Bandwidths for Different Access Solutions
Table 10.2 Traffic Patterns and Bandwidth Requirements of Broadband Applications

List of Figures

Introduction

FIGURE 1 Overview of different CWDM-based applications.
FIGURE 2 Metro CWDM wavelength grid defined by ITU-T G.694.2.

Chapter 1

FIGURE 1.1 Laser wavelength drift and filter passband requirements for a CWDM channel.
FIGURE 1.2 G.695 application code example.
FIGURE 1.3 Black box transmitter and receiver.
FIGURE 1.4 Black link concept.
FIGURE 1.5 Point-to-point CWDM links.
FIGURE 1.6 Linear add-drop CWDM black link. OM/OD, optical multiplexer/demultiplexer.
FIGURE 1.7 CWDM black link ring (hubbed).
FIGURE 1.8 CWDM black link ring (nonhubbed).
FIGURE 1.9 Assumed attenuation coefficients for non-low-water-peak fibers.
FIGURE 1.10 Assumed attenuation coefficients for low or zero water-peak fibers.

Chapter 2

FIGURE 2.1 Comparison of spectral attenuation of ZWP fiber (ITU-T G.652D) with standard water peak single mode fiber (SSMF, ITU-T G.652). The ITU-proposed CWDM channels are overlaid, whereas those highlighted in red indicate channels gained by using ZWP fiber.
FIGURE 2.2 Refractive index profile of (a) a step index fiber bent at constant radius of curvature and (b) the equivalent straight fiber with "tilted" index profile. Dotted line is effective index of the fundamental mode.
FIGURE 2.3 Predicted macrobend loss for SSMF across the CWDM channel spectrum.
FIGURE 2.4 Model of the core of a fiber in the vicinity of a microbend. Power carried in the fundamental mode before the microbend is coupled into the fundamental and into lossy higher order modes by the microbend.
FIGURE 2.5 Predicted microbending loss as a function of wavelength according to Marcuse microbend model. The four curves show the change in wavelength dependence and the change in the magnitude of the attenuation as the correlation length, L_c, of the microbending power spectrum increases from 1 to 500 µm.
FIGURE 2.6 Dispersion versus wavelength for common transmission fibers. SSMF: standard single mode fiber — ITU-T G.652; DSF: dispersion shifted fiber — ITU-T G.653; LD NZDF: low dispersion nonzero dispersion fiber — ITU-T G.655; and MD NZDF: medium dispersion nonzero dispersion fiber — ITU-T G.656.

FIGURE 2.7 Calculated pulse broadening for an initial Gaussian pulse with $T_0 = 30$ ps, dynamic chip only, and a wavelength $\lambda = 1550$ nm as a function of the accumulated dispersion: D·L.

FIGURE 2.8 Impact of accumulated dispersion on the receiver sensitivity for a 10-Gb/s NRZ signal at 1551 nm. (Taken from Thiele, H.J., Nelson, L.E., and Das, S.K., *Electron. Lett.*, 39, 1264–1266, 2003. With permission.)

FIGURE 2.9 Power transmitted through a short length of single mode fiber in the vicinity of the effective cut-off wavelength.

FIGURE 2.10 Core alignment of the optical fiber.

FIGURE 2.11 RIC-ODD and the conventional RIT process.

FIGURE 2.12 Histograms of fiber loss distribution at 1385 and 1550 nm for the RIC-ODD fibers.

FIGURE 2.13 A deuterium-treated AllWave® fiber shows virtually no hydrogen aging loss for the full CWDM spectrum from 1270 to 1610 nm. Its permanent low loss over the service lifetime is virtually assured.

FIGURE 2.14 Postcompensated transmission line using DCF at the receiver.

FIGURE 2.15 Performance of a DCF module for 8-channel CWDM by OFS. (*Top*) Residual dispersion from 1470 to 1610 nm; (*bottom*) total attenuation, including splices and connectors.

Chapter 3

FIGURE 3.1 Wavelength used in an optical communications link. (a) Single wavelength transmitter; (b) extension to a four-channel WDM system/ colored transmitter.

FIGURE 3.2 Comparison of performance and cost of FP and DFB lasers.

FIGURE 3.3 Interaction of laser chirp and dispersion for DML. (*Left*) Prechirped DML pulses; (*top*) dispersive pulse broadening for $D > 0$; (*bottom*) pulse broadening for $D < 0$; (*right*) resulting output signal which is further broadened ($D > 0$) or recompressed ($D < 0$).

FIGURE 3.4 Responsivity of an InGaAs/InP PIN photodetector. (From Lee, T., Burrus, C.A., and Dentai, A.G., *J. Quantum Electronics JQE*, 15, 1979. With permission.)

FIGURE 3.5 (a) Early transmitters and receivers; (b) principle of a basic bidirectional link using two separate fibers and pairs of transmitter and receiver (TX, RX).

FIGURE 3.6 Laser transmitter using a temperature control circuit for wavelength stabilization.

FIGURE 3.7 Transmitter without temperature control as commonly used for CWDM.

FIGURE 3.8 Wavelength drift of an uncooled CWDM laser within the 1550-nm CWDM passband, the actual wavelength is a function of temperature.

FIGURE 3.9 Definition of CWDM passband center for uncooled lasers, shown for channel at 1550 nm.

FIGURE 3.10 Statistical distribution of laser wavelengths around 1550 nm at room temperature. (*Center*) Nominal laser wavelength at $T = 25°C$; gray lines: production yield, ±2 nm, also at $T = 25°C$.

FIGURE 3.11 Temperature variations and production yield with lasers distributed around the passband center.

FIGURE 3.12 Building blocks of a typical receiver used in transceiver applications.

FIGURE 3.13 A 16-channel standard CWDM transmission link using 16 pairs of transmitters (TX) and receivers (RX).

FIGURE 3.14 Set transceivers for use in CWDM; each device has an optical plug for the receiver and transmitter.

FIGURE 3.15 (a) Bidirectional CWDM link consisting of two separate unidirectional links and 2 × 16 transceivers leading to 16-channel CWDM transmission in each direction; (b) bidirectional CWDM link over a single fiber with 2 × 8 transceivers and a total of 16 wavelengths; and (c) bidirectional link with 16 transceivers at both sides. Each CWDM wavelength is used twice in this setup and combined into the same multiplexer port.

FIGURE 3.16 A 1 × 9 transceiver.

FIGURE 3.17 An SFF transceiver.

FIGURE 3.18 A GBIC transceiver for 1610-nm CWDM wavelength.

FIGURE 3.19 An SFP transceiver for 1570-nm CWDM wavelength.

FIGURE 3.20 Integrated transceiver units using parallel optics and ribbon fiber.

FIGURE 3.21 (a) Concept of highly integrated 16-channel CWDM transceiver as repeater for extended reach CWDM and networking functions; (b) concept of highly integrated 16-channel CWDM transceiver as add/drop multiplexer for extended reach CWDM and networking functions.

Chapter 4

FIGURE 4.1 Basic concepts of multiplexing. Combining four different channels A, B, C, and D in (a) SDM, (b) TDM, and (c) WDM.

FIGURE 4.2 Principle of chirped fiber Bragg grating reflecting different wavelengths. (From Ref. 7. With permission.)

FIGURE 4.3 Two methods to write fiber Bragg gratings. (From Ref. 7. With permission.)

FIGURE 4.4 Optical add drop multiplexer using one grating and two circulators. (From Ref. 7. With permission.)

FIGURE 4.5 Fiber Bragg grating with athermalized package design. (From Ref. 7. With permission.)

FIGURE 4.6 Fiber Bragg gratings of the center wavelength with temperature and the results of athermalization. (From Ref. 7. With permission.)

FIGURE 4.7 Relative strength of the refractive index modulation in (a) a uniform grating and (b) an apodized grating. (From Ref. 7. With permission.)

FIGURE 4.8 Calculated response function of an (a) uniform grating and (b) an apodized grating. (From Ref. 7. With permission.)

FIGURE 4.9 Measured spectral loss for a fiber Bragg grating written in single mode fiber. Shorter wavelengths experience additional loss. (From Ref. 7. With permission.)

FIGURE 4.10 Measured spectral loss for an apodized fiber Bragg grating in nonstandard fiber with high numerical aperture, transmitted and reflected signal. (From Ref. 7. With permission.)

FIGURE 4.11 Structural elements of an arrayed waveguide demultiplexer. (From Ref. 8. With permission.)

FIGURE 4.12 Response function of an arrayed waveguide. The spectrum shows the typical Gaussian profile. (From Ref. 8. With permission.)

FIGURE 4.13 Definition of some basic performance parameters.

FIGURE 4.14 Coating layer structure of a 3-cavity thin-film filter.

FIGURE 4.15 An increasing number of cavities results in a smaller stop-band or improved adjacent channel isolation.

FIGURE 4.16 Definitions for input and output beams, IN, Transmit and reflect.

FIGURE 4.17 Edge pass filter (long pass) splitting the 1310-nm window from 1550-nm window.

FIGURE 4.18 Example for a CWDM bandpass filter. (Reference in the table was changed to min of 25 dB.) The filter element transmits the light corresponding to the 1551-nm channel while all other wavelength appear on the reflect port.
FIGURE 4.19 Example for a special type of bandpass filter called "4-skip-0."
FIGURE 4.20 Process flowchart for thin-film dielectric coatings.
FIGURE 4.21 *In situ* monitoring capability of modern coating chambers. A laser beam is directed through the spinning substrate to monitor the progress of the deposition process.
FIGURE 4.22 Transmittance and reflectance spectrum (changed to 25 dB) of a CWDM 1551-nm filter; the light trace shows the calculated reflectance.
FIGURE 4.23 A typical 3-port package.
FIGURE 4.24 Key optical elements of a 3-port package; details are discussed in the text.
FIGURE 4.25 Two basic concepts to align optical fiber: (a) through physical contact to a passive alignment structure and (b) via active alignment.
FIGURE 4.26 Conceptual drawing of a 6-port device; the same filter is used twice in two independent 3-ports sharing the same package.
FIGURE 4.27 Cascaded 3-port devices in a CWDM 8-channel module. A secondary housing (blue) needs to protect the fiber management.
FIGURE 4.28 Conceptual drawing of a free-space multiple bounce architecture. Individual thin-film filter elements are attached to an optical block to form the core of the device.
FIGURE 4.29 Size comparison of a standard module based on 3-port technology versus a new miniature design concept.
FIGURE 4.30 A new generation of integrated devices for optoelectronic interfaces will enable more compact and cost-effective system solutions.

Chapter 5

FIGURE 5.1 Measured AllWave fiber (ZWPF) spectral attenuation as a function of wavelength for all considered 16 CWDM channels compared to conventional single mode fiber.
FIGURE 5.2 A 16-channel CWDM n-node hubbed ring network.
FIGURE 5.3 Thin film filter implementation of OADM to add/drop wavelength band B(1) = $\{\lambda_1, \lambda_2, \lambda_3, \lambda_4\}$ and node loss model.
FIGURE 5.4 A 16-channel CWDM n-node hubbed ring network with variable span lengths.
FIGURE 5.5 (a) Logical star network of four local exchanges (COs) and a tandem exchange and (b) equivalent hubbed ring configuration.
FIGURE 5.6 Application of wavelength assignment in 4-node ring with unidirectional traffic.
FIGURE 5.7 Hub loss model and TFF implementation of OADM to add/drop wavelength band B(1) to B(4); channels $\{\lambda_1,...,\lambda_{16}\}$.
FIGURE 5.8 (a) Node loss model and TFF implementation of OADM to add/drop wavelength band B(1)=$\{\lambda_1, \lambda_2, \lambda_3, \lambda_4\}$. (b) Node loss model and TFF implementation of OADM to add/drop wavelength band B(2); channels $\{\lambda_5...\lambda_8\}$. (c) Hub loss model and TFF implementation of OADM to add/drop wavelength band B(3); channels $\{\lambda_9...\lambda_{12}\}$. (d) Node loss model and TFF implementation of OADM to add/drop wavelength band B(4); channels $\{\lambda_{13}...\lambda_{16}\}$.
FIGURE 5.9 Filter (including connectors and splices), fiber, and ring losses plotted as functions of the nodes N_1 to N_4 and their assigned wavelength bands B(1) to B(4) for nonattenuation slope compensating wavelength assignment.

FIGURE 5.10 Calculated ring perimeter for both wavelength assignment approaches across the CWDM band, 4-node hubbed ring.

FIGURE 5.11 Filter, fiber, and ring losses plotted as functions of the nodes N_1 to N_4 and their assigned wavelength bands B(1) to B(4) for attenuation slope compensating wavelength assignment.

FIGURE 5.12 Wavelength assignment in 8-node unidirectional ring. (a) configuration with bands assigned to nodes with no preference and (b) configuration with attenuation slope compensating wavelength assignment.

FIGURE 5.13 Calculated ring perimeter for both wavelength assignment approaches across the CWDM band, 8-node hubbed ring.

FIGURE 5.14 (a) Topology of a meshed network and (b) equivalent N-node unidirectional ring.

FIGURE 5.15 Attainable ring perimeters calculated as function of the CWDM center wavelengths for the CWDM-based 6-node logical mesh/unidirectional ring; simple wavelength assignment versus attenuation slope compensating assignment.

Chapter 6

FIGURE 6.1 CWDM architecture with semiconductor-based amplifiers using band (de)multiplexers. (Taken from Thiele, H.J., Nelson, L.E., Thomas, J., Eichenbaum, B., Spiekman, L.H., and van den Hoven, G.N., Proceedings OFC 2003, vol. 1, pp. 23–24, 2003. With permission.)

FIGURE 6.2 Measured receiver sensitivity at 10^{-9} BER of an optically preamplified pin-diode receiver using a clean intensity-modulated signal at 10 Gb/s. As the optical filter bandwidth is varied, the decrease in sensitivity due to increased spontaneous-spontaneous beat noise can clearly be observed.

FIGURE 6.3 Hybrid EDFA–TDFA amplifier configuration: the four shorter wavelengths are amplified by the TDFA, whereas the remaining four CWDM wavelengths are amplified by the EDFA. (Taken from Sakamoto, T., Mori, A., and Shimizu, M., Proceedings OFC 2004, vol. 2, paper ThJ5. With permission.)

FIGURE 6.4 Gain after equalization and NF characteristics of hybrid EDFA–TDFA amplifier. (Taken from Sakamoto, T., Mori, A., and Shimizu, M., Proceedings OFC 2004, vol. 2, paper ThJ5. With permission.)

FIGURE 6.5 Amplified spontaneous emission spectrum of an SOA, which typically has a smooth parabola-like shape. (Taken from Spiekman, L.H., *Optical Fiber Telecommunications*, Academic Press, 2002. With permission.)

FIGURE 6.6 Measured TE and TM gains and PDG of an SOA chip using a coupled quantum well structure with broadband gain. (Taken from Park, S., Leavitt, R., Enck, R., Luciani, V., Hu, Y., Heim, P.J.S., Bowler, D., and Dagenais, M., *Photon. Technol. Lett.*, 17(5), 980–982, 2005. With permission.)

FIGURE 6.7 Dependence of gain, saturation output power, and NF on signal wavelength. Note that these parameters are referenced to the chip facets. (Taken from Morito, K., Tanaka, S., Tomabechi, S., and Kuramata, A., *Photon. Technol. Lett.*, 17(5), 974–976, 2005. With permission.)

FIGURE 6.8 Typical configuration of a packaged SOA chip. The device has two fiber-chip couplings. Optional items like isolators may be placed in the light path. (Taken from Spiekman, L.H., *Optical Fiber Telecommunications*, Academic Press, 2002. With permission.)

FIGURE 6.9 Gain versus output power curve of an SOA. The gain is quasi-linear in the small signal regime, but saturates at higher power levels. (Taken from Spiekman, L.H., *Optical Fiber Telecommunications*, Academic Press, 2002. With permission.)

FIGURE 6.10 (a) Ultrafast gain compression and slower recovery of an SOA by an intense optical pulse. The gain recovery time is determined by the carrier lifetime. (b) Pump-probe setup with which the curve was measured. (Taken from Spiekman, L.H., *Optical Fiber Telecommunications*, Academic Press, 2002. With permission.)

FIGURE 6.11 Input (inset) and output spectra of an SOA showing an 8-channel CWDM spectrum upgraded in one of the CWDM channels with eight DWDM channels. (Taken from Iannone, P., Reichmann, K., and Spiekman, L., Proceedings OFC 2003, vol. 2, pp. 548–549, 2003. With permission.)

FIGURE 6.12 Transmission of eight WDM channels modulated at 10 Gb/s across 6 × 40 km spans of standard fiber using a total of nine SOAs. External modulation is used here to investigate the transmission over several cascaded SOA-amplified spans. (Taken from Spiekman, L.H., *Optical Fiber Telecommunications*, Academic Press, 2002. With permission.)

FIGURE 6.13 Access architecture with SOAs in the end user nodes. The SOAs provide the amplification needed in the metro ring, and double as modulators for the upstream data. (Taken from Spiekman, L.H., *Optical Fiber Telecommunications*, Academic Press, 2002. With permission.)

FIGURE 6.14 Flat gain profile (100 nm centered on C- and L-band) obtained by Raman-pumping fiber using a multi-wavelength source. (Taken from Namiki, S. and Emori, Y., *IEEE J. Sel. Topics Quantum Electron.*, 7(1), 3–16 (invited paper), 2001. With permission.)

FIGURE 6.15 A 12-wavelength-channel WDM high-power pump LD unit. (Taken from Namiki, S. and Emori, Y., *IEEE J. Sel. Topics Quantum Electron.*, 7(1), 3–16 (invited paper), 2001. With permission.)

FIGURE 6.16 Two SOA-Raman hybrid schemes: (a) SOA + Raman and (b) Raman + SOA. (Taken from Chen, Y., Pavlik, R., Visone, C., Pan, F., Gonzales, E., Turukhin, A., Lunardi, L., Al-Salameh, D., and Lumish, S., Proceedings OFC 2002, pp. 390–391. With permission.)

FIGURE 6.17 Gain versus wavelength for a hybrid Raman-SOA amplifier and its constituent SOA and Raman stages. (Taken from Iannone, P.P., Reichmann, K.C., Zhou, X., and Frigo, N.J., Proceedings OFC 2005, paper OthG3. With permission.)

FIGURE 6.18 Various configurations of an amplified CWDM link.

Chapter 7

FIGURE 7.1 Attenuation coefficient of legacy standard single-mode fiber and advanced low water peak single-mode fiber. (Taken from Winzer, P.J., Fidler, F., Thiele, H.J., Matthews, M., Nelson, L.E., Sinsky, J.H., Chandrasekhar, S., Winter, M., Castagnozzi, D., Stulz, L.W., and Buhl, L.L., *J. Lightwave Technol.*, 23(1), 203–210, 2005. With permission.)

FIGURE 7.2 Typical CWDM system, as specified by ITU-T G.695, for $N = 16$ channels: (a) unidirectional and (b) bidirectional. (Taken from Winzer, P.J., Fidler, F., Matthews, M., Nelson, L.E., Thiele, H.J., Sinsky, J.H., Chandrasekhar, S., Winter, M., Castagnozzi, D., Stulz, L.W., and Buhl, L.L., *J. Lightwave Technol.*, 23(1), 203–210, 2005. With permission.)

FIGURE 7.3 Insertion loss of typical thin-film CWDM multiplexers. (Taken from Thiele, H.J., Winzer, P.J., Sinsky, J.H., Stulz, L.W., Nelson, L.E., and Fidler, F., *Photon. Technol. Lett.*, 16(10), 2004. With permission.)

FIGURE 7.4 Upgrade paths for an existing WDM system, either by increasing the bit-rate (option 1) or additional wavelength channels (option 2a–c).

FIGURE 7.5 Chromatic dispersion of standard-dispersion fiber (SSMF and standard LWPF).

FIGURE 7.6 A DWDM overlay substituting a single CWDM channel.

FIGURE 7.7 Correction curves of the enhanced FEC scheme used in our experiment. For comparison, standard RS(255,239) FEC and the FEC-free (uncorrected) case are also shown. (Taken from Winzer, P.J., Fidler, F., Thiele, H.J., Matthews, M., Nelson, L.E., Sinsky, J.H., Chandrasekhar, S., Winter, M., Castagnozzi, D., Stulz, L.W., and Buhl, L.L., *J. Lightwave Technol.*, 23(1), 203–210, 2005. With permission.)

FIGURE 7.8 BER vs. OSNR for NRZ-OOK at 10.7 Gb/s. *Squares*: back-to-back; *circles*: after 60 km of SSMF; *arrows*: BER-dependent dispersion penalties.

FIGURE 7.9 Measured variation of output power for uncooled CWDM laser.

FIGURE 7.10 Widely varying static characteristics of the nonselected DMLs covering the full 16-channel CWDM band from 1310 to 1610 nm. (Taken from Thiele, H.J., Winzer, P.J., Sinsky, J.H., Stulz, L.W., Nelson, L.E., and Fidler, F., *Photon. Technol. Lett.*, 16(10), 2004. With permission.)

FIGURE 7.11 Frequency-chirp and intensity waveforms measured for two nonselected CWDM lasers at (a) 1510 nm and (b) 1530 nm.

FIGURE 7.12 Principle of wavelength detuning in DFB lasers; variation of α-factor, differential gain and gain curve of the device.

FIGURE 7.13 Signal power and receiver sensitivity as a function of transmission distance.

FIGURE 7.14 Power margin and sensitivity penalty for 16 CWDM channels at 10 Gb/s for 40-km NZDSF transmission link. (Taken from Thiele, H.J., Winzer, P.J., Sinsky, J.H., Stulz, L.W., Nelson, L.E., and Fidler, F., *Photon. Technol. Lett.*, 16(10), 2004. With permission.)

FIGURE 7.15 Power penalty as a function of ER (100-km link, 4 Gb/s, laser active region $w = 1.75$ μm wide). (Taken from reference Corvini, P.J. and Koch, T.L., *J. Lightwave Technol.*, 5(11), 1591–1595, 1987. With permission.)

FIGURE 7.16 CWDM laser output signal at 2.5 Gb/s. *Left*: detected with 20-GHz PIN diode; *right*: with additional electrical filtering (cut-off at 2.488 GHz).

FIGURE 7.17 To improve modulation performance, each laser in a coaxial package was mounted at the end of a 50-Ω microstrip line, in series with a 47-Ω surface mount chip resistor.

FIGURE 7.18 *Left*: Modulation characteristics of the DMLs using the improved mounting of Figure 7.17. The characteristics of a 10-Gb/s rated DML is shown for reference. *Right*: Two eye diagrams of typical lasers at 10 Gb/s. (Taken from Thiele, H.J., Winzer, P.J., Sinsky, J.H., Stulz, L.W., Nelson, L.E., and Fidler, F., *Photon. Technol. Lett.*, 16(10), 2004. With permission.)

FIGURE 7.19 DWDM upgrade with four DWDM channels replacing CWDM channel number 13 at 1550 nm in a combined DWDM/CWDM transmission link. *Inset*: example of overlap between CWDM and DWDM multiplexer spectrum for the 1550-nm channel. (Taken from Thiele, H.J., Nelson, L.E., and Das, S.K., *Electron. Lett.*, 39(17), 2003. With permission.)

FIGURE 7.20 Measured power penalties (at 10^{-9} BER) for the 2.5-Gb/s reference CWDM system, channels modulated with 8-dB ER and $2^{23} - 1$ PRBS word length, transmission over 60 km of: AllWave fiber (*squares*) and low water peak NZDSF (*circles*).

FIGURE 7.21 Impact of accumulated dispersion on performance of the 1550-nm 10-Gb/s DWDM channel over a post-compensated 60-km AllWave fiber span: *circles*: 6-dB ER; *squares*: 8-dB ER; *crosses*: back-to-back. (Taken from Thiele, H.J., Nelson, L.E., and Das, S.K., *Electron. Lett.*, 39(17), 2003. With permission.)

FIGURE 7.22 Variation of accumulated dispersion for the 1550-nm DWDM channels in the CWDM-DWDM transmission link.

FIGURE 7.23 Penalty measurement of 1310-nm CWDM channel after 60 km, upgraded with a 10-Gb/s cooled DML. (Taken from Thiele, H.J., Nelson, L.E., and Das, S.K., *Electron. Lett.*, 39(17), 2003. With permission.)

FIGURE 7.24 *Left*: Optical spectrum of the 100-Gb/s capacity mixed bit-rate CWDM system over 60 km AllWave fiber. *Right*: 4×10-Gb/s channels of the DWDM overlay.

FIGURE 7.25 Experimental setup for 10-Gb/s uncompensated transmission over a mixed fiber plant.

FIGURE 7.26 Dispersion of SSMF (*dashed*) and NZDSF (*solid*) across the CWDM spectrum. Dotted vertical lines indicate nominal CWDM wavelengths.

FIGURE 7.27 Optical spectrum of 10-Gb/s EA-EML and DML overlay channels after 50-km TrueWave-RS fiber. *Inset*: back-to-back filtered eye diagram of center DML. (Taken from Thiele, H.J., Das, S.K., Boncek, R., and Nelson, L.E., Proceedings OECC 2003, paper 16E1-4, pp. 717–718. With permission.)

FIGURE 7.28 *Left*: 16×10-Gb/s CWDM spectrum measured after multiplexing; *right*: BER measurements for all 10-Gb/s channels at $T = 25°C$ after 40-km of simultaneous transmission over low water peak NZDSF. (Taken from Thiele, H.J., Winzer, P.J., Sinsky, J.H., Stulz, L.W., Nelson, L.E., and Fidler, F., *Photon. Technol. Lett.*, 16(10), 2004. With permission.)

FIGURE 7.29 (a) Receiver sensitivity for all CWDM channels (■), maximum available receiver power (5), and power for BER= 10^{-9} after 40 km (O). (b) Calculated power margin after 40-km fiber propagation (■) and sensitivity penalty with respect to the back-to-back case (O). (Taken from Thiele, H.J., Winzer, P.J., Sinsky, J.H., Stulz, L.W., Nelson, L.E., and Fidler, F., *Photon. Technol. Lett.*, 16(10), 2004. With permission.)

FIGURE 7.30 *Top*: Maximum achievable laser case temperature allowing BER = 10^{-9} after 40-km fiber propagation. *Below*: back-to-back eye diagrams for all CWDM channels at temperature corresponding to BER = 10^{-9}. (Taken from Thiele, H.J., Winzer, P.J., Sinsky, J.H., Stulz, L.W., Nelson, L.E., and Fidler, F., *Photon. Technol. Lett.*, 16(10), 2004. With permission.)

FIGURE 7.31 Block diagram of the integrated electronic equalization and FEC chip set. This is the device for the receiver.

FIGURE 7.32 Back-to-back sensitivities of all 16 CWDM channels (channel BER = 10^{-3}; corrected BER < 10^{-16}). Drive conditions optimized for maximum transmission distance. *Circles*: without equalization; *squares*: with equalization. (Taken from Winzer, P.J., Fidler, F., Matthews, M., Nelson, L.E., Sinsky, J.H., Chandrasekhar, S., Winter, M., Castagnozzi, D., Stulz, L.W., and Buhl, L.L., *J. Lightwave Technol.*, 23(1), 203–210, 2005. With permission.)

FIGURE 7.33 Receiver sensitivity (*solid*: BER = 10^{-3}; *dashed*: BER = $6 \cdot 10^{-5}$) vs. transmission distance and accumulated CD for the 1550-nm CWDM laser. *Circles*: unequalized receiver; *squares*: equalized receiver; *hatched area*: equalization gain. Optical eye diagrams are shown at various distances. (Taken from Winzer, P.J., Fidler, F., Thiele, H.J., Matthews, M., Nelson, L.E., Sinsky, J.H., Chandrasekhar, S., Winter, M., Castagnozzi, D., Stulz, L.W., and Buhl, L.L., *J. Lightwave Technol.*, 23(1), 203–210, 2005. With permission.)

FIGURE 7.34 Receiver sensitivity at BER = 10^{-3} vs. transmission distance for the 1550-nm CWDM laser (*circles*: unequalized receiver; *squares*: equalized receiver). The dashed line represents the optical power evolution as a function of transmission distance, while the dotted line includes any lumped optical loss and represents the available optical power at the receiver. The hatched area indicates unattainable transmission distances caused by high penalty around 20 km in the unequalized case. (Taken from Winzer, P.J., Fidler, F., Thiele, H.J., Matthews, M., Nelson, L.E., Sinsky, J.H., Chandrasekhar, S., Winter, M., Castagnozzi, D., Stulz, L.W., and Buhl, L.L., *J. Lightwave Technol.*, 23(1), 203–210, 2005. With permission.)

FIGURE 7.35 (a) Maximum transmission distance for all CWDM channels and (b) CD at maximum distance shown in (a). (Taken from Winzer, P.J., Fidler, F., Theil, H.J., Matthews, M., Nelson, L.E., Sinsky, J.H., Chandrasekhar, S., Winter, M., Castagnozzi, D., Stulz, L.W., and Buhl, L.L., *J. Lightwave Technol.*, 23(1), 203–210, 2005. With permission.)

FIGURE 7.36 Fully bidirectional fiber communication system.

FIGURE 7.37 (a) Measured BER as a function of received signal power for different SIRs using a 10-Gb/s rated DML at 1550 nm. (b) Back-to-back eye diagram of the 10-Gb/s laser without interferer. (c) Measurement setup, incorporating two attenuators to set SIR and received power, a polarization controller to align signal and interferer, and 10 km of NZDSF to decorrelate signal and interferer. (Taken from Winzer, P.J., Fidler, F., Matthews, M., Nelson, L.E., Sinsky, J.H., Chandrasekhar, S., Winter, M., Castagnozzi, D., Stulz, L.W., and Buhl, L.L., *J. Lightwave Technol.*, 23(1), 203–210, 2005. With permission.)

FIGURE 7.38 10-Gb/s receiver sensitivity penalties vs. SIR at different target BERs, measured at 1550 nm using the setup of Figure 7.37c. (a) 10-Gb/s rated DML with ~7.5-dB extinction and (b) 2.5-Gb/s rated DML with ~3-dB extinction.

FIGURE 7.39 Experimental setup to generate backward traffic and emulate the worst-case scenario for a fully bidirectional CWDM system.

Chapter 8

FIGURE 8.1 A fiber optics network with long-haul, metro regional and metro access.

FIGURE 8.2 CWDM ring topology using four nodes $N_1...N_4$ interconnected via links.

FIGURE 8.3 CWDM MUX at Tx side of the link shown with three different channels having a center wavelength $\lambda_1...\lambda_3$.

FIGURE 8.4 CWDM DEMUX at Rx side of the link shown with three different channels having a center wavelength $\lambda_1...\lambda_3$.

FIGURE 8.5 CWDM point-to-point connection between two buildings.

FIGURE 8.6 (a) OADM schematic with a single add/drop channel plus express channels, the OADM is functionally composed of a DEMUX-MUX pair; (b) usage of an OADM in a protected ring configuration.

FIGURE 8.7 Add/drop multiplexer, multiplexer and transceiver used to create a CWDM ring with four nodes and protection.

FIGURE 8.8 Add/drop multiplexer, multiplexer and primary transceiver used to create a CWDM ring with four nodes and no protection.

FIGURE 8.9 MUX/DEMUX scaling from 1 to 16 wavelengths for full-spectrum CWDM.

FIGURE 8.10 Node with OADM connected Ethernet switch.

FIGURE 8.11 OADM connected to Ethernet with network fiber cut, compared to Figure 8.10.

FIGURE 8.12 Transparency between signals with different protocols. At the far end, not shown in this figure, is a DEMUX that separates the wavelengths after transmission with the individual protocol traffic unaffected by the other signals in the same path.
FIGURE 8.13 Path to extended storage networks with CWDM: (a) single fiber connection, (b) two channels, and (c) three channels multiplexed.
FIGURE 8.14 SONET link with a single wavelength at 1310 nm.
FIGURE 8.15 Adding CWDM wavelengths to a single wavelength link. Here, a common solution is shown for adding IP (Ethernet router) or fiber channel data services to a SONET link without a major expensive upgrade.
FIGURE 8.16 Using DWDM (lower part of figure) multiplexed on CWDM to provide a capacity upgrade for the network of Figure 8.15.

Chapter 9

FIGURE 9.1 Three-step evolution to today's HFC networks with (a) a small local office serving a few homes via coax lines, (b) to larger numbers of subscribers by use of more and more RF amplifiers, and (c) very large areas with hundreds of thousands of subscribers served by a combination of fiber and coax distribution.
FIGURE 9.2 Primary distribution for CATV networks based on an optical star topology distributing the RF signals $f_1...f_n$ via fiber to remote nodes.
FIGURE 9.3 Impact of laser threshold on signal output of directly modulated laser. The large driving signal causes clipping at the output.
FIGURE 9.4 Drive current and laser output for a nonlinear P-I characteristic.
FIGURE 9.5 Trunk network with externally modulated 1550-nm transmitter, amplifier for reach extension, and subsequent splitting network distributing the RF signals to the nodes.
FIGURE 9.6 A narrowcast implementation with only eight CWDM narrowcast optical channels λ_1 to λ_8. At each hub, a portion of the λ_{trunk} is split off and λ_n is dropped completely. These are combined and sent to the HFC node in the neighborhood.
FIGURE 9.7 Return path aggregation using different CWDM wavelengths (1...N) for the upstream.
FIGURE 9.8 Interaction of two channels spaced 100 nm apart, for example, 1470 and 1570 nm CWDM channels by SRS. The short wavelength channel acts as a pump for the longer wavelength, resulting in nonlinear crosstalk; g_{12} is the Raman gain coefficient and reaches its maximum at wavelengths that are 100 nm longer than the pump.
FIGURE 9.9 Measured SRS crosstalk as a function of modulation frequency for the 1570-nm channel in the presence of 1470-nm channel, 14 km SSMF, 3 dBm/channel, according to results in Ref. 8.
FIGURE 9.10 Measured SRS crosstalk as a function of CWDM channel spacing and launch power, high powers and wide channel spacing degrade the transmission performance, according to Ref. 8.

Chapter 10

FIGURE 10.1 FTTx approaches — from FTTCab/C to FTTH/U.
FIGURE 10.2 P2P and P2MP concept for FTTx.
FIGURE 10.3 Active and passive access networks.
FIGURE 10.4 A PON model.

FIGURE 10.5 TDM downstream and TDMA upstream in TDM-PON.
FIGURE 10.6 Single fiber bidirectional transmission using one single fiber.
FIGURE 10.7 Network overlay combining CWDM and the TDM-PON.
FIGURE 10.8 Central office design using the single-fiber and double-fiber approach.
FIGURE 10.9 Outside plant for distribution using demultiplexer and power splitter.
FIGURE 10.10 Two different ONU designs.
FIGURE 10.11 Spectrum of an upgrade for a 2-channel legacy network using a CWDM multi-wavelength extension.
FIGURE 10.12 Possible upgrade designed over a legacy infrastructure.
FIGURE 10.13 Deployment of a CWDM solution for G/EPON and P2P links.
FIGURE 10.14 Modular ONU design.
FIGURE 10.15 Proposed service overlay infrastructure of the TDM-PON.
FIGURE 10.16 Details of a multi-service ONU design.
FIGURE 10.17 Typical bandwidth required for broadband applications.
FIGURE 10.18 Multi-carrier model and service provider selection.
FIGURE 10.19 ONU for single and multiple service reception.
FIGURE 10.20 CWDM overlay on a legacy EPON/GPON.
FIGURE 10.21 CWDM ring for network overlay.
FIGURE 10.22 CWDM to enhance connectivity in RoF networks.
FIGURE 10.23 CWDM combined with D/UDWDM.
FIGURE 10.24 Routing profile of DWDM + CWDM using the FSR of the DWDM AWG router.

Preface

Although CWDM technology is reaching broad acceptance in the telecommunications industry, with the exception of some review articles and white papers, most material currently available on CWDM is isolated to device-related or specific system topics in journal articles and conference papers. As the technology progressed and the concept of CWDM moved toward commercialization, the idea emerged to combine the various isolated aspects of CWDM into a comprehensive book. In 2003, Ray Bonceck of OFS started preparations to collect the various aspects of CWDM in a single place and put them into perspective. When he was not able to continue this project, we were asked to step in and continue it, since we had been working on various aspects of CWDM previously. Taking the lead in this unique project was both challenging and interesting. Marcus had been working on CWDM-related systems since 1985 as a systems and design engineer at Quante Corp, Raynet Corporation, and Ericsson Telecom. For the past 8 years, he has been president/CEO of Fiber Network Engineering, a company that designs and develops optical transmission products for Datacom, Telecom, and CATV applications. After earning a Ph.D. in Electrical Engineering from University College London in 2000, Hans started as a member of technical staff at Bell Laboratories, Crawford Hill, New Jersey, working on optical transmission and soon moved into the area of metro/CWDM. Here, he has performed several groundbreaking experiments on CWDM, and is now with Siemens where he is leading efforts in optical transmission link and network optimization.

It soon became clear that there was a tremendous amount of material waiting to be organized into a comprehensive description of the technology and its applications in networking. We had to decide the focus of the book, because certain topics such as transceivers for CWDM and metro turned out to be extensive topics by themselves. We decided on a book structured to cover both CWDM technologies and applications, integrating component issues, system architectures, and concepts for extensions and upgrades as well as practical applications of CWDM. Through the participation of the contributing authors, this project has finally been possible. For this we offer our heartfelt thanks.

Although the book is intended to be self-contained, starting with standards, components for CWDM, and then moving to applications, each chapter provides sufficient information to be used independently and is followed by references to relevant papers and articles for further study. The intended readers of this book range from graduate students to development engineers seeking a broad overview of systems as well as some detail about components and applications related to CWDM. The concept followed in the chapters assumes a certain familiarity of the reader with the fundamentals of optical communications. We hope that this book will provide the reader with practical information and serve for a better understanding of the recent existing developments in the dynamic field of CWDM.

Acknowledgments

This work would not have been possible without the help and support of many people. We would like to thank the publishers of Taylor & Francis for making this project possible and providing help and support throughout the creation process. Taisuke Soda and Jessica Vakili were extremely helpful in the process of publishing. The original idea for this book came from Ray Bonceck while he was at OFS in Norcross, GA. The basic structure of this book is based on his ideas, which we developed further and invited the authors to make contributions to these topics. Our authors also deserve credit for their excellent work and their willingness to devote their time to this project despite sometimes tight deadlines. Without their strong commitment to this project this work would never have been possible. They are Mike Hudson (CWDM Standards), Lars Grüner-Nielsen, David Peckham, and Kai Chang (Optical Fibers to Support CWDM), Ralf Lohrmann (WDM Filters for CWDM), Charles Ufongene (Optimizing CWDM for Nonamplified Networks), Leo Spiekman (Amplifiers for CWDM), Peter Winzer (CWDM Upgrade Paths and Toward 10 Gb/s), Jim Aldridge (CWDM in Metropolitan Networks), Jim Farina (CWDM in CATV/HFC Networks), and Josep Prat and Carlos Bock (CWDM for Fiber Access Solutions). In addition to our authors, we want to thank all the technical reviewers who offered us their help when compiling the book: Bob Norwood (University of Arizona), Dan Parsons (Broadlight), Robert Blum (Gemfire), Tran Muoi (OCP, Inc.), Michael Eiselt (ADVA), Joachim Vathke (HHI), Larry Folzer (Turin), Frank Levinson (Finisar), Alka Swanson (Princeton Lightwave), Chien Yu Kuo (ADI), and Ken Ahmad and Dave Piehler.

Last but not least we would like to thank our families for their understanding and continued support throughout this busy time.

1 CWDM Standards

Mike Hudson

CONTENTS

1.1 Introduction 1
1.2 ITU-T Recommendation G.694.2 2
1.3 ITU-T Recommendation G.695 3
    1.3.1 Application Code Nomenclature 3
    1.3.2 Black Box and Black Link Approaches 5
        1.3.2.1 Black Box 5
        1.3.2.2 Black Link 7
    1.3.3 Unidirectional and Bidirectional Transmission 8
    1.3.4 Topologies 8
    1.3.5 Power Budget 10
    1.3.6 Distance 11
    1.3.7 Path Penalty 13
    1.3.8 Channel Plans 13
    1.3.9 Center Wavelength Deviation 13
    1.3.10 Application Code Summary 15
1.4 ITU-T Recommendation G.671 15
References 17

1.1 INTRODUCTION

Agreed technical standards, in many industries, benefit both producers and consumers by stimulating the uptake of a new technology, reducing costs, and increasing the overall market size for the technology. Coarse wavelength division multiplexing (CWDM) standards have allowed the manufacturers of CWDM lasers and optical filters to commit to mass production of a small range of standardized components that they know will be used by all of their customers, and thereby achieve economies of scale. Manufacturers of telecommunication systems equipment who incorporate standardized components into their designs can reduce their design effort and materials costs. Purchasers of systems that comply with the standards directly benefit from the flow-on of component and system cost reductions, but they also benefit from knowing that there is a reduced technical risk when adopting the new technology, and that there is a higher likelihood of interoperability of equipment supplied by different vendors. Interoperability is important for network operators because they can avoid becoming locked in to buying from just one supplier. This freedom to purchase from any supplier stimulates price competition among the vendors and reduces long-term support risks as suppliers discontinue product lines or go out of business. Interoperability benefits the suppliers because they know that even if they do not win a contract to supply equipment for a new network, they can still offer their products for network expansion, upgrades, and ongoing maintenance.

There are several international standards for CWDM components, systems, and applications. The most important are:

- ITU-T Recommendation G.694.2 (2003), Spectral Grids for WDM Applications: CWDM Wavelength Grid.
- ITU-T Recommendation G.695 (2004), Optical Interfaces for Coarse Wavelength Division Multiplexing Applications.
- ITU-T Recommendation G.671 (2002), Transmission Characteristics of Optical Components and Subsystems.

This chapter is by no means a substitute for a thorough review of the individual standards. Rather, it is intended to provide an overview of the main features of each standard and some historical perspective of decisions made at the time of their development. These standards, the concepts behind them, and their implications for component vendors, system vendors, and network operators will be examined here.

1.2 ITU-T RECOMMENDATION G.694.2

ITU-T Recommendation G.694.2, *Spectral Grids for WDM Applications: CWDM Wavelength Grid,* specifies the wavelengths to be used for CWDM. As it did for dense wavelength division multiplexing (DWDM), standardizing the set of wavelengths to be used for CWDM was a very important first step in the proliferation of CWDM technology into the marketplace because it allowed suppliers to avoid having to make a vast range of components with slightly different specifications and it allowed network operators to proceed with confidence that interoperability would be achievable. The original version of G.694.2, published in June 2002, defined a CWDM wavelength grid with 20-nm spacing beginning at 1270 nm and ending at 1610 nm. At that time, and still today, most manufacturers of CWDM lasers were making lasers that complied with this wavelength grid when operated at a temperature of 25°C with a typical manufacturing variation of ±3 nm, or ±2 nm for reduced tolerance parts. When this manufacturing tolerance is combined with a typical CWDM laser temperature variation of +0.1 nm/°C over an operating temperature range of 0 to 70°C, this gives CWDM lasers a central wavelength variation of fully 13 nm since from each of the wavelengths within the ±3 nm interval the laser wavelength will shift by 7 nm towards longer wavelength with increasing temperature. Since the laser wavelengths are specified for 25°C, the actual wavelength at 0°C (shortest wavelength) is shifted by −2.5 nm so that the window becomes asymmetric, typically –5.5 to +7.5 nm, with respect to the CWDM channel. For example, in the 1550-nm CWDM channel, the laser wavelength would be located between 1544.5 and 1557.5 nm at any temperature. As a consequence, CWDM filters were being specified with a 1-nm offset from the G.694.2 wavelengths and a ±6.5 nm passband, although filters with slightly narrower and wider passbands were also available and remain available today. Figure 1.1 shows the laser wavelength drift with respect to the passband. Therefore, one has to distinguish carefully between the CWDM channel, for example, 1550-nm channel, and the actual center wavelength of the equipment used for that particular channel.

In recognition of the fact that CWDM components and systems were effectively operating with a +1 nm offset and with a desire to avoid specifying an asymmetric central wavelength deviation value in G.695, in December 2003 the ITU-T published a revision of G.694.2 that specified the set of 18 CWDM channels shown in Table 1.1. Note that it may be possible to have CWDM systems that use wavelengths outside the range defined in G.694.2, although their use is somewhat impractical. Wavelength channels shorter than 1270 nm will suffer from dramatically increasing attenuation as one approaches the cable cut-off wavelength for G.652 fiber, 1260 nm. Wavelengths longer than 1610 nm will suffer from increased attenuation and bending loss of the fiber. In addition, the responsivity of the photodetectors falls off dramatically beyond 1610 nm so that both effects severely decrease the available link budget for the CWDM system.

1.3 ITU-T RECOMMENDATION G.695

ITU-T Recommendation G.695, *Optical Interfaces for Coarse Wavelength Division Multiplexing Applications*, was first published in February 2004. G.695 builds upon the basic wavelength grid that was agreed 2 years earlier in G.694.2 adding the detailed information necessary for application interoperability that could be agreed at the time G.694.2 was produced. It specifies the optical interface parameters for several applications, each having different numbers of channels, bit rates, distances, and fiber types. In particular, applications were specified, for example, transceivers, line cards, switch blades running SONET, Ethernet, Fiber Channel, and PSDH rates plus the sub-groupings of channel plans with SX, LX, LH for Ethernet and SR, and IR and LR for SONET systems. G.695 allows for two different approaches to the engineering of CWDM links, the so-called black box and black link approaches.

1.3.1 APPLICATION CODE NOMENCLATURE

In all of the ITU-T's optical interface recommendations, "application codes" are used as a kind of shorthand to unambiguously describe a particular kind of interface. This is analogous to Ethernet, where 10BASE-T, 100BASE-FX, and 1000BASE-SX all have very particular meanings that are well known by everyone in the business. An application code is a special alphanumeric code that allows a vendor, systems integrator, or network operator to unambiguously specify, in a

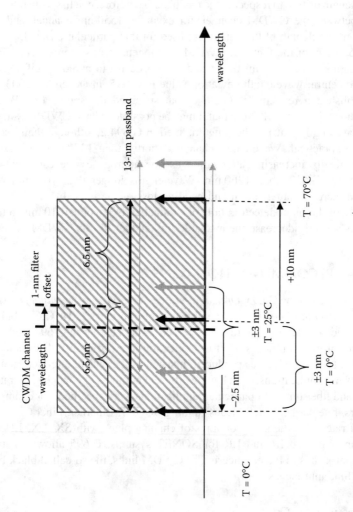

FIGURE 1.1 Laser wavelength drift and filter passband requirements for a CWDM channel.

TABLE 1.1
G.694.2 CWDM Wavelength Grid (Defined by the Center Filter Wavelengths)

Channel Number	Nominal Central Wavelength (nm)
1	1271
2	1291
3	1311
4	1331
5	1351
6	1371
7	1391
8	1411
9	1431
10	1451
11	1471
12	1491
13	1511
14	1531
15	1551
16	1571
17	1591
18	1611

compact, well-known form, the optical parameters for a CWDM link or interface. G.695 defines application codes for CWDM interfaces and these application codes are then used throughout G.695 when specifying the optical parameters for each kind of CWDM interface. The application code contains digits and letters that describe the application, as illustrated in the example of Figure 1.2, Table 1.3. Finally, a list of all of the application codes (i.e., the different kinds of CWDM interfaces for different applications) that have been standardized in G.695 is provided at the end of this chapter.

1.3.2 BLACK BOX AND BLACK LINK APPROACHES

There are two fundamental transmission interface reference models used in G.695. They are called the "black box" approach and the "black link" approach.

1.3.2.1 Black Box

In the black box model, shown in Figure 1.3, the CWDM lasers and the multiplexing filters are lumped together in a notional or actual black box. Similarly, the demultiplexing filters and receivers are lumped together in another black box. Physical realizations of the black box approach may put the CWDM lasers and optical multiplexer into a single physical unit with internal spliced fiber connections

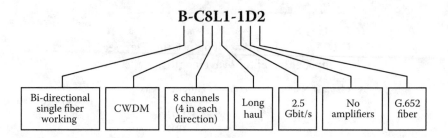

FIGURE 1.2 G.695 application code example.

between the CWDM lasers and the multiplexer, or they may have individual transponder cards and a separate optical multiplexer unit with patch cords joining all the components together. In either case, the G.695 black box interface is always taken to be the multichannel signal at the output of the optical multiplexer or at the input of the demultiplexer. The black box's internal optical signals between the CWDM lasers and the multiplexer and between the demultiplexer and the receivers are not standardized in G.695 for black box applications.

The black box approach is important because it allows system designers to make the trade-offs with attenuation at different wavelengths combined with the multiplexer/demultiplexer loss at each wavelength. This enables designers to optimize the system for maximum reach, compact size, and improved thermal performance. Because of its "plug-and-play" nature, a system that uses the black link approach can eliminate the need for detailed optical engineering by the network operator, making it easy to reliably deploy and use CWDM technology.

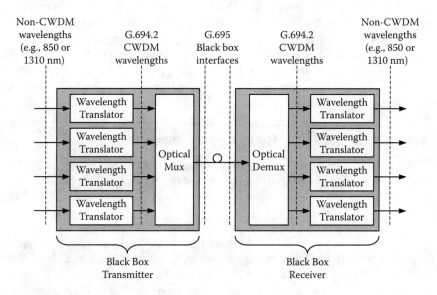

FIGURE 1.3 Black box transmitter and receiver.

CWDM Standards

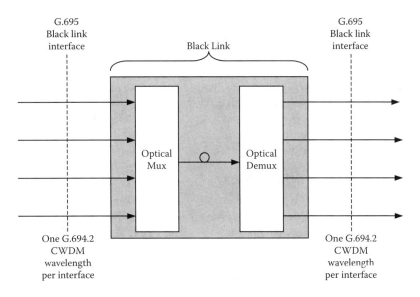

FIGURE 1.4 Black link concept.

1.3.2.2 Black Link

The concept of a black link, shown in Figure 1.4, was introduced into G.695 to describe the concatenation of an optical multiplexer, an optical fiber cable, and an optical demultiplexer, along with all the associated optical connectors, patch cords, and splices. The black link model allows the optical characteristics to be defined at the single channel interface between a CWDM transceiver, typically a CWDM small form-factor pluggable (SFP) or gigabit interface converter (GBIC) transceiver* and the tributary inputs of an optical multiplexer or the tributary outputs of an optical demultiplexer. Because G.695 does not specify details about the multiplexer or the demultiplexer, the black link approach allows the combined insertion loss of the multiplexer/demultiplexer pair to be optimized for each CWDM wavelength. For example, the combined insertion loss for a 1470-nm signal can be made to be less than the insertion loss for a 1550-nm signal, thereby compensating for the fact that the attenuation of the optical fiber is greater at 1470 nm than it is at 1550 nm, and so the overall insertion loss for each wavelength on the black link can be minimized to provide longer reach. The downside of the black link approach is that in order to achieve maximum performance like maximum transmission distance for every expected suppliers' components over the expected temperature range for the lifetime of the system, the user must do the necessary optical engineering calculations for every CWDM link in their network.

A black link can be used to provide virtual dark fiber services for several users, with each service operating at a different CWDM wavelength. Application codes

---

* For transceiver standards, See Chapter 3, Section 3.5.2.

for black links are prefixed with "S-" (Single Channel Interface Specifications). Bidirectional black link applications have not yet been standardized in G.695.

1.3.3 Unidirectional and Bidirectional Transmission

G.695 standardizes unidirectional applications requiring two fibers for both-way transmission (dual fiber working) and bidirectional applications that use a single fiber for both directions of transmission (single fiber working). Application codes for bidirectional applications are prefixed with "B-" (Bi-directional application).

1.3.4 Topologies

The 2004 version of G.695 only addresses point-to-point CWDM links, as shown in Figure 1.5, that is, no optical ring topologies with add/drop multiplexing (OADM) or optical networking. The second version of G.695, due for publication in early 2005, will extend the black link concept to cater for linear CWDM chains with OADMs (see Figure 1.6) and CWDM rings with OADMs (see Figure 1.7 and Figure 1.8). This extension of the standard will be accomplished by adding an interferometric cross-talk specification for the demultiplexed CWDM channel signals. Interferometric crosstalk is a measure of how much the remaining signal power from an imperfectly added CWDM channel (wavelength) interferes with a dropped signal on the same CWDM channel. Currently, it is thought that a value of 45 dB is sufficient to account for inteferometric crosstalk since the optical power penalty is less than 0.25 dB and is therefore negligible when engineering black links. All other black link parameters, like transmit and receive power levels, dispersion tolerance, and so on, will remain the same as in the first version of G.695, to ensure

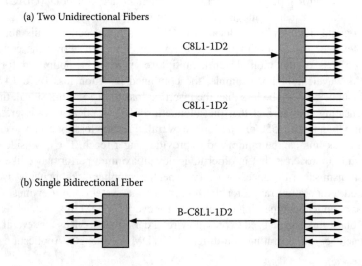

FIGURE 1.5 Point-to-point CWDM links.

CWDM Standards

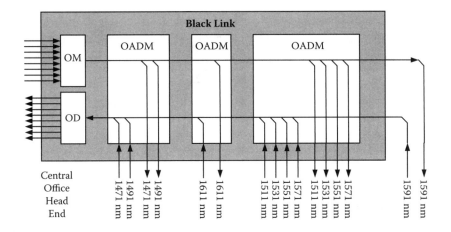

FIGURE 1.6 Linear add-drop CWDM black link. OM/OD, optical multiplexer/demultiplexer.

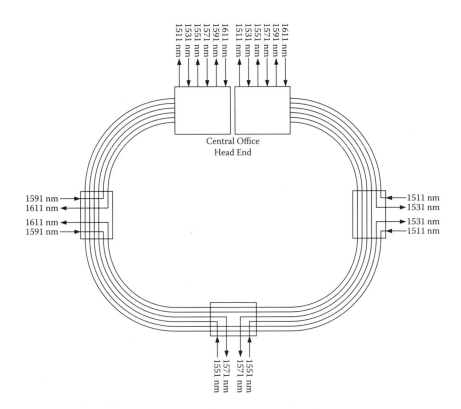

FIGURE 1.7 CWDM black link ring (hubbed).

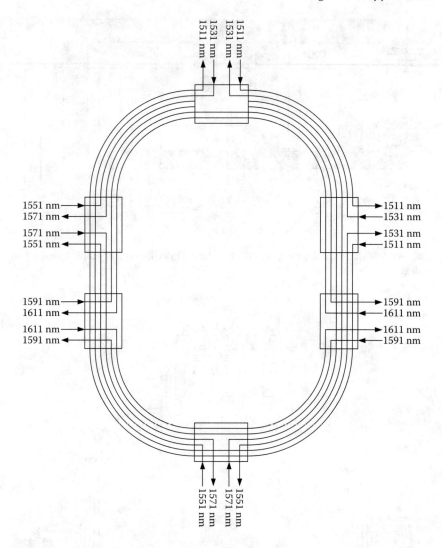

FIGURE 1.8 CWDM black link ring (nonhubbed).

backward compatibility with components and equipment that complies with the first version of G.695. Engineering considerations for CWDM linear chains and rings will also be addressed in future revisions of the standard so that engineers are aware of the issues involved in designing networks with these new topologies.

1.3.5 Power Budget

Power budgets are generally considered the starting point of any optical link or network design. The power budgets used during the development of each application

CWDM Standards

code were based on surveys of commonly available 1.25- and 2.5-Gb/sec CWDM lasers, receivers, SFF and SFP transceivers, and CWDM multiplexers and demultiplexers. The following characteristics were agreed as being representative of the surveyed components, and therefore a suitable basis for standardization:

- Laser output power: 0 to +5 dBm, to allow for the use of uncooled, direct modulated lasers
- 8 Channel CWDM multiplexer insertion loss: 1 to 3.5 dB, to allow for a variety of multiplexer designs and potential future improvements in filter design and fabrication technology
- 8 Channel CWDM demultiplexer insertion loss: 1 to 4.0 dB, again to allow for a variety of demultiplexer designs and advances in filters.
- 2.5 Gb/s P-intrinsic-n (PIN) receiver input power: −18 to 0 dBm, to cater for the typical range of sensitivity and overload values seen in low cost PIN receivers
- 2.5 Gb/s avalance photodiode (APD) receiver input power: −28 to −9 dBm
- 1.25 Gb/s APD receiver input power: −30 to −9 dBm

1.3.6 DISTANCE

The achievable transmission distance for each application code depends on the maximum link attenuation specified for the application code and the expected attenuation of the fiber at each wavelength. In the case of black link applications, the insertion loss of the optical multiplexer and demultiplexer at each wavelength also determine the achievable reach. Black box applications include the optical multiplexer and demultiplexer within the black box and so the allowed attenuation between the output of one black box (not the laser!) and the input of another black box (not the optical receiver) is lower than for black link applications which measure the link attenuation between the laser and the optical receiver. However, if the same lasers, multiplexer, fiber, demultiplexer and receivers were used in either a black box implementation or in a black link deployment, then the reach would be the same for both the black box and the black link even though the allowed link attenuation for the black link may be, for example, 7.5 dB, more than for the black boxes.

In some deployments, for example, those with lower than average fiber loss, the reach may be dispersion limited rather than being attenuation limited and may significantly exceed the target distances that are stated in G.695. The insertion losses of the optical multiplexers and demultiplexers that are used in black link applications are not standardized and so it is not possible for G.695 to state a target distance for black links. Instead, G.695 provides an appendix that shows the reach that can be expected for various insertion loss values of the multiplexer/demultiplexer pair. To assist component, system and network designers, G.695 has an appendix that shows the assumed minimum and maximum attenuation coefficient values at each wavelength that were used while specifying the parameter values

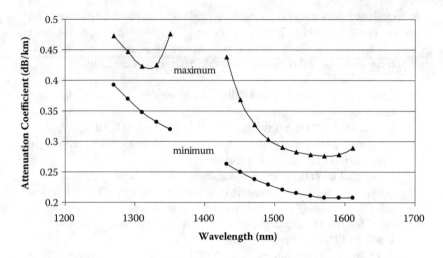

FIGURE 1.9 Assumed attenuation coefficients for non-low-water-peak fibers.

for each application code in G.695. Separate attenuation coefficient values were used for "wet" and "dry" fibers, as shown in Figure 1.9 and Figure 1.10.

Note that optical fiber contains even tiny traces of OH⁻ ions (i.e., water) and has quite high attenuation at or near 1385 nm. This type of fiber is commonly called "wet" fiber. Most of the older optical fiber, typically, fiber installed before 2000, will have this "water peak" in its attenuation profile, and therefore will not support longer transmission spans of all of the G.694.2 wavelengths. Optical fiber that is specifically manufactured to have a low or completely absent water peak is often called "dry" fiber. Most manufacturers of optical fiber now produce "dry"

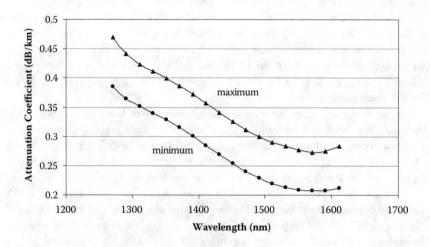

FIGURE 1.10 Assumed attenuation coefficients for low or zero water-peak fibers.

CWDM Standards 13

fiber that is suitable for use with all wavelengths in the CWDM wavelength grid. More about ITU specified fiber types can be found in Chapter 2, Section 2.3.

1.3.7 PATH PENALTY

CWDM has long been promoted on the basis of the inherent low cost that can be achieved by combining uncooled directly modulated lasers and optical filters with a wide passband and inter-channel guard-band. In most of the ITU-T optical transmission recommendations that were developed before G.695, it was customary to specify a 1-dB path penalty for low dispersion systems and 2 dB for high dispersion systems. The chromatic dispersion of optical fiber is significantly different at each CWDM wavelength, as shown in more detail in Chapter 2. At the longest CWDM wavelengths, chromatic dispersion is quite large, typically up to 21.2 ps/nm/km at 1610 nm. Based on technical contributions from a number of the participating experts involved in the development of G.695, it was decided that for a bit-error ratio (BER) of 1×10^{-12}, the benchmark path penalties would be 1.5 dB for 2.5 Gb/sec applications with chromatic dispersion up to about 1000 ps/nm (corresponding to approximately 60 km distance at 1550 nm) and 2.5 dB for 2.5 Gb/sec applications with chromatic dispersion greater than about 1000 ps/nm. In addition, for long reach 1.25 Gb/sec, a 1.5 dB path penalty value was agreed. In G.695, the path penalty includes the effects of chromatic dispersion, polarization mode dispersion (PMD), and optical crosstalk between the CWDM channels. Signal-to-noise ratio reduction due to optical amplification is not considered a path penalty.

1.3.8 CHANNEL PLANS

Different sets of G.694.2 wavelengths are used for applications that use 4, 8, 12, or 16 CWDM channels. The wavelengths were chosen to maximize the reach. In the case of 12-channel applications, the wavelengths in the E-band near the fiber's OH⁻ absorption peak at 1385 nm, which experience relatively high attenuation, were avoided so that both older "wet" and newer "dry" fibers can be supported by the same set of 12-channel application codes. The channel plans used in G.695 are shown in Table 1.2.

1.3.9 CENTER WAVELENGTH DEVIATION

During the development of G.695, there was difficulty in reaching agreement on a single value for the center wavelength deviation parameter as already presented in Figure 1.1. Two values were proposed: ±6.5 nm (i.e., 13-nm passband) and ±7 nm (i.e., 14-nm passband). Support for a ±6.5 nm passband was based on lasers with a manufacturing variation of ±3 nm, an operating temperature range of 0 to 70°C and a wavelength drift of 0.1 nm/°C, and the expectation that optical filters with a 13 nm passband would be less expensive than 14 nm or wider filters. The main argument in support of ±7 nm was that it would allow for an extra 5°C of operating

TABLE 1.2
G.695 Channel Plans

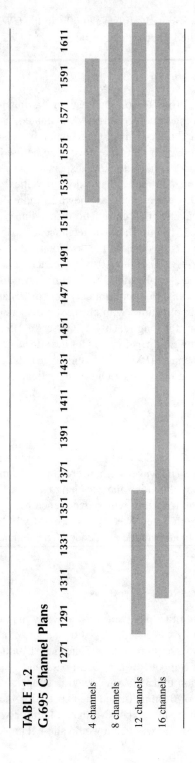

CWDM Standards 15

temperature range for the CWDM lasers in support of an extended temperature range or, alternatively, to allow for manufacturing variations when incorporating lasers into manufactured systems. Eventually an "11th hour" compromise was reached where the standard would specify ±6.5 nm but with a note that ±7 nm equipment and components may exist and that special care must be taken when deploying a mixture of standard ±6.5 and ±7 nm equipment or components.

1.3.10 Application Code Summary

Table 1.3 summarizes the application codes that have been standardized in G.695. A variety of 4-, 8-, and 12-channel CWDM applications have been fully specified in G.695. Sixteen channel CWDM applications have been tentatively included in an appendix to G.695, but these parameter values have not yet been fully ratified by the ITU-T because it was felt that not enough practical component specification data, measurements, test results, and real-life experience had been presented to the ITU-T working group from component vendors, system vendors, and users. The ITU-T is awaiting contributions from relevant experts who can confirm that the tentative values are practical before moving them out of an informative appendix into the normative (ratified) body of the G.695 recommendation.

1.4 ITU-T RECOMMENDATION G.671

ITU-T Recommendation G.671, *Transmission Characteristics of Optical Components and Subsystems,* identifies optical transmission-related parameters for the optical components that are commonly used in optical telecommunications systems. It also specifies values for these parameters and specifies testing methods, usually by reference to IEC standards. The optical components addressed by G.671 include attenuators, splitters, connectors, filters, isolators, splices, switches, dispersion compensators, multiplexers and demultiplexers, and OADMs.

For CWDM multiplexers and demultiplexers, the following parameters are defined:

- Channel insertion loss: The reduction in optical power between an input and an output of a CWDM device.
- Channel insertion loss deviation: The maximum variation of insertion loss across the wavelength range of a CWDM channel.
- Reflectance: The ratio of reflected power to incident power at a port.
- Polarization dependent loss (PDL): The maximum variation of insertion loss due to the polarization state.
- Polarization dependent reflectance: The maximum variation in reflectance due to a variation of the state of polarization.
- Allowable input power: The maximum optical power that can be applied to an optical component without causing damage.
- PMD: Is usually described in terms of a differential group delay (DGD), which is the time difference between different polarization

TABLE 1.3
G.695 Application Code Summary

Application Code	Number of Channels[a]	Bit Rate Class[b]	Fiber Type[c]	Target Distance[d] (km)
Unidirectional Black Box				
C4S1-1D2	4	2.5G	G.652	?
C4S1-1D3	4	2.5G	G.653	?
C4S1-1D5	4	2.5G	G.655	?
C4L1-1D2	4	2.5G	G.652	?
C4L1-1D3	4	2.5G	G.653	?
C4L1-1D5	4	2.5G	G.655	?
C8S1-1D2	8	2.5G	G.652	27
C8L1-1D2	8	2.5G	G.652	55
C16S1-1D2	16	2.5G	G.652 C or D	20
C16L1-1D2	16	2.5G	G.652 C or D	42
Bidirectional Black Box				
B-C4L1-0D2	2 +2	1.25G	G.652	90
B-C4L1-0D3	2 +2	1.25G	G.653	90
B-C4L1-1D2	2 +2	2.5G	G.652	80
B-C4L1-1D3	2 +2	2.5G	G.653	83
B-C8S1-1D2	4 +4	2.5G	G.652	27
B-C8L1-0D2	4 +4	1.25G	G.652	64
B-C8L1-0D3	4 +4	1.25G	G.653	64
B-C8L1-1D2	4 +4	2.5G	G.652	55
B-C8L1-1D3	4 +4	2.5G	G.653	58
B-C12L1-0D2	6 +6	1.25G	G.652	42
B-C12L1-1D2	6 +6	2.5G	G.652	38
B-C16S1-1D2	8 +8	2.5G	G.652 C or D	20
B-C16L1-1D2	8 +8	2.5G	G.652 C or D	42
Unidirectional Black Link				
S-C8S1-1D2	8	2.5G	G.652	e
S-C8S1-1D3	8	2.5G	G.653	e
S-C8S1-1D5	8	2.5G	G.655	e
S-C8L1-1D2	8	2.5G	G.652	e
S-C8L1-1D3	8	2.5G	G.653	e
S-C8L1-1D5	8	2.5G	G.655	e

[a]The number of channels for bidirectional applications is expressed as $N + N$, where N is the number of channels in each direction. For example, a bidirectional application with four channels in each direction (4 +4) uses eight different wavelengths.

[b]A bit rate class of 1.25G means any bit rate from nominally 622 Mb/s to nominally 1.25 Gb/s. A bit rate class of 2.5G means any bit rate from nominally 622 Mb/s to nominally 2.67 Gb/s.

[c]G.652 fiber is standard single mode optical fiber (e.g., SMF-28). It may be either "wet" fiber (i.e., G.652 A or G.652 B) or "dry," "low water peak," or "zero water peak" fiber (i.e., G.652 C or G.652 D). G.653 fiber is dispersion-shifted single mode fiber. G.655 fiber is nonzero dispersion-shifted single mode fiber. "Dry" fiber must be used for the 16-channel applications.

[d]Target distances are not specified in G.695, they are only provided as indicative values.

[e]The expected reach for black link applications depends on the insertion losses of the optical multiplexer and optical demultiplexer at each wavelength.

states of an optical signal. DGD is usually not important for signals up to 2.5 Gb/sec, but becomes quite important at 10 Gb/sec and faster.
- Channel wavelength range: The range of wavelengths that belong to a CWDM channel. Usually expressed as the nominal central wavelength plus/minus the central wavelength deviation. For example, the channel wavelength range for the 1551-nm CWDM channel is 1551 ± 6.5 nm, which means any wavelengths from 1544.5 to 1557.5 nm.
- Ripple: The variation in insertion loss within a CWDM channel's wavelength range.
- Adjacent channel isolation: The difference in insertion loss between a channel that is passed by a CWDM filter and the CWDM channels that use either the next lower or next higher nominal central wavelength.
- Nonadjacent channel isolation: The difference in insertion loss between a channel that is passed by a CWDM filter and any CWDM channel other than the adjacent channels.
- Bidirectional (near-end) isolation: Because bidirectional CWDM multiplexer/demultiplexer devices have both input channels and output channels at the same side of the device, input light for one direction can appear on the output port for the other direction.
- Unidirectional (far-end) crosstalk attenuation: A measure of the part of the optical power of each wavelength exiting from a CWDM demultiplexer port at wavelengths different from the nominal wavelength range for that port.
- Bidirectional (near-end) crosstalk attenuation: In a bidirectional CWDM multiplexer/demultiplexer, this is a measure of the optical power of each wavelength entering a multiplexer input port that exits a demultiplexer output port on the same bidirectional multiplexer/demultiplexer.

Unfortunately, the currently published version of G.671 (June 2002) does not specify the values for any of these parameters, all of them being "for further study," and so there is some way to go before CWDM optical components are fully standardized.

REFERENCES

1. ITU-T Recommendation G.694.2 (2003), *Spectral Grids for WDM Applications: CWDM Wavelength Grid.*
2. ITU-T Recommendation G.695 (2004), *Optical Interfaces for Coarse Wavelength Division Multiplexing Applications.*
3. ITU-T Recommendation G.671 (2002), *Transmission Characteristics of Optical Components and Subsystems.*

2 Optical Fibers to Support CWDM

Kai H. Chang, Lars Grüner-Nielsen, and David W. Peckham

CONTENTS

2.1 Introduction to Optical Fibers 20
2.2 Fiber Properties and Effects on CWDM System Performance 21
    2.2.1 Fiber Attenuation 21
    2.2.2 Bending-Induced Loss 22
        2.2.2.1 Macrobending Loss 23
        2.2.2.2 Microbending Loss 24
    2.2.3 Chromatic Dispersion 27
        2.2.3.1 Dispersion Effects 28
        2.2.3.2 Dispersion Effect when Transmitter Has Chirp 29
    2.2.4 Polarization Mode Dispersion 31
    2.2.5 Cut-Off Wavelength 31
        2.2.5.1 Theoretical Cut-Off Wavelength 32
        2.2.5.2 Effective Cut-Off Wavelength 32
    2.2.6 Nonlinear Effects 34
    2.2.7 Geometric Properties 35
2.3 Overview of Common Transmission Fibers Used for CWDM 37
    2.3.1 Standard Single Mode Fiber (ITU-T G.652) 37
        2.3.1.1 Low Water Peak and Zero Water Peak Fiber 37
        2.3.1.2 Low Bend Loss Fiber Designs 38
    2.3.2 Nonzero Dispersion-Shifted Fiber (ITU-T G.655) 39
    2.3.3 NZDSF for Wideband Optical Transport (ITU-T G.656) 41
2.4 Zero-OH$^-$ Single Mode Fibers for CWDM Applications 41
    2.4.1 Introduction 41
    2.4.2 Manufacturing Process for Zero-OH$^-$ AllWave® Fiber 43
    2.4.3 The RIC-ODD Process and AllWave Fiber Performance 44
        2.4.3.1 Fiber Loss 44
        2.4.3.2 Interface Quality and Fiber Strength 46
        2.4.3.3 Fiber PMD 46
        2.4.3.4 Fiber Geometry 46

            2.4.3.5 Hydrogen Aging Losses 47
            2.4.3.6 Conclusions 50
2.5 Dispersion-Compensating Fibers 50
      2.5.1 Basic Principles 50
      2.5.2 DCF for CWDM 51
      2.5.3 Raman-Pumped DCF 53
References 53

2.1 INTRODUCTION TO OPTICAL FIBERS

Much of the single mode optical fiber development effort during the 1990s was focused on optimizing fiber performance in long-haul transmission systems that carry OC-192 (10 Gb/s) signals and utilize dense wavelength division multiplexing (DWDM) and erbium-doped optical fiber technologies. Terrestrial long-haul DWDM systems initially operated within the 35-nm wide spectral band known as the C-band (1530 to 1565 nm) and electrical regeneration was typically required at intervals of 1000 km or less. Wavelength division multiplexing (WDM) technologies have currently advanced to support wider optical bandwidths across the S-, C-, and L-bands, increased signal rates of 40 Gb/s or higher and ultra long-haul, unregenerated transmission distances of greater than 1500 km.

The fiber design problem for wideband, high bit rate, long-haul systems is an optimization of the tradeoffs of fiber properties with the goals of (i) mitigating the deleterious effects of fiber nonlinearity, (ii) supporting the ability for wideband dispersion compensation, and (iii) supporting wideband, low noise, distributed Raman-gain. Of course, the optimization also required that the resulting fibers could be manufactured at reasonable costs and that the fiber could be spliced, cabled, and installed in the telecom environment with low, stable attenuation. The current generation of medium dispersion fibers (MDFs), such as TrueWave REACH, represents the state-of-the-art in long-haul fiber design.

As the telecommunications network capacity bottleneck shifted from long-haul networks to metro and access networks, the focus in network investments shifted to metro and access systems. Likewise, the fiber development focus shifted from supporting high ultimate capacity DWDM systems to coarse wavelength division multiplexing (CWDM) systems that are cost effective for carrying metro and access traffic. This work led to the widespread acceptance of full spectrum fibers that provide low loss over a broad wavelength range by eliminating the OH$^-$ absorption peak at 1385 nm and the sensitivity of fiber loss to exposure to molecular hydrogen.

Section 2.2 reviews fiber transmission properties and discusses how they relate to CWDM system performance. In Section 2.3, an overview of common transmission fibers used for CWDM systems will be given. In the following, Section 2.4 goes through development, properties, and manufacturing of the first zero-OH$^-$ transmission fiber, the OFS AllWave® fiber. Finally, Section 2.5 introduces dispersion-compensating fiber and possible application for CWDM.

2.2 FIBER PROPERTIES AND EFFECTS ON CWDM SYSTEM PERFORMANCE

2.2.1 Fiber Attenuation

Optical amplification technology is not generally utilized with CWDM systems because of the need for amplification over the broad wavelength band from 1310 to 1610 nm. Without low-cost, broadband amplification technology readily available and the power limitations imposed by low-cost directly modulated lasers (DMLs), the span budget may be limited to less than 30 dB, the span loss will limit the span distance to between 40 to 80 km [1]. Since for bit rates up to 2.5 Gb/s the power budgets of the higher loss channels will determine the reach of a CWDM system*, the spectral shape of fiber attenuation has a determining role in CWDM system design. More information on amplification in CWDM systems can be found in Chapter 6.

A typical spectral attenuation curve from 1000 to 1700 nm of a modern germano-silicate-based optical fiber on a 150 mm diameter bobbin is shown in Figure 2.1. The spectral shape is dominated by the λ^{-4} dependence of the Rayleigh scattering loss. In the vicinity of 1550 nm, the loss reaches the lowest value of around 0.185 dB/km. At 1310 nm, the loss curve has a local minimum of approximately 0.325 dB/km. At wavelengths longer than 1550 nm, the loss level rises due to packaging effects, such as the macrobending loss associated with the 150 mm diameter bends from spooling. There may be additional attenuation at long wavelength resulting from the "short wavelength tails" of infrared (IR) absorption loss, such as contamination from OH^-.

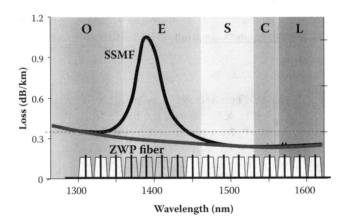

FIGURE 2.1 Comparison of spectral attenuation of ZWP fiber (ITU-T G.652D) with standard water peak single mode fiber (SSMF, ITU-T G.652). The ITU-proposed CWDM channels are overlaid, whereas those highlighted in grey indicate channels gained by using ZWP fiber.

---

* The impact of dispersion on CWDM lasers will be explored in Section 2.2.3 and also in Chapter 3.

At 1385 nm, overtones of the OH⁻ contamination-related molecular absorption manifest themselves in the well known "water peak." The 1385-nm loss can be further degraded by ambient temperature exposure of the fiber to H_2. The molecular hydrogen diffuses into the interstitial spaces in the glass matrix, reacting at chemically reactive glass defect sites to form absorbing species, such as SiOH, resulting in growth of the water peak with time. As will be discussed in detail later in this chapter, the growth in water peak is permanent and even at low levels of H_2 exposure at room temperature the magnitude can grow to 1 dB/km or more when certain processing-related atomic defects are present in the glass structure.

Progress in glass processing techniques during the late 1990s has led to reduction in OH⁻ contamination and the magnitude of the absorption peak. Figure 2.1 also illustrates the effects of these processing improvements. The attenuation curve labeled SSMF (standard single mode fiber) shows the increase in loss above the Rayleigh scattering level in the region around 1385 nm that is typical of SSMF. The curve labeled ZWP (zero water peak) shows the attenuation of a modern fiber with reduced OH⁻ contamination and the water peak effectively removed. Additional discussion of the topic of fiber attenuation and attenuation aging effects will be presented more fully in Section 2.4.

Furthermore, Figure 2.1 shows an overlay of the CWDM signal wavelengths on the spectral attenuation curve. The current international telecommunications union (ITU-T) proposal for CWDM systems includes placing 18 CDWM channels on a 20-nm spaced grid that spans the wavelength range from 1270 to 1610 nm, discussed in Chapter 1 in detail. ZWP fibers have low attenuation over the 1360 to 1440 nm spectral range that comprises the E-band, allowing efficient transmission on the four CWDM channels located within the E-band. Thus, the modern full spectrum fibers enable transmission on the 16 low-loss CDWM channels located over the wavelength range from 1310 to 1610 nm.

2.2.2 Bending-Induced Loss

Optical fiber loss can also be affected by the presence of axial bends or deformation. Bends that result in deflections of the fiber axis that are large compared to the fiber core diameter, for example, loops or the bending resulting from spooling, are referred to as macrobends. The smaller the radius of the macrobend, the higher is the resulting attenuation. In addition, small-scale deformations of the axis (small relative to fiber core diameter), for example, caused by pressing the fiber against a rough surface within a cable, result in attenuation through mode coupling effects and are referred to as microbends. Whereas macrobends are typically characterized by their radius of curvature, for example, 10 mm radius, the axis deformation of microbends is typically stocastic in nature and therefore characterized by the power spectrum of the axis deformation. Both the magnitude and the shape of the power spectrum of the microbend deformations affect the magnitude and the spectral shape of microbending loss.

2.2.2.1 Macrobending Loss

The following is a brief heuristic explanation of the macrobending loss phenomenon based on the tilted profile bending loss model proposed by Heiblum [2]. A mode of an optical fiber is a bound mode, that is, has a radial evanescent field in the cladding, when the longitudinal propagation constant, β, is greater than the plane-wave propagation constant of the cladding, $n_{clad} \cdot k$, where k is the free space propagation constant and nclad is the index of refraction of the cladding material. In other words, the inequality

$$\beta > n_{clad} \cdot k \quad (2.1)$$

must hold. By dividing both sides of Equation 2.1 by k, we have

$$\beta/k > n_{clad} \quad (2.2)$$

The ratio β/k is the fundamental mode effective index, which must be greater than the cladding index for the field in the cladding to decay radially. If the effective index of the mode is less than the cladding index, then the radial field solution in the cladding changes becomes oscillatory, resulting in radial propagation of energy.

Now consider a step index optical fiber with index profile shown in curve (a) of Figure 2.2 that is bent at a constant radius of curvature. The effective index

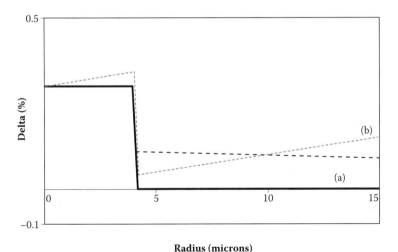

FIGURE 2.2 Refractive index profile of (a) a step index fiber bent at constant radius of curvature and (b) the equivalent straight fiber with "tilted" index profile. Dotted line is effective index of the fundamental mode.

of the fundamental mode is shown with the dotted line. Heiblum showed, by a coordinate system transformation using a conformal map, that a bent fiber with the index profile shown in Figure 2.2, curve (a), has equivalent behavior to a straight fiber with the tilted index profile shown in Figure 2.2, curve (b). The slope of the tilted profile is inversely proportional to the radius of curvature.

For the straight fiber with the tilted index profile at radii greater than Rc, the effective index is less than the cladding index. Therefore, the field becomes oscillatory in the radial direction and energy is carried away from the axis of the fiber, resulting in excess loss as energy leaks radially away from the fiber axis. The rate of energy loss depends on the fraction of the mode power that is outside the radiation caustic at radius Rc. The loss formula derived by Marcuse [3] for the rate of energy loss per unit length of fiber is

$$2\alpha = \frac{4}{\sqrt{\pi^3}} \frac{1}{\sqrt{\gamma R^3}} \frac{|I_1|^2}{I_2} \left| H_{\beta R}\left(\frac{2\pi n_{\text{clad}}}{\lambda}(R+a)\right) \right|^{-2}, \qquad (2.3)$$

where a is the core radius, R is the bend radius, $I_1 = \int_0^\infty E(z)\,dz$, $I_2 = \int_0^{2\pi} d\phi \int_0^\infty |E|^2\, r\, dr$ and $\gamma = \sqrt{\beta^2 - n_{\text{clad}}^2 k^2}$. Figure 2.3 (top) shows the predicted macrobend loss over the CWDM band of a typical SSMF for a 16-mm radius. The macrobend loss increases rapidly with wavelength. Figure 2.3 (bottom) shows the strong dependence of macrobending loss of a typical SSMF as the bend radius is varied.

2.2.2.2 Microbending Loss

Microbends are small deflections of the fiber axis such as those that would be imposed when a fiber is pressed against a rough surface. Coupled mode theory can be applied to the problem of transmission through a single axis deflection [4] to derive an expression for the power loss. For a single mode fiber, the problem becomes expressing the fundamental mode of the undeflected fiber (at the input to the microbend) in terms of an expansion of the modes of the deflected fiber (at the output of the microbend), as illustrated in Figure 2.4. The set of guided modes and radiation modes of the fiber make a complete, orthogonal set of basis functions for the expansion.

The coefficients of the expansion give the coupling strength between the modes. The fundamental-mode coupling coefficient determines the loss of the microbend, since it can be assumed that energy coupled into the leaky, radiation or cladding modes is lost. For a length of fiber with microbends of random amplitude that are distributed randomly along the fiber axis, as when a fiber is pressed against a rough surface, the microbending can be modeled as an ensemble of individual deflections whose characteristics are described by the power spectral density of the axis deflections as a function of axial position. Marcuse derived a useful approximate microbend loss formula for step index fibers by using a set of linearly polarized (LP) modes and cladding modes as basis functions for the

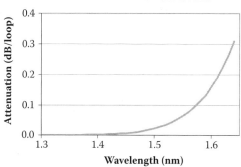
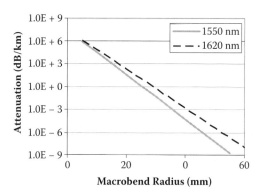

FIGURE 2.3 Predicted macrobend loss for SSMF across the CWDM channel spectrum.

field expansion and by assuming that the size of the random axis deflections are described by a zero mean Gaussian distribution function [5]. The length-normalized microbending loss is:

$$2\alpha = \sqrt{\pi}\sigma^2 L_c \left(\frac{2nka\Delta}{b\omega}\right)^2 \sum_s \exp\left\{-\left[(\beta_g - \beta_{1s})\frac{L_c}{2}\right]^2\right\} \frac{J_1^2(j_{1s}(a/b))}{J_0^2(j_{1s})} \exp\left(-2\frac{a^2}{b^2}\right), \quad (2.4)$$

where L_c is the correlation length of the microbends, σ is the root mean square (RMS) amplitude of the microbends, a is the core radius, b is the cladding radius, Δ is the core delta, ω is the mode field radius, $(\beta_g - \beta_{1s})$ is the difference between the fundamental mode and cladding mode propagation constants, $J_\nu(x)$ is the Bessel function of the first kind of order ν, and $j_{\nu s}$ are the roots of the Bessel functions ($J_\nu(j_{\nu s}) = 0$). Since the basis functions in the coupled mode analysis

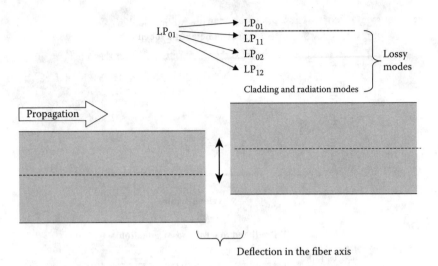

FIGURE 2.4 Model of the core of a fiber in the vicinity of a microbend. Power carried in the fundamental mode before the microbend is coupled into the fundamental and into lossy higher order modes by the microbend.

used to derive Equation 2.4 are the discrete set of cladding modes, the loss formula contains a summation over the cladding modes of radial order s, which is computationally simple to implement. Figure 2.5 shows the predicted loss spectrum for a typical SSMF with various correlation length microbends. For microbends of correlation length less than about 100 μm, the excess loss is relatively large and varies slowly with wavelength. For the same RMS magnitude (σ) microbends,

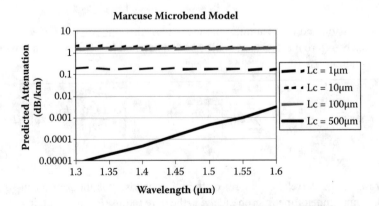

FIGURE 2.5 Predicted microbending loss as a function of wavelength according to Marcuse microbend model. The four curves show the change in wavelength dependence and the change in the magnitude of the attenuation as the correlation length, L_c, of the microbending power spectrum increases from 1 to 500 μm.

Optical Fibers to Support CWDM

but with the correlation length greater than about 350 µm, the excess loss is considerably smaller in magnitude and increases rapidly with wavelength.

2.2.3 CHROMATIC DISPERSION

The chromatic dispersion plays a critical role in the propagation of optical signals since the spectral components associated with the individual pulses travel with different speeds along the fiber. Even if fiber attenuation is overcome with a sufficiently large link budget and amplification, dispersive pulse broadening remains a limiting factor in CWDM systems. Dispersion in a single mode fiber is defined as:

$$D = \frac{1}{L}\frac{d\tau}{d\lambda}, \tag{2.5}$$

where τ is the group delay time, λ is the wavelength, and L is the length of the fiber. The dispersion is closely related to the second derivative β_2 of the propagation constant β,

$$D = -\frac{2\pi c}{\lambda^2}\beta_2 \quad \text{and} \quad \beta_2 = \frac{d^2\beta}{d\lambda^2}, \tag{2.6}$$

where c is the speed of light in vacuum. Theoretical textbooks often define the second derivative of β as the dispersion, but in practical use the definition in Equation 2.5 is utilized.

The dispersion in single mode fibers can be divided into three terms [6]:

- Material dispersion
- Waveguide dispersion
- Profile dispersion

The *material dispersion* is due to the change of the refractive index of the glass versus wavelength. The *waveguide dispersion* is caused by the change of the mode field distribution in the fiber versus wavelength and consequently the change in effective index versus wavelength. The *profile dispersion*, which is due to the different material dispersion of the different dopants of the fiber, is very small in normal fibers and can, therefore, be neglected. The material dispersion is the dominant dispersion factor in ITU-T G.652 standard nonshifted telecommunication fibers (SSMF). In dispersion-shifted fibers (DSFs), the material and waveguide dispersion are equal in magnitude, while the waveguide dispersion is dominant in, for example, dispersion-compensating fibers.

In Figure 2.6, typical dispersion curves for some common transmission fiber types are shown in the CWDM wavelength band (1270 to 1610 nm). Today, the most common fiber type for CWDM is the SSMF, with typical dispersion of –4 ps/(nm·km) at 1270 nm and +20 ps/(nm·km) at 1610 nm.

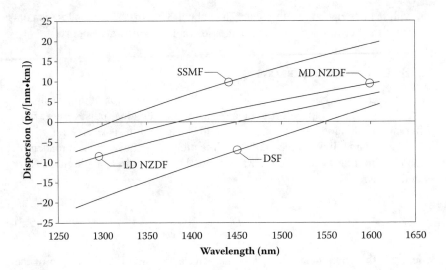

FIGURE 2.6 Dispersion versus wavelength for common transmission fibers. SSMF: standard single mode fiber — ITU-T G.652; DSF: dispersion shifted fiber — ITU-T G.653; LD NZDF: low dispersion nonzero dispersion fiber — ITU-T G.655; and MD NZDF: medium dispersion nonzero dispersion fiber — ITU-T G.656.

2.2.3.1 Dispersion Effects

Because of the nonzero bandwidth of an optical signal, the dispersion leads to pulse broadening due to different group delays of the different spectral components of the signal. For narrowband externally modulated (chirp-free) lasers, the bandwidth of the signal is governed by the modulation. Increased modulation speed leads to increased bandwidth. The tolerable pulse broadening on the other hand decreases with the modulation frequency. Therefore, the dispersion tolerance decreases with the square of the bit rate (B) and linearly with the transmitted length [7]. The maximum transmission length for dispersion-limited transmission is therefore given as:

$$L_{max} = \frac{K}{B^2 |D|}, \qquad (2.7)$$

where K is a constant depending on details of the transmitter and receiver, the modulation format, and the allowed penalty. For nonreturn to zero (NRZ) modulation and a 1-dB eye-closure penalty, $K \approx 104{,}000$ km/((Gb/s)2·(ps/(nm km)) [8]. Maximum dispersion limited transmission distances for bit rates of 2.5, 10, and 40 Gb/s are shown in Table 2.1 for a conventional SSMF for the two outermost CWDM channels at 1270 and 1610 nm. NRZ modulation, a 1-dB eye-closure penalty, and chirp-free transmitters are assumed. It is clear from Table 2.1 that dispersion can be a limiting factor that for bit rates of 10 Gb/s and above even for the relative short distances used in CWDM systems.

TABLE 2.1
Transmission Distances on SSMF of NRZ Signal for 1-dB Eye-Closure Penalty Due to Dispersion Using Transmitter without Chirp

Bit Rate (Gb/s)	1270 nm Reach (km)	1610 nm Reach (km)
2.5	4160	830
10	260	52
40	16	3.3

2.2.3.2 Dispersion Effect when Transmitter Has Chirp

Directly modulated lasers (DMLs) are the preferred choice for CWDM systems of today: this laser type is attractive due to low cost and high output power. However, DMLs have a significant frequency chirp, which can significantly decrease the dispersion-limited distance. The frequency chirp $\Delta \upsilon$ of a DML can be modeled as [9]:

$$\Delta \upsilon = \frac{1}{2\pi} \frac{d\phi}{dt} = \frac{\alpha}{4\pi} \left(\frac{1}{P(t)} \frac{dP(t)}{dt} + \kappa \Delta P(t) \right), \qquad (2.8)$$

where ϕ is the phase, $P(t)$ is the output power as a function of time (t), α is the linewidth enhancement factor, and κ is the adiabatic chirp coefficient. The first term in Equation 2.8 represents the dynamic chirp, whereas the second term represents the adiabatic (static) chirp. The value of the chirp parameters α and κ will depend on the actual laser design, and even quite some variation due to production tolerances must be expected. Typical values quoted in the literature is $\alpha \sim 3$ and $\kappa \sim 20$ THz/W.

In general, numerical simulation will be required to model the exact influence of the interplay of chirp and dispersion on transmission performance. However, for the special case of Gaussian pulses given by:

$$P(t) = P_0 \exp\left[-\frac{t^2}{T_0^2} \right], \qquad (2.9)$$

and further ignoring the adiabatic chirp ($\kappa = 0$), Agrawal [7] has found that the pulse after transmission over a fiber with dispersion D and length L will still be Gaussian with the new pulse with T_1. Hence, the pulse broadening is

$$\frac{T_1}{T_0} = \sqrt{\left(1 - \frac{\alpha \beta_2 L}{T_0^2}\right)^2 + \left(\frac{\beta_2 L}{T_0^2}\right)^2}, \qquad (2.10)$$

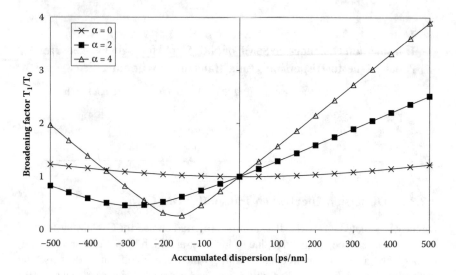

FIGURE 2.7 Calculated pulse broadening for an initial Gaussian pulse with $T_0 = 30$ ps, dynamic chirp only, and a wavelength $\lambda = 1550$ nm as a function of the accumulated dispersion: D·L.

where β_2 is given by D from Equation 2.6. In Figure 2.7, the calculated pulse broadening for a Gaussian pulse with $T_0 = 30$ ps and different chirp values is shown as a function of the accumulated dispersion. $T_0 = 30$ ps is equivalent to a full width half maximum (FWHM) pulse width of 50 ps, which is again half the bit period for a 10-Gb/s signal.

When comparing the result for the chirped pulses with the result for the unchirped pulse ($\alpha = 0$), it is observed that for $D > 0$, the presence of chirp leads to significantly increased dispersion-induced pulse broadening, which increase with increased chirp. However, for up to a certain amount of negative accumulated dispersion, the chirp is actually beneficial and leads to pulse compression so that one can avoid the use of dispersion-compensating fibers in some cases. Increased chirp increases the amount of possible pulse compression but also decreases the amount of tolerable negative accumulated dispersion, as seen in Figure 2.7 for accumulated dispersion below −350 ps/nm.

The conclusions drawn on the interplay between dispersion and chirp were presented here for the special case of a Gaussian pulse with only dynamic chirp; however, the validity has also been shown for more general cases. Thiele et al. [10] measured the impact of accumulated dispersion on a 1550-nm CWDM channel, directly modulated with a 10-Gb/s NRZ signal. The measured receiver sensitivity for extension ratios (ERs) of 6 and 8 dB is shown in Figure 2.8. Higher ER will give more laser chirp. It is noted from Figure 2.8 that the pulse compression obtained for negative accumulated dispersion actually leads to improved receiver sensitivity compared to back to back. In general, with DMLs, it is possible to transmit up to a 200 km at 2.5 Gb/s [9]. However, when using 10-Gb/s transmission

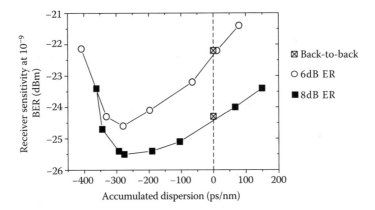

FIGURE 2.8 Impact of accumulated dispersion on the receiver sensitivity for a 10-Gb/s NRZ signal at 1551 nm. (Taken from Thiele, H.J., Nelson, L.E., and Das, S.K., *Electron. Lett.*, 39, 1264–1266, 2003. With permission.)

with DMLs, the above analysis shows that dispersion has to be dealt with for distances just above ~10 km, for example, by the use of transmission fibers with negative dispersion or by dispersion-compensating fiber.

2.2.4 Polarization Mode Dispersion

In practice, optical fibers are not perfectly cylindrical symmetric. The derivation from perfect cylindrical symmetry will lead to a break-up of the fundamental mode in two polarization modes, which again can have different propagation constants. This leads to another source of pulse broadening: polarization mode dispersion (PMD) [11]. PMD can be shown to increase with the square root of the length of the fiber and is, therefore, measured in ps/$\sqrt{km}$

Modern transmission fibers typically have PMD in the range 0.02 to 0.03 ps/$\sqrt{km}$, whereas old fibers might have considerable larger PMD values. For older fibers, PMD as high as 2 to 3 ps/$\sqrt{km}$ has been reported [12]. A common used rule of thumb is that a transmission system can tolerate a total PMD equal to approximately 10% of the bit period. For example, a 10-Gb/s system can tolerate a total PMD of ~10 ps. Due to the typical, relatively short length of CWDM systems and bit rates of 2.5 Gb/s, PMD will typically not be an issue in CWDM and is not discussed further in this chapter.

2.2.5 Cut-Off Wavelength

The cut-off wavelength of a single mode optical fiber is defined as the wavelength above which only a single bound mode propagates. To avoid signal distortion or noise generation from the effects of multi-path interference or modal noise, it is desirable to operate single mode fibers above their cut-off wavelength. Therefore, single mode optical fibers are designed so that the cut-off wavelength is shorter than the lowest signal wavelength in the intended application space. For example, the single mode fiber developed for use in the 1310-nm window during the early 1980s, now generally referred to as SSMF and specified in the ITU-T G.652 standard, typically has the maximum effective cable cut-off wavelength specified to be 1260 nm. The cut-off wavelength is particularly important for CWDM, where the wavelengths can be as short as 1270 nm for the 18-channel grid. In this section, we discuss the theoretical and effective cut-off wavelengths of step index, single mode fibers.

2.2.5.1 Theoretical Cut-Off Wavelength

The weakly guided mode analysis by Gloge [13] introduced the approximate LP modes and showed that a step index optical fiber with infinite cladding supports the propagation of only the fundamental LP_{01} mode when the normalized frequency or V-number of the waveguide is less than 2.405. Therefore, the theoretical cut-off wavelength for a step index fiber λ_c^{th} can be defined as

$$\lambda_c^{th} \approx \frac{2\pi n_1 a}{2.405} \sqrt{2 \cdot \Delta}, \tag{2.11}$$

where n_1 is the refractive index of the core, a is the core radius, and Δ is the relative index difference between the core and cladding. At wavelengths greater than λ_c^{th}, all modes other than the LP_{01} are leaky or radiation modes with energy propagation in the radial direction and are, therefore, extremely lossy.

2.2.5.2 Effective Cut-Off Wavelength

Since the *theoretical* cut-off wavelength λ_c^{th} is not directly measurable, the *effective* cut-off wavelength was proposed and has been universally adopted. The effective cut-off wavelength λ_c^{eff} is a phenomenologically determined parameter that can be directly measured by observing the total optical power that is transmitted through a sample of fiber as a function of the source wavelength.

To illustrate the concept of the effective cut-off wavelength, let us consider the behavior of the LP_{11}, which is usually the first higher order mode, at wavelengths shorter than λ_c^{th}. Far below λ_c^{th}, the LP_{11} mode is tightly confined within the core region and its losses will generally be comparable to those of the fundamental mode. As the wavelength increases, the LP_{11} mode becomes less tightly confined to the core. The decreasing mode confinement gives rise to excess LP_{11} mode loss in the presence of axial imperfections, such as microbends or macrobends. Generally, at wavelengths about 100 nm or so below λ_c^{th}, the LP_{11} mode has become loosely confined to the fiber core and LP_{11} mode losses of tens of dBs per meter readily occur when the fiber axis is not perfectly straight. At this wavelength and higher, the LP_{11} attenuation is so high that it does not *effectively* transmit energy over distances of more than a few meters. This has

Optical Fibers to Support CWDM

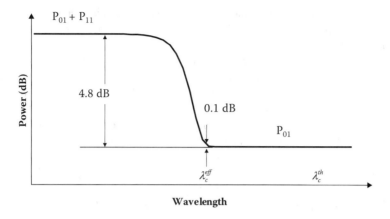

FIGURE 2.9 Power transmitted through a short length of single mode fiber in the vicinity of the effective cut-off wavelength.

lead to the phenomenologically defined effective cutoff wavelength λ_c^{eff} of a single mode fiber.

Figure 2.9 shows the power as a function of wavelength that is observed at the output of a 2-m length of fiber over the range of wavelengths where the transition from single mode behavior to two mode behavior occurs. The power launched into the LP_{01} and LP_{11} modes at the input of the fiber are referred to here as P_{01} and P_{11}, respectively. If we assume that the loss of the LP_{01} mode in the 2-m length of fiber is negligible, then the total power at the output of the fiber as a function of wavelength can be written as:

$$P_{out}(\lambda) = P_{01} + P_{11} \cdot e^{-\alpha_{11}(\lambda) \cdot L}, \tag{2.12}$$

where $\alpha_{11}(\lambda)$ is the LP_{11} attenuation as a function of wavelength. Considering wavelengths much shorter than the LP_{11} mode effective cut-off, then the LP_{11} mode is well confined and the excess attenuation is low, so that $P_{out}(\lambda) \sim P_{01} + P_{11}$. If we assume that $P_{11} = 2 \cdot P_{01}$ (the LP_{01} mode is a singly degenerate set of modes and the LP_{11} mode is a doubly degenerate set of modes), then $P_{out}(\lambda) \sim 3 \cdot P_{01}$ at these short wavelengths. At wavelengths greater than the LP_{11} mode effective cut-off, the LP_{11} mode is highly lossy and $P_{out}(\lambda) \sim P_{01}$. Between these two extremes, the power exiting the fiber falls between $3 \cdot P_{01}$ and P_{01}, as the fiber transitions from two-moded behavior to single-moded behavior.

Note that in Figure 2.9, the shortest wavelength where *only* LP_{01} power, P_{01}, is observed and is considerably below λ_c^{th}. In other words, at wavelengths considerably below λ_c^{th}, the LP_{11} mode has become effectively cut-off. By convention, the effective cut-off wavelength has been defined to be the wavelength where $P_{out}(\lambda)$ has risen by 0.1 dB above P_{01}. It can be shown from Equation 2.12 that the attenuation of the LP_{11} mode at λ_c^{eff} is 19.4 dB. It is important to note that since the length and layout of the fiber sample will determine the level of excess LP_{01} attenuation, the location of the transition from two-moded to single-moded behavior will vary with fiber length and layout.

The *fiber* effective cut-off wavelength has been defined by international standards groups to be measured on a 2-m length of fiber that is deployed in a "nominally" straight configuration except for a single 28-cm diameter loop [14]. This fiber configuration was defined so that fiber manufacturers could easily implement the procedure in factories on readily available spectral attenuation test benches [15]. Since this factory-friendly measurement configuration may not represent field-deployed conditions, a need arose to relate the fiber effective cut-off wavelength to the effective cutoff wavelength of the fiber when it was deployed as a cable section or jumper in an operating transmission system.

Many groups [16,17] studied how the LP_{11} mode cut-off scales in wavelength as the length and bending configuration of fiber under test is varied. The studies showed that length and bending scaling of cut-off varied significantly for different fiber designs. For example, the change in the cut-off wavelength with variation in length of the fiber under test was significantly different for the matched-cladding, depressed-cladding, and dispersion-shifted fiber designs manufactured during the late 1980s [18].

To ensure that fibers are effectively single-moded in the various deployed configurations, the *cable* effective cut-off wavelength has been defined. The industry defined the worst-case deployment scenarios for outside plant, building and interconnection cables. For outside plant cables, the concern is that a short section of restoration cable, as short as 20 m in length, may be spliced into the transmission path to replace a damaged section of cable. If the fiber in the short restoration cable is not effectively single-moded at the system operating wavelength, then modal noise can be generated by the paired splices [19]. Thus the outside plant cable cut-off wavelength deployment configuration, designed to mimic a 20-m restoration cable and the associated splice closures at the end of its ends, is defined as a 22-m length of fiber, coiled with minimum bend diameter of 28 cm, with one 75-mm diameter loop at each end of the fiber. Many suppliers of fiber for use in the outside plant specify that the cable effective cut-off wavelength of their fiber is ≤1260 nm.

2.2.6 Nonlinear Effects

Nonlinear effects will normally not be of importance in CWDM systems due to the relatively short transmission lengths and moderate power levels without amplifiers. Therefore, only a brief introduction will be given here. There are three important types of nonlinear effects in optical fibers: the optical Kerr effect, stimulated Raman scattering (SRS), and stimulated Brillouin scattering (SBS) [7].

The optical Kerr effect is represented as an intensity-dependent change of the refractive index n:

$$n = n_0 + n_2 \frac{P}{A_{\text{eff}}}, \tag{2.13}$$

where n_0 is the linear refractive index, n_2 is the nonlinear refractive index, P is the power, and A_{eff} is the effective area of the mode defined as:

$$A_{\text{eff}} = \frac{2\pi \left[\int_0^\infty p(r)r \, dr \right]^2}{\int_0^\infty p(r)^2 r \, dr}, \qquad (2.14)$$

where $p(r)$ is the near-field power at radius r. The Kerr effect is the origin for three different effects on optical transmission: self-phase modulation (SPM), cross-phase modulation (XPM), and four-wave mixing (FWM). SPM is a power-induced phase shift of the signal, induced by the signal itself. Similarly, XPM results in a phase shift induced on the signal by the presence of another signal with a different wavelength. In a dispersive fiber, the phase shift can be converted by phase to amplitude conversion into signal distortion and hence lead to transmission impairments.

FWM is a process where new spectral components are generated in a parametric process. When those nonlinear mixing products fall within the filter bandwidth of existing channels, nonlinear crosstalk will degrade the transmission performance. FWM can get reduced by high local dispersion as found in SSMF and — similar to XPM — is not significant in CWDM systems due to the wide channel spacing of 20 nm.

In comparison to the Kerr-effect-based nonlinearities, SRS and SBS are both inelastic scattering processes where a photon of the propagating signal interacts with the silica fiber medium and during this process a new photon with lower energy is emitted. SRS described the interaction with molecular vibrations (phonons), while SBS represents the interaction with acoustic waves in the silica fiber. SRS results in both forward and backward scattering and is a quite broadband process with a bandwidth of ~70 nm in the CWDM band and can potentially degrade CWDM channels, as discussed in Chapter 9. However, the Raman effect can also be advantageous for designing broadband CWDM amplification, as discussed in Chapter 6. SBS occurs only in the backward direction and is a very narrowband process with a bandwidth of ~10 MHz. Since SBS is also power-dependent with a typical threshold-like behavior, the onset of this effect defines a limit to how much power can be launched into a channel.

2.2.7 GEOMETRIC PROPERTIES

The geometric properties of an optical fiber [core-cladding concentricity, core and cladding ovality, outside diameter (OD), and curl] are primarily a concern when considering the loss that occurs at joints between fibers or the efficiency of coupling energy into or out of a fiber. The geometric properties can effect the core alignment at a splice or within a connector, which impacts the joint loss. Similarly, the geometric properties can effect alignment of the fiber core to light sources or optical detectors and thus impact coupling efficiency. By assuming that the radial dependence of the LP_{01} mode has a Gaussian shape, the loss in dB of a joint that has lateral axis offset of d is

$$\alpha = 4.34 \cdot \left(\frac{d}{\omega_0} \right)^2, \qquad (2.15)$$

where ω_0 is the mode field radius [20]. Using Equation 2.15 and assuming that $\omega_0 = 4.6$ microns, that is, the typical mode field radius for SSMF at 1310 nm, then the attenuation of a joint with lateral offset of 1.0 microns is 0.21 dB. This is roughly equivalent to the 1310-nm attenuation of 600 m of fiber.

Figure 2.10a and 2.10b illustrate how variations in the OD and core-cladding concentricity, respectively, affect core alignment when a simple V-groove splice technique is used to align the outer diameter of the fibers at a joint. Connector and splicing techniques are generally more elaborate than the V-groove splice

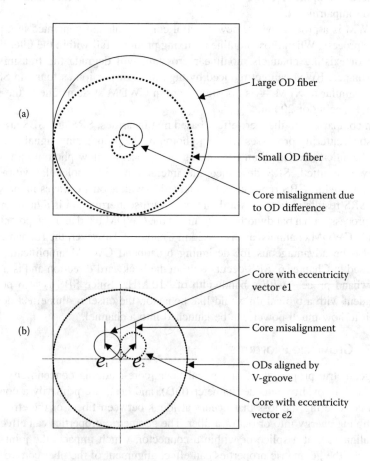

FIGURE 2.10 Core alignment of the optical fiber.

TABLE 2.2
ITU-T G.652 Fiber Geometry Requirements

Parameter	Requirement	Tolerance
Core concentricity error	<0.6 μm	n/a
Fiber diameter	125.0 μm	±1.0
Cladding ovality	<1.0%	n/a

shown in Figure 2.10 and will introduce additional sources of alignment error beyond those directly resulting from fiber geometry effects. Therefore, to ensure the lowest joint losses possible, tight control of fiber geometry is required. The ITU-T G.652 recommendations for SSMF include the requirements for geometric properties shown in Table 2.2.

2.3 OVERVIEW OF COMMON TRANSMISSION FIBERS USED FOR CWDM

In this section, we provide a brief overview of the various commercially available fiber types that are used in CWDM systems.

2.3.1 STANDARD SINGLE MODE FIBER (ITU-T G.652)

The most widely available single mode fiber used in telecommunications systems is specified in the ITU-T G.652 recommendation and is generally referred to as SSMF. The ITU-T G.652 recommendation is quite broad and fibers of several designs, manufactured by all of the principal fabrication techniques [outside vapor deposition (OVD), vapor axial deposition (VAD), modified chemical vapor deposition (MCVD), and plasma chemical vapor deposition (PCVD)], fall within that specification. The majority of the commercial SSMF falls into a narrow range of radial index of refraction profiles and are referred to as matched-cladding fiber because the index of refraction of the low loss cladding region adjacent to the core matches that of the outer cladding region, resulting in approximately constant cladding index across the entire cladding region. These fibers typically have approximately step index core shapes, that is, constant index of refraction within the core region with an abrupt change in index of refraction at the core cladding boundary. Table 2.3 shows a list of the nominal transmission properties of typical matched cladding fibers.

2.3.1.1 Low Water Peak and Zero Water Peak Fiber

As will be later described in Section 2.4, the optical fiber fabrication technology has been developed to reduce the level of OH⁻ contamination to the point that

TABLE 2.3
Fiber Properties of Typical Matched Cladding and Depressed Cladding Fiber Meeting G.652B Recommendation

Fiber Property	Typical Value for Matched Cladding Fiber	Typical Value for Depressed Cladding Fiber
Mode field diameter at 1310 nm (μm)	9.2	8.8
Nominal zero dispersion wavelength (nm)	1312	1308
Dispersion at 1550 nm (ps/[nm km])	17	17
Cable cut-off wavelength (nm)	<1260	<1260
Relative microbending loss at 1550 nm	1.8	1
Macrobending loss for 32-mm radius bend (1 turn)	<0.1 at 1625 nm	<0.05 at 1625 nm
Polarization mode dispersion (ps/km$^{0.5}$)	<0.15	<0.15
Macrobending loss for 16-mm radius bend	<0.05 at 1550 nm	<0.05 at 1625 nm

the absorption peak centered about 1385 nm, known as the water peak, can be dramatically reduced (LWP fiber) or virtually eliminated (ZWP fiber), as shown in Figure 2.1. The benefits of the improvement in fabrication technology and associated attenuation reduction are reflected in the loss requirements contained in the ITU-T G.652D fiber specification. Table 2.4 compares the loss requirements of the various fiber classifications covered by the ITU-T G.652A–D recommendation.

2.3.1.2 Low Bend Loss Fiber Designs

With the current focus of telecommunications investment in the access and metro portion of the network, there is a renewed interest by carriers in fibers with low sensitivity to bending loss. The primary factors that are driving the interest in

TABLE 2.4
Cabled Fiber Loss Requirements of the ITU-T G.652 Recommendation

	G.652 A	G.652 B	G.652C/ D	G.652 D	Commercially Available ZWP Fiber
1310-nm attenuation (dB/km)	0.5	0.4	0.4	0.4	<0.34
1385-nm attenuation (dB/km)	n/a	n/a	*	*	<0.31
1550-nm attenuation (dB/km)	0.4	0.35	0.3	0.3	<0.21
1625-nm attenuation (dB/km)	n/a	0.4	n/a	n/a	<0.24

*The sampled attenuation average at 1385 nm shall be less than or equal to the maximum value specified for the range, 1310 to 1625 nm, after hydrogen aging according to IEC 60793-2-50 regarding the B1.3 fiber category.

Optical Fibers to Support CWDM

low bend loss fibers are generally related to the desire to extend fiber through the network all the way to the customer premises and the unique requirements of this new operating environment leading to the fiber to the home (FTTH) concept (see Chapter 10). It is desirable for the distribution cables and drop cables used in the access network to be as small and low cost as possible. This generally means the cable will provide less protection from the external forces that will result in macro- and micro-bending. End users also desire or require that the hardware installed at their premises be unobtrusive and small in size, resulting in fibers being routed in small bend diameters within the hardware. In addition, to decrease the installation cost of cable and the associated hardware, crafts people with lower training and skill levels may be performing the installations. Therefore, carriers desire fiber that will perform well if a lower level of skill and care is exercised during fiber installation and increased bending results. Compounded with all of the above factors is the fact that the full spectrum CWDM systems utilize wavelengths at the long wavelength end of the L-band where it is particularly challenging to design fiber with low micro- and macro-bend sensitivity (see Figure 2.3). Table 2.5 summarizes the properties of low bend loss fibers that are currently commercially available.

2.3.2 Nonzero Dispersion-Shifted Fiber (ITU-T G.655)

Nonzero dispersion-shifted fiber (NZDSF) has primarily been developed for applications where erbium-doped fiber amplifier (EDFA) and DWDM technologies are deployed, such as long-haul transmission systems. The ITU-T G.655 and ITU-T G.656 fiber specifications were developed to cover the wide range of NZDSFs that are commercially available. At 1550 nm, the magnitude of the

TABLE 2.5
Macrobend Loss Performance (in dB) of SSMF (60, 50, and 15 mm Diameters are for 100 turns, 30 and 32 mm for 1 turn)

	Commercial Matched Clad Spec. (MFD = 9.2)	Typical Depressed Clad G.652 (MFD = 8.8)	G.652D Fiber with Enhanced Bending	Non G.652 Compliant Fiber w/MFD = 8.6	Non G.652 Compliant Fiber w/MFD = 6.3
60 mm diameter at 1625 nm	<0.05	<0.05	<0.05	—	—
50 mm diameter at 1550 nm	<0.05	<0.05	<0.05 at 1625	<0.005	—
30 mm diameter at 1550 nm	<0.05	<0.05	—	<0.005	—
32 mm diameter at 1625 nm	—	—	<0.1	—	—
15 mm diameter at 1625 nm	—	—	—	—	<0.5

Note: MFD = mode field diameter (μm).

dispersion of NZDSF is typically within the range of about 4 to 8 ps/(nm km), compared to about 17 ps/(nm km) for SSMF. Therefore, the 10-Gb/s system reach before dispersion compensation is required can be 2 to 4 times longer for transmission over NZDSF compared to SSMF. The extended dispersion limited reach of NZDSF, compared to that of SSMF, is attractive for applications in DWDM metropolitan ring networks where the circumference of the rings may extend to a few hundred kilometers. Since CDWM systems will be deployed in metropolitan networks, it is likely that CWDM systems will operate on NZDSF as well as SSMF. We therefore briefly discuss NZDSF properties in this section.

In Figure 2.6, the dispersion was shown as a function of wavelength for typical ITU-T G.652 fiber and ITU-T G.655 NZDSF. For NZDSF, the index of refraction profile is tailored so that the magnitude of the total dispersion (the sum of the material, waveguide, and profile dispersions) in the C-band is (i) large enough to disrupt the phase-matched interaction across DWDM channels necessary for inter-channel nonlinear impairments, such as FWM, but (ii) lower than SSMF so that the amount of dispersion-compensating fiber is reduced or the uncompensated dispersion-limited transmission distance is increased. For closely spaced 10-Gb/s channels in DWDM systems, about 2 ps/(nm km) is sufficient to suppress FWM. Two widely deployed NZDSFs are Corning's LEAF® [21] and Lucent/OFS's TrueWave® RS [22]. The design philosophies chosen by the developers of these two NZDSFs were slightly different: the LEAF designers chose to have large nonlinear effective area of ~72 μm^2, defined in Equation 2.14, while maintaining acceptable micro- and macro-bending properties which necessarily led to large dispersion slope and therefore large variation in dispersion across the C- and L-bands. On the other hand, the TrueWave RS designers chose to reduce dispersion slope and the resulting dispersion variation across the transmission bands; however, this choice requires that the effective area be maintained at ~55 μm^2 in order for bending properties to be at acceptable levels. With low dispersion slope transmission fiber, it is generally easier to achieve acceptable properties of the matching dispersion-compensating modules, so that the properties of the overall transmission line (attenuation, residual dispersion, PMD, etc.) can be better optimized when deployed in long-haul applications. Most NZDSF fibers deployed to date have properties that are within the range defined by these two pioneering NZDSF designs. This group of fibers has become referred to as low dispersion NZDSF (LDF), referring to the relatively low level of dispersion in the C- and L-bands. Recently, fiber fabrication improvements have been implemented by several NZDSF suppliers that provide ITU-T G.655 fiber with reduced attenuation at 1385 nm, for example, Sumitomo Electric's PureAccess fiber [23] has an attenuation ≤0.37 dB/km at 1385 nm. The Corning MetroCor® [24] is a somewhat unique NZDSF (outside of ultra long-haul submarine applications) because it has a zero dispersion wavelength longer than the transmission band (C-band), leading to dispersion values of approximately -7 ps/(nm km) at 1550 nm. Transmission through the negative dispersion fiber results in pulse compression of the positively chirped pulses that result when DMLs are used as the light source.

Optical Fibers to Support CWDM   41

Consequently, an increase in the uncompensated reach of C-band transmission at 10 Gb/s is seen with negative dispersion fiber over more typical positive dispersion NZDSF and with SSMF [24].

2.3.3 NZDSF FOR WIDEBAND OPTICAL TRANSPORT (ITU-T G.656)

Recent interest in 40-Gb/s transmission rates and distributed Raman amplification has led to the development of yet another group of NZDSF, referred to as MDFs. The ITU-T G.656 recommendation was developed for MDFs, which are primarily intended for use in ultra long-haul transmission systems. MDFs typically have shorter zero dispersion wavelength than ITU-T G.655 fibers so that there is sufficient dispersion in the 1450-nm wavelength region where Raman pump lasers for C-band amplification are located in order to minimize noise associated with pump–pump or signal–pump FWM. In addition, optimum performance at 40 Gb/s generally requires higher levels of transmission fiber dispersion to mitigate intrachannel nonlinear effects. Since MDFs are most likely to be deployed in 40-Gb/s ultra long-haul networks, the probability that CDWM transmission will take place over these fibers may be quite low.

2.4 ZERO-OH⁻ SINGLE MODE FIBERS FOR CWDM APPLICATIONS

2.4.1 INTRODUCTION

Since the beginning of silica optical fiber development, a ubiquitous loss peak centered around 1385 nm had been recognized as the vibrational second-overtone absorption of the hydroxyl group OH⁻. This OH⁻ or "water" peak causes increased optical loss from about 1360 to 1460 nm, a wavelength region now designated as the E-band by the standards bodies. Because of high OH⁻ loss in typical commercial single mode fibers, most telecommunication systems until recently have avoided this E-band and instead used the two windows at either side of the OH⁻ peak — namely, the O- and C-bands centered at 1310 and 1550 nm, respectively. Although the benefits of completely eliminating the OH⁻ loss in optical fibers had been obvious for more than two decades, the enormous technical and commercial challenges of realizing a viable zero-OH⁻ manufacturing process led many to believe its impracticality. Consequently, the presence of OH⁻ peak in optical fiber was assumed to be unavoidable and system applications had to be designed around it. In this chapter, we will discuss the development of zero-OH⁻ AllWave fiber that essentially eliminated the OH⁻ loss. It was in fact the realization of such ZWP fiber or, in less demanding applications, a low water peak (LWP) fiber that provided the key enabling component and the impetus for the development of CWDM. In other words, it was the advent of ZWP/LWP fibers that opened up additional 100 nm of the E-band and made low-cost, full spectrum (from 1270 to 1610 nm) CWDM systems with wide channel spacing possible. In the medium and short distance local and access networks, low-cost CWDM with ZWP/LWP SSMF is ideally suited to aggregate traffic from the rapidly growing number of FTTH and fiber to the building (FTTB) broadband users and deliver it to the long-haul network in a very cost effective way. The availability of the full spectrum is the key that allows carriers to gain the most leverage from low-cost 10-Gb/s CWDM technology.

While it has always been clear that to achieve LWP or ZWP fibers one has to keep the OH^- contamination in silica to an extremely low level — less than 0.1 ppb of OH^- in the core of ZWP fiber or less than 0.005 dB/km of added loss at 1385 nm, for instance — the biggest challenge was to devise a practical, high yield and low-cost manufacturing process. To put this challenge in perspective, even to this day despite all the improvements in fiber manufacturing technologies over the past decades, some commercial single mode fibers still have a 1385-nm loss as high as 2 dB/km, which is equivalent to about 40 ppb of OH^- in the fiber core. But as it turned out in the course of ZWP fiber development, eliminating OH^- contamination was only part of the challenge, as important new hydrogen aging loss mechanisms were discovered where the initially low OH^- peak in ZWP/LWP fibers could grow by as much as a few tenths of dB/km or more [25]. This happened when certain extremely reactive atomic defects were present in the fibers and they rapidly reacted with a trace amount of molecular hydrogen at ambient temperature over a short period of only a few days. Because of the ubiquitous nature of these reactive defects in commercial silica fibers and the inevitable presence of small amounts of molecular hydrogen in fiber installations, a solution to the hydrogen aging loss problem must be found for ZWP/LWP fibers so that the loss, particularly at the OH^- peak, remains permanently low throughout their service lifetime. Hence, in the following sections, we will include a discussion on the solution to the hydrogen aging loss problem and the development of the standards for hydrogen aging test.

Although the elimination or reduction of the OH^- loss is the key to ZWP/LWP fibers and low-cost CWDM applications, other fiber performance parameters are also important for the remaining application areas. In particular, to support the future migration to 40 Gb/s and the use of L-band, high performance in fiber PMD and macrobending, respectively, are required. In fact, the aim was to develop the highest performing SSMF with zero-OH^- peak, meeting or exceeding the most stringent ITU-T G652D standards. But, perhaps even more important than the technical performance, the cost for manufacturing ZWP/LWP fibers must be low as it is the critical driver for any mass fiber deployment toward the end users. These different aspects of ZWP/LWP fiber performance, manufacturing process, and cost will be discussed subsequently. Finally, it should be noted that although the focus of ZWP/LWP fiber development and applications have been primarily on SSMF of the step-index design with a zero-dispersion near 1310 nm, much of the zero-OH^- technology described here could also benefit fibers of other designs and for different applications. For example, ZWP/LWP NZDFs could have increased bandwidth for DWDM applications as well as improved Raman pumping at 1450 nm due to the lower loss fiber loss at that wavelength.

2.4.2 MANUFACTURING PROCESS FOR ZERO-OH⁻ ALLWAVE® FIBER

Although the original zero-OH⁻ AllWave fiber was developed in 1997 with the Rod-in-Tube (RIT) process [26], here instead we will describe the new and improved Rod-in-Cylinder (RIC) and Overclad-during-Draw (ODD) manufacturing process [27]. RIC-ODD increases the preform size to >5000 fiber km and achieves low-cost and ZWP fiber performance exceeding the most demanding ITU-T 652D standards.

Essentially, the RIC-ODD process entails a totally mechanical RIC assembly and drawing this large assembly directly into fiber with the ODD process [28–30]. The RIC assembly is constructed by inserting a VAD core rod and an optional thin-walled first overclad tube into a large, hollow overclad cylinder of up to 170 mm OD, 60 mm inside diameter (ID), and 3 m in length to form a preform capable of yielding over 5000 km of fiber. The bottom end of the cylinder is machined into a conical taper and a hole is drilled through its walls so that a quartz plug-and-pin assembly can be inserted. The plug-and-pin is used to hold-up the VAD core rod and the first overclad tube inside the cylinder. The conical taper of the cylinder bottom facilitates the initial seal and glass drop during the RIC-ODD draw process. The cylinder also has a quartz handle attached to the top end for handling by a robotic manipulator. When this completely mechanical RIC assembly is lowered into the draw furnace and a vacuum is applied through the hollow handle at the top, the overclad cylinder and the first overclad tube collapse onto the core rod and form a seal at the tapered end of the cylinder. The plug-and-pin and the rest of the cylinder taper will then be melted and dropped off to begin the high-speed fiber draw, as the RIC assembly is further lowered into the furnace and the vacuum-assisted ODD collapse continues.

The RIC-ODD process has several obvious cost advantages over the conventional RIT process because, for one thing, it bypasses several costly high temperature thermal processing steps (see Figure 2.11). First of all, RIC uses the large overclad cylinder as the overclad material directly without the need and added expense of drawing the cylinder into smaller overclad tubes. Secondly, the RIC-ODD process collapses the preform during fiber draw and therefore it completely eliminates the need for a separate overclad process to form a consolidated preform prior to draw. Furthermore, because of its large preform size, RIC-ODD preforms can have a significantly higher fiber yield and be more cost effective at high draw speeds due to fewer setups and less waste at the ends of preform. All these cost savings were realized in the development of the complete RIC-ODD process, which included the abovementioned low-cost RIC-ODD and plug-and-pin mechanical assembly, machined surface finish for the overclad cylinder, surface cleaning procedures, preform handling, and fiber drawing for the large RIC-ODD preforms. Each of the process steps was proven in terms of yield and fiber quality. Subsequently, some of the key G.652C/D fiber performance parameters are discussed with respect to this newly developed RIC-ODD process.

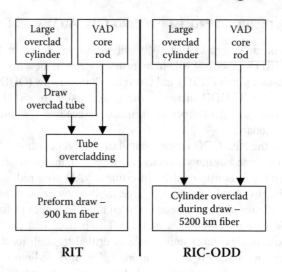

FIGURE 2.11 RIC-ODD and the conventional RIT process.

2.4.3 THE RIC-ODD PROCESS AND ALLWAVE FIBER PERFORMANCE

2.4.3.1 Fiber Loss

To achieve the ultimate zero-OH^- performance, the VAD core soot body is carefully dehydrated with chlorine and consolidated into a core rod. The VAD core rod is then etched with a hydrogen-free (i.e., "dry") plasma torch to remove its surface OH^-. The first overclad tube and the overclad cylinder, both made by the OVD process with an OH^- level typically less than 0.3 ppm, are also lightly etched by acids to remove surface contaminants and OH^-. Using Fourier Transform Infrared (FTIR) spectroscopy and fiber loss analyses, it is found that in the ODD process less than 2 ppm OH^- remains at the interfaces (assuming a width of about 1 µm in a 125 µm OD fiber). Such a small amount of interface OH^- will not adversely affect the fiber's 1385-nm loss to a significant degree as long as the interfaces are placed at a sufficiently large distance from the fiber core (e.g., a clad-to-core ratio "D/d" or "b/a" > 3). The median 1385 and 1550-nm losses for the RIC-ODD process achieved world-class levels of 0.276 and 0.187 dB/km, respectively (see Figure 2.12). The essentially zero-OH^- loss (i.e., no observable OH^- peak) at 1385 nm in particular far exceeds the requirements for G.652C/D fibers.

It should be noted that this RIC-ODD process has some inherent advantages for low 1385-nm loss:

1. The VAD core bodies have no open holes or center-lines like those appearing in the MCVD, PCVD, and OVD processes so it is relatively easy to keep OH^- completely out of core bodies with a separate dehydration and consolidation process.

Optical Fibers to Support CWDM

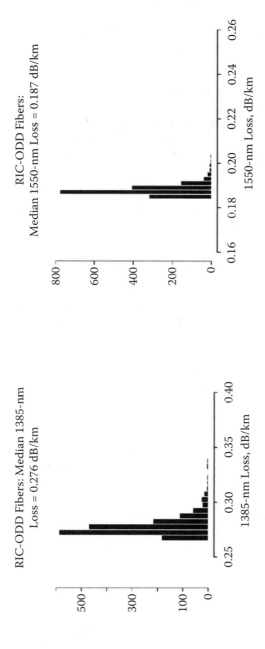

FIGURE 2.12 Histograms of fiber loss distribution at 1385 and 1550 nm for the RIC-ODD fibers.

2. The RIT or RIC overclad material, made by a separate OVD process, can also be completely dehydrated.
3. The surface OH⁻ on the core rods and overclad tubes or cylinders can be easily removed by plasma or acid etch.
4. It is fairly easy to keep the interfaces in RIT or RIC dry with a vacuum ODD process or with a more aggressive dehydration or etching procedure, utilizing Cl⁻ or F⁻ containing gases during overclad.

In contrast, the complete dehydration of soot-on-glass interfaces for the alternative OVD and VAD overclad processes can be difficult, especially for large preforms. For MCVD and PCVD processes where the glass layers are typically deposited without forming soot layers first, a complete dehydration of consolidated glass layers also can be difficult. However, even in these cases, it is possible to greatly reduce the 1385-nm OH⁻ loss peak down to hundredths of dB/km (i.e., to the LWP level) through vigilant practice of eliminating any leakage and avoiding any moisture contamination throughout the chemical delivery systems.

2.4.3.2 Interface Quality and Fiber Strength

Initially, there were concerns about the interface quality and fiber strength for the RIC-ODD fibers because the machined surfaces for the RIC cylinders are rougher than the smooth, drawn overclad tubes used in RIT. But a cylinder surface preparation and cleaning procedures were developed that resulted in no interface problems, such as airlines or bubbles, and achieved a fiber break rate of <5/fMm at 100 kpsi (0.7 GPa) proof test.

2.4.3.3 Fiber PMD

For ultra low fiber PMD, a patented draw process [31] was implemented in conjunction with RIC-ODD and it imparts frozen-in spins in the core of the fiber. To guard against any localized high PMD problems, possible airlines in fiber were detected and cut out during fiber draw and proof test. In addition, the avoidance of mechanical twist for fiber wound onto spools, the development of accurate low-mode-coupled (LMC) method for PMD measurement, and the low core and overclad ovality in the VAD core rod and overclad cylinder all contributed toward the reduction of fiber PMD. The result is a typical LMC PMD ≤ 0.02 ps/√km for RIC-ODD fiber, which again far surpasses the G.652C/D requirements.

2.4.3.4 Fiber Geometry

The RIC-ODD fiber performance in core eccentricity, clad noncircularity, and fiber curl is excellent, normally comparable to or even better than that for the RIT fibers. This is no surprise since the machined RIC overclad cylinder has virtually no ovality and bow and its relative geometry tolerances in siding, ID,

OD, and cross-sectional area are much tighter than those for the drawn RIT overclad tubes. The good geometry performance of RIC-ODD fiber also results in an excellent typical splice loss of <0.02 dB.

2.4.3.5 Hydrogen Aging Losses

To reliably support many decades of revenue generating services, the optical loss in fibers should not degrade with time. It is therefore a challenge to the fiber manufacturers to minimize the risk that the optical loss could increase due to the chemical reaction between the inevitable atomic defects in fibers and the trace amount of molecular hydrogen normally present in or around optical cables. This hydrogen aging loss is a particularly important problem for the ZWP/LWP fibers since all known hydrogen reactions with silica fibers will at least result in the loss increase at the 1385-nm OH⁻ peak. Additional loss increases at other wavelengths are possible, depending on the types of defects present in silica fibers and the hydrogen reactions involved. As it turned out, during the course of the ZWP fiber development, a very common but previously unknown silica defect was discovered: it is extremely reactive and can cause significant hydrogen aging loss at the OH⁻ peak even upon a brief exposure to a trace amount of hydrogen at room temperature. Next we will discuss this and two other types of hydrogen aging losses relevant to ZWP/LWP fibers, as well as the countermeasures against them.

Basically, there are three types of hydrogen aging losses that must be avoided in germanium-doped silica fibers to ensure reliability in optical transmission over the service lifetime of 25 years or more. These hydrogen aging losses are caused by different types of atomic defects or impurities present in the silica fibers. The severity of hydrogen aging loss degradation is entirely dependent on the fiber manufacturing process and the purity of the silica material used.

The first two types of hydrogen aging losses involve two species of extremely reactive silica defects: (i) a pair of NBOHCs (i.e., two nonbridging oxygen hole centers, $\equiv$Si-O• •O-Si$\equiv$) and (ii) the peroxy radical ($\equiv$Si-O-O•) plus Si E' center (•Si$\equiv$) defect (i.e., the $\equiv$Si-O-O• •Si$\equiv$ defect) where • denotes an unpaired electron at the broken chemical bond. These two types of silica defects (involving Si and O atoms only and no Ge) are extremely reactive. Even at room temperature, they can react almost instantaneously with trace amounts of hydrogen and cause significant loss increases of up to a few tenths of dB/km or more. The two hydrogen reaction mechanisms can be described as follows:

$$\equiv\text{Si-O}\bullet \;\bullet\text{O-Si}\equiv + \text{H}_2 \rightarrow \equiv\text{Si-O-H} + \text{H-O-Si}\equiv$$

NBOHCs **1385 nm** (A)

$$\equiv\text{Si-O-O}\bullet \;\bullet\text{Si}\equiv + \text{H}_2 \rightarrow \equiv\text{Si-O-O-H} + \text{H-Si}\equiv \rightarrow \equiv\text{Si-O-O-Si}\equiv + \text{H}_2$$

peroxy radical + Si E' **1385 nm + 1530 nm** (B)

Hydrogen reaction (A) with NBOHCs results in an OH⁻ peak at 1385 nm and nothing else. Hydrogen reaction (B) with peroxy radical and Si E' center results in an OH⁻ peak at 1385 nm and an "SiH" peak at 1530 nm, which are metastable and can decay at room temperature, although a significant fraction of the loss increases will remain even after several months.

Room temperature hydrogen tests have shown that the two hydrogen reactions mentioned earlier typically reach saturation in less than 4 days in 0.01 atmospheres of hydrogen. Because the hydrogen-induced losses in general is a function of the product of hydrogen partial pressure and elapsed time for hydrogen reaction, this implies that the partial pressure of hydrogen in the cable installation needs to be much less than 4 ppm over the 25-year lifetime in order to avoid these two hydrogen aging losses, which is clearly an impractical solution since the measured hydrogen partial pressure in a typical cable installation is of the order of 400 ppm or more. So, to reduce the risk of hydrogen aging loss, it is critical to minimize these two types of silica defects in the fiber manufacturing process. Because of the importance of hydrogen aging loss for the ZWP/LWP fibers, a room temperature test for the hydrogen aging loss is now required by all standards bodies, including ITU-T G.652C/D.

Now, the two types of reactive silica defects have in their origin the same oxygen-rich ≡Si-O-O-Si≡ precursors. Their difference comes from the fact that the chemical bonds are broken (and frozen in) at different places during the thermal processing of silica (e.g., high temperature fiber draw): NBOHCs arise from a broken bond between the two oxygen atoms, whereas peroxy radical and Si E' are the result of a broken bond between oxygen and silicon in ≡Si-O-O-Si≡. So, there are two ways to minimize the two types of reactive silica defects: (C) by adjusting the oxidation and reduction conditions in dehydration and consolidation in the preform-making process to reduce the oxygen-rich ≡Si-O-O-Si≡ precursors and (D) by optimizing the fiber draw conditions to reduce the number of broken and frozen-in chemical bonds. Furthermore, it is possible to passivate any remaining reactive silica defects in fiber by treating the drawn fiber spools with a small amount of deuterium at ambient temperature [32,33]. In fact, this patented deuterium treatment process for AllWave fiber has been used to completely eliminate these egregious hydrogen aging losses caused by the reactive silica defects.

The *deuterium* reactions work in a similar way as the hydrogen reactions (A) and (B) mentioned above:

≡Si-O• •O-Si≡ + D2 → ≡Si-O-D + D-O-Si≡ (C)
NBOHCs **1900 nm**

≡Si-O-O• •Si≡ + D2 → ≡Si-O-O-D + D-Si≡ → ≡Si-O-O-Si≡ + D2
peroxy radical + Si E′ **1900 nm + 2100 nm** (D)

The resulting OD⁻ and SiD vibrational absorption occur at wavelengths approximately a factor of √2 higher than those corresponding to OH⁻ and SiH (or more precisely, the square root of the ratio of reduced mass of deuterium to that of hydrogen in OD⁻/OH⁻ or SiD/SiH vibrations). So, now the OD⁻ and SiD absorption losses are harmless because they occur at much longer wavelengths (>>1625 nm)

Optical Fibers to Support CWDM

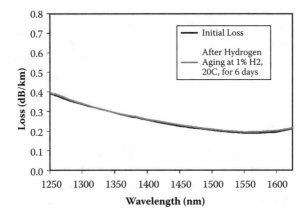

FIGURE 2.13 A deuterium-treated AllWave fiber shows virtually no hydrogen aging loss for the full CWDM spectrum from 1270 to 1610 nm. Its permanent low loss over the service lifetime is virtually assured.

and are completely outside the normal operating wavelength windows even for CWDM. Furthermore, the reactive silica defects after being passivated by the deuterium reaction are no longer available to cause additional hydrogen aging loss in the field. In other words, the deuterium-treated AllWave fiber will essentially maintain its zero-OH$^-$ peak and suffer no hydrogen aging loss increase throughout its service lifetime (see Figure 2.13).

Although less reactive, the third type of hydrogen aging loss that could be of concern is when there is alkali (Na, Li, K, etc.) contamination in the germanium-doped silica fiber [34]. Alkali contamination can be as low as a fraction of ppma (parts per million atomic) and still results in significant hydrogen aging loss over time. This is because the activation energy for hydrogen reaction is greatly reduced when the normal, high-activation energy Ge defects (which are inevitably generated in the fiber manufacturing process) interact with alkali impurities. The hydrogen reaction can be described as follows:

$$\begin{array}{ccc} Na^+ & Na & Na \\ \equiv Si\text{-}O\bullet \;\; \bullet O\text{-}Ge\equiv + H_2 & \rightarrow & Si\text{-}O\text{-}H + H\text{-}O\text{-}Ge \\ Na^+ & & \mathbf{1385\ nm + Long\ \lambda\ loss} \end{array}$$

Alkali contamination can arise from the use of natural quartz material, which inevitably contains alkali impurities, insufficient purification or contamination in preform processing. When there is alkali contamination, the hydrogen aging loss has an OH$^-$ peak as well as a "long wavelength loss" that increases with wavelength beyond 1360 nm. An accurate quantitative model was developed based on extensive hydrogen studies and it predicts a hydrogen aging loss of the order of 0.02 to 0.04 dB/km from 1360 to 1625 nm (which includes the OH$^-$ peak) for a germanium-doped fiber with 1 ppm alkali contamination after 25 years under typical cable operating conditions of 20°C and 400 ppm H$_2$.

The zero-OH⁻ AllWave fiber made with the RIC-ODD process is free of alkali contamination because both the VAD core rod and the cylinder overclad are made by high purity synthetic silica process and care is taken in the fiber manufacturing process to avoid alkali impurities. Thus, the AllWave fiber is also free of the risk of increased hydrogen aging loss in the OH⁻ peak and longer wavelengths due to alkali contamination.

2.4.3.6 Conclusions

The development of the zero-OH⁻ AllWave fiber and its low-cost, large preform (>5000 fiber km) manufacturing process are described. The key elements of this low-cost manufacturing process are a totally mechanical RIC assembly, a high-speed, low-PMD ODD fiber draw process, and a fiber treatment process with deuterium that essentially eliminates the risk for hydrogen aging loss. The fiber performance even exceeds the highest ITU-T G.652C/D standards. In particular, the ultimate performance in loss and hydrogen aging loss that essentially guarantees AllWave fiber's zero-OH⁻ and permanent low loss for the full spectrum from 1260 to 1625 nm over its service lifetime plus its extremely low PMD are ideally suited to CWDM applications. With further development of high purity low OH⁻ tubing, including up- and down-doped tubes, zero-OH⁻ high performance fibers with more complicated index profiles can also be made with the RIC-ODD process. In addition, variations or elements of the RIC-ODD process can be adapted to make low-cost, high performance preforms for fiber-draw operations.

2.5 DISPERSION-COMPENSATING FIBERS

2.5.1 BASIC PRINCIPLES

The basic principle for use of dispersion-compensating fiber (DCF) is quite simple [35,36]. Although there is a lot of literature discussing the impact of different dispersion maps on transmission performance, we want to focus here on simple, single span transmission links that are typically associated with CWDM. A DCF with a dispersion sign opposite to the transmission fiber, for example, SSMF, is inserted into the link. In general, we distinguish two different cases: DCF at the beginning of the link is known as precompensation, whereas compensation at the receiver is known as postcompensation, as shown in Figure 2.14.

Depending on the amount of added DCF, the transmission fiber may not be exactly compensated. When more DCF than transmission fiber is used, the link is over-compensated; for less DCF, the link is under-compensated. In this case, the residual link dispersion D_{res} in Equation 2.16 is either negative (more DCF) or positive (less DCF). It is defined as

$$D_{res}(\lambda) = D_{TF}(\lambda)L_{TF} + D_{DCF}(\lambda)L_{DCF} , \qquad (2.16)$$

where $D_{TF}(\lambda)$ and $D_{DCF}(\lambda)$ are the dispersion coefficients of the transmission fiber and DCF versus wavelength λ, and L_{TF} and L_{DCF} are the length of the transmission

Optical Fibers to Support CWDM

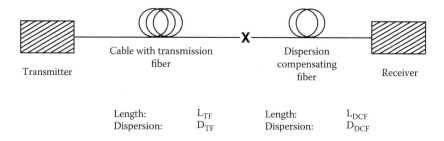

FIGURE 2.14 Postcompensated transmission line using DCF at the receiver.

fiber and DCF, respectively. In practical systems, the residual dispersion is often required to have a value different from zero to suppress penalties from, for example, XPM or frequency chirp from the transmitter, as already discussed in Section 2.3.

A major drawback of DCF for unamplified transmission links is that the total link attenuation is increased by the DCF insertion loss degrading the signal-to-noise ratio (SNR). To overcome this problem, the DCF should have as high a negative dispersion D_{DCF} as possible and, at the same time, as low attenuation α_{DCF} as possible. Therefore, a figure of merit (FOM) for dispersion-compensating fibers can be defined as:

$$\text{FOM} = -\frac{D_{DCF}}{\alpha_{DCF}} \quad (2.17)$$

For discrete DCF modules equipped with SSMF connectors, the module loss is given as:

$$\alpha_M = \frac{|D_{tot}|}{\text{FOM}} + 2\alpha_S + 2\alpha_C, \quad (2.18)$$

where D_{tot} is the total dispersion of the module, α_s is the splice loss between DCF and SSMF, and α_C is the connector-connector loss. The splice loss depends strongly on the DCF type and splice method. A typical connector-connector loss is 0.1 dB. Optimization of the loss of a DCF module is a matter of maximizing FOM and minimizing the splice loss. Depending on the type of DCF, the FOM can vary between 200 and 400 ps/(nm · dB) and the splice loss between 0.05 and 0.4 dB [36].

2.5.2 DCF for CWDM

In order to have a similar performance for all channels, $D_{res}(\lambda)$ should not vary too much across the wavelength range. The variation of dispersion with wavelength is described by the higher order terms of Equation 2.6 when an expansion in powers of the wavelength λ is performed. The first derivative of *D* in λ is known as the *dispersion slope* and is sufficient for describing the approximate variation of the fiber dispersion with wavelength. When designing a DCF for CWDM applications, the challenge is the very wide required operation bandwidth. Record bandwidths of 165 nm with a residual dispersion of ±0.18 ps/(nm km) was reported by Gorlier et al. [37] using optimized DCF for a special low dispersion slope NZDF. Miyamoto et al. [38] have reported on a DCF for SSMF with a bandwidth of 140 nm (1480 to 1620 nm) with a residual dispersion of ±0.15 ps/(nm · km). Residual dispersion for a commercially available DCF for compensation of the eight uppermost CWDM channels (1470 to 1610 nm) is shown in Figure 2.15 (top). The attenuation, including splices and connectors, for a module for compensation of 75 km SSMF using this DCF is shown in Figure 2.15 (bottom).

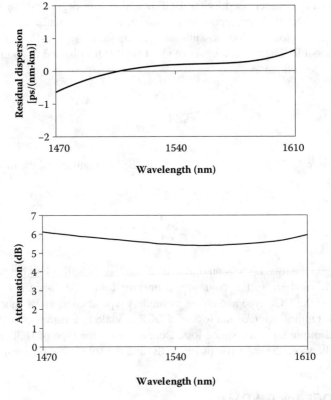

FIGURE 2.15 Performance of a DCF module for 8-channel CWDM by OFS. (*Top*) Residual dispersion from 1470 to 1610 nm; (*bottom*) total attenuation, including splices and connectors.

2.5.3 RAMAN-PUMPED DCF

Raman amplification is an attractive amplifier technology for CWDM systems, as will be discussed in more detail in Chapter 6. Quite different from, for example, rare-earth-doped fiber amplifiers, with Raman amplification it is possible to get gain at any wavelength. Another advantage with Raman amplification is the possibility of very broad gain bandwidth by use of multiple pump wavelengths. Miyamoto et al. [38] have reported on a Raman-pumped wideband DCF with a 103 nm (1485 to 1588 nm) continuous gain without any interleave between pump and signal wavelength. In CWDM systems, it will, however, be possible to interleave pump and signal wavelengths to get even higher gain bandwidths. Tanaka et al. [39] have reported on a 211-nm gain bandwidth (1450 to 1661 nm) using interleaving of pump and signals. For CWDM systems, it should not be a problem to interleave pump and signal channels due to the wide 20-nm CWDM spacing; it should therefore in principle be possible to obtain gain in the full 340-nm CWDM band.

DCF is a good Raman gain medium [35,36,38], due to a relatively high germanium doping level and a small effective area. To get sufficient gain with a reasonable pump power, a discrete Raman amplifier has to contain several kilometers of fiber, adding extra dispersion to the system that must be handled in the overall dispersion management. Dispersion-compensating Raman amplifiers integrate two key functions: dispersion compensation and discrete Raman amplification into a single component.

REFERENCES

1. Miyamoto, T. and Lindsay, R., Raman amplification extends CWDM's reach, *LightWave*, January 2005.
2. Heiblum, M. and Harris, J., Analysis of curved optical waveguides by conformal transformation, *IEEE J. Quantum Electron.*, 11, 75–83, 1975
3. Marcuse, D., Influence of curvature on the losses of doubly clad fibers, *Appl. Optics*, 21, 4208, 1982.
4. Marcuse, D., *Theory of Dielectric Optical Waveguides,* Academic Press, New York, 1974.
5. Marcuse, D., Microdeformation losses in single mode optical fibers, *Appl. Optics,* 23, 1984.
6. Petermann, K., Constraints for fundamental-mode spot size for broadband dispersion-compensated single mode fibres, *Electron. Lett.,* 19, 712–714, 1983.
7. Agrawal, G.P., *Nonlinear Fiber Optics*, 4th ed., Academic Press, New York, 2006.
8. Nelson, L. and Zhu, B., *Raman Amplifiers for Telecommunications*, Springer-Verlag, New York, 2004, chap. 19.
9. Ackerman, D.A., Johnson, J.E., Ketelsen, L.J.P., Eng, L.E., Kiely, P.A., and Mason, T.G.B., *Optical Fiber Telecommunications IVA*, Academic Press, San Diego, CA, 2002, chap. 12.
10. Thiele, H.J., Nelson, L.E., and Das, S.K., Capacity-enhanced coarse WDM transmission using 10 Gbit/s sources and DWDM overlay, *Electron. Lett.*, 39, 1264–1266, 2003.

11. Kogelnik, H., Nelson, L.E., and Jopson, R.M., *Optical Fiber Telecommunications IVB*, Academic Press, San Diego, CA, 2002, chap. 15.
12. Barcelos, S., Serra, T., Rando, R.F., Rigon, E.L., Sasaki, N., Arellano, W., Alkschbirs, A., Lima, D., and Oliveira, R., Polarization Mode Dispersion (PMD) Field Measurements: Audit of Newly Installed Fiber Plants, Proceedings OFC/ NFOEC 2005, paper NThC3.
13. Gloge, D., Weakly guiding fibers, *Appl. Optics*, 10, 2252–2258, 1971.
14. TIA Fiber Optic Test Procedure FOTP-455-80.
15. Wang, C., Villarrud, C., and Burns, W., Comparison of Cut-off Wavelength Measurements for Single Mode Fibers, Proceedings NBS Symposium on Optical Fiber Measurements, September 1982.
16. Nijnus, H. and Van LeeuWen, K., Length and Curvature Dependence of Effective Cut-off Wavelength and LP11 Mode Attenuation of Single Mode Fibers, Proceedings NBS Symposium on Optical Fiber Measurements, September 1984.
17. Peckham, D. and Sears, F., Relationship between Fiber and Cable Cut-off Wavelength of Depressed-cladding Fiber, Proceedings OFC 1988, paper WE5.
18. Wei, L., Saravanos, C., and Lowe, R., Practical Upper Limits to Cut-off Wavelength for Different Single Mode Fiber Designs, Proceedings of the NBS Symposium on Optical Fiber Measurements, September, 1986.
19. Isser, A., Matthews, J., Hopiavouri, E., and Dorland, C., Dependence of Modal Noise on Cable Repair Section Deployment Conditions for Single-mode Fibers, Proceedings OFC 1988, paper WE2.
20. Marcuse, D., Loss analysis of single mode fiber splices. *Bell Syst. Tech. J.*, 56, 703–718, 1977.
21. Lui, Y., Antos, J., and Newhouse, M., Large Area Dispersion-Shifted Fibers with Dual-Ring Index Profiles, Proceedings OFC 1996, paper WK15.
22. Peckham, D.W., Judy, A.F., and Kummer, R.B., Reduced Dispersion Slope, Non-Zero Dispersion Fiber, Proceedings ECOC 1998, Madrid, p139, 1998.
23. Sakabe, I., Ishikawa, H., Tanji, H., Terasawa, Y., Ueda, T., and Ito, M., Bend-insensitive SM fiber and its application to access network systems, *IEICE Trans. Electron.*, E88-C, 896–903, 2005.
24. Woodfin, A., Tomkos, I., and Filios, A., New negative-dispersion fiber in metropolitan networks, *Lightwave*, January 2002.
25. Chang, K.H., Kalish, D., and Pearsall, M.L., New Hydrogen Aging Loss Mechanism in the 1400-nm Window, Proceedings OFC 1999, paper PD22.
26. Chang, K.H., Kalish, D., Miller, T.J., and Pearsall, M.L., Method of Making a Fiber having Low Loss at 1385 nm by Cladding a VAD Preform with a d/D<7.5, U.S. Patent 6,131,415, October 17, 2000.
27. Chang, K.H., Fletcher, J.P., Rennell, J., Nakajima, A., Vydra, J., and Sattmann, R., Next Generation Fiber Manufacturing for the Highest Performing Conventional Single Mode Fiber, Proceedings OFC/NFOEC 2005, Anaheim, CA, paper JWA5, March, 2005.
28. Dong, X., Hong, S.-P., Miller, T.J., and Smith, D.H., Collapsing a Multitube Assembly and Subsequent Optical Fiber Drawing in the Same Furnace, U.S. Patent 6,460,378B1, October 8, 2002.
29. Fabian, H. and Miller, T.J., Method for Producing an Optical Fiber, WO Patent application 03/080522A1, 2003.

30. Fletcher, J.P. III, Miller, T., Rennell, J.A., Smith, D.H., Bauer, P., Norbert, C., Sattmann, R., and Sowa, R., Rod-in-Tube Optical Fiber Preform and Method, U.S. Patent application 2004/0,107,735A1, 2004.
31. Hart, A.C., Jr., Huff, R.G., and Walker, K.L., Method of Making a Fiber having Low Polarization Mode Dispersion due to a Permanent Spin, U.S. Patent 5,298,047, March 29, 1994; Hart A.C. Jr., Huff, R.G. and Walker, K.L., Article Comprising Optical Fiber having Low Polarization Mode Dispersion, Due to Permanent Spin, U.S. Patent 5,418,881, May 23, 1995.
32. Lemaire, P.J. and Walker, K.L., Glass Optical Waveguides Passivated against Hydrogen-induced Loss Increases, U.S. Patent 6,499,318B1, December 31, 2002.
33. Chang, K.H., Kalish, D., and Miller, T.J., Method of Making an Optical Fiber using Preform Dehydration in an Environment of Chlorine-containing Gas, Fluorine-Containing Gases and Carbon Monoxide, U.S. Patent 6,776,012B2, August 17, 2004.
34. Chang, K.H., Alkali Impurities and the Long-Wavelength Hydrogen-Induced Aging Loss in Ge-Doped Silica Fibers, Proceedings OFC/NFOEC 2005, Anaheim, CA, paper OThQ3, March 2005.
35. Ramachandran, S., Ed., *Fiber-based Dispersion Compensation*, Series on Optical and Fiber Communication Reports, Springer-Verlag, New York, 2007.
36. Grüner-Nielsen, L., Wandel, M., Kristensen, P., Jørgensen, C., Jørgensen, L.V., Edvold, B., Pálsdóttir, B., and Jakobsen, D., Dispersion compensating fibers, *J. Lightwave Technol.*, 23, 3566–3579, 2005.
37. Gorlier, A., Sillard, P., Beaumont, F., de Montmorillon, L.-A., Fleury, L., Guénot, Ph., Bertaina, A., and Nouchi, P., Optimized NZDF-based Link for Wide-Band Seamless Terrestrial Transmissions, Proceedings OFC 2002, paper ThGG7.
38. Miyamoto, T., Tsuzaki, T., Okuno, T., Kakui, M., Hirano, M., Onishi, M., and Shigematsu, M., Raman Amplification over 100-nm Bandwidth with Dispersion and Dispersion Slope Compensation for Conventional Single Mode Fiber, Proceedings OFC 2002, paper TuJ7.
39. Tanaka, T., Torii, K., Yuki, M., Nakamoto, H., Naito, T., and Yokota, I., 200-nm Bandwidth WDM Transmission around 1.55 µm using Distributed Raman Amplifier, Proceedings ECOC 2002; paper PD4.6.

3 CWDM Transceivers

Marcus Nebeling

CONTENTS

- 3.1 Introduction 58
- 3.2 Sources for CWDM 61
  - 3.2.1 Laser Types and Their Properties 62
    - 3.2.1.1 Fabry Perot Laser 62
    - 3.2.1.2 Vertical Cavity Surface Emitting Laser 63
    - 3.2.1.3 Distributed Feedback Laser 63
    - 3.2.1.4 Fiber Gating Laser 64
    - 3.2.1.5 Externally Modulated Laser 65
  - 3.2.2 Application to CWDM 66
    - 3.2.2.1 Dispersion Tolerance 66
- 3.3 Detectors for CWDM 68
  - 3.3.1 PIN Diodes 69
  - 3.3.2 APD Diodes 70
- 3.4 Transmitters and Receivers 70
  - 3.4.1 Evolution of The Optical Transceiver 70
  - 3.4.2 Laser Transmitters 72
  - 3.4.3 Uncooled Transmitters for CWDM 73
    - 3.4.3.1 Wavelength Drift 74
    - 3.4.3.2 Power Variation 76
  - 3.4.4 Optical Receivers 76
  - 3.4.5 CWDM Transmission Link 78
- 3.5 Transceivers 78
  - 3.5.1 Building Blocks 79
  - 3.5.2 Standards 82
    - 3.5.2.1 Small Form Factor (SFF) 83
    - 3.5.2.2 Gigabit Interface Converter (GBIC) 84
    - 3.5.2.3 Small Form Pluggable (SFP) 85
  - 3.5.3 Transceivers for 10 Gb/s 85
  - 3.5.4 Trends for Future CWDM Transceivers 86
    - 3.5.4.1 Capacity/Performance 86
    - 3.5.4.2 Modularity 86
    - 3.5.4.3 Integration 88
- References 90

3.1 INTRODUCTION

With the need for greater capacity of optical communication systems, wavelength division multiplexing (WDM) technology was developed. WDM basically puts multiple signals on a single fiber, each one represented by a different wavelength. While early systems utilized two or three wavelengths with relatively wide separation, we would today characterize those as wide wavelength division multiplexing (WWDM) systems, and primarily they employed dual window (Fused Biconic Taper) FBT couplers. These devices typically operated in the range of 810/870/1310 nm or 1310/1550 nm, with light sources such as Fabry Perot (FP) lasers or light emitting diodes (LEDs) operating at those wavelengths. Those early systems [1] eventually evolved in what we today know as coarse wavelength division multiplexing (CWDM) and dense wavelength division multiplexing (DWDM) systems. While DWDM is the technology of choice for long-haul or ultra long-haul (ULH) applications across the network backbones, in metro applications CWDM rapidly filled the gap between DWDM and WWDM systems, yet less expensive and complex than DWDM. Today, CWDM is becoming more widely accepted as an important transport architecture, where up to 18 wavelengths are available for transmitting information over a fiber link in metro or access networks.

In this chapter, we are going to focus primarily on the lasers, transmitters, and transceivers that form the building blocks of a wavelength division multiplexed system. As shown in Figure 3.1a, a simple communication system uses just a single laser transmitting information on any wavelength λ which meets the criteria that it can be detected by an appropriately designed receiver and falls within the spectral low-loss region of the optical fiber, bridging the distance between transmitter and receiver unit. Using the concept of wavelength multiplexing, however, results in several changes with respect to the aforementioned simple single-channel communications link. Figure 3.1b depicts an extended version of the simple communications link where we now use four different wavelengths λ_1 to λ_4 to transmit the information over the fiber.

The multiplexing of different wavelengths with the multiplexer (MUX)/demultiplexer (DEMUX) pair in Figure 3.1b adds further complexity to the system since more transmitter and receiver units are needed and additional components are required for multiplexing and demultiplexing of the four wavelengths. The components used for multiplexing are covered in more detail in Chapter 4. The biggest advantage of WDM is the increase of capacity achieved on a single fiber. The link in Figure 3.1b has four times the capacity of the reference link assuming is the same bit-rate is used for each link. Having a sufficient number of ports at the multiplexer devices, we can also perform in-service upgrade without traffic interruption. The existing wavelengths are not affected when additional channels are added at new wavelengths. Moreover, the lasers at the different wavelengths can operate at independent bit-rates, based on the capacity demand as well as physical properties of the link. The upgrade paths for a CWDM system and the requirements for lasers in such capacity and wavelength-upgraded systems are

CWDM Transceivers

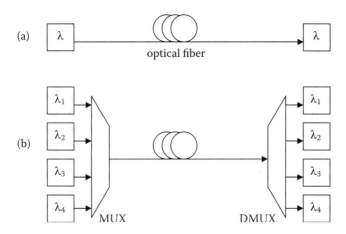

FIGURE 3.1 Wavelength used in an optical communications link. (a) Single wavelength transmitter; (b) extension to a four-channel WDM system/colored transmitter.

discussed in Chapter 7 and therefore complement the information covered here in this chapter on CWDM transceivers.

WDM, or more specifically, CWDM, requires a control over the laser wavelengths used in the transmission link. Unlike single channel systems with a wide tolerance to laser wavelength variation, the presence of neighboring channels dictates the wavelength range of each of the lasers. By exceeding the limits, crosstalk to other channels or excess loss due to the presence of wavelength selective multiplexers can occur. The wavelength spacing between channels in early DWDM systems was typically 200 GHz, eventually dropping down to 100 or 50 GHz. Even narrower spacing is made move difficult by the onset of linear and nonlinear crosstalk due to component performance and spectral bandwidth of the modulated laser sources. The wavelengths for the CWDM are defined according to a standardized frequency/wavelength grid developed by the International Telecommunication Union (ITU) (http://www.itu.int). This is commonly referred to as the ITU grid, extending from 1270 to 1610 nm for CWDM with a spacing of 20 nm between the 18 CWDM channels. More information on CWDM standards can be found in Chapter 1. In the following, Table 3.1 summarizes the nominal wavelength as well as the wavelength range of CWDM lasers under operating conditions. The min and max boundaries take into account the passband of the CWDM multiplexers, the temperature drift of uncooled CWDM lasers of 0.1 nm/°C, and the operating temperature range of 0 to 70°C. The lasers must remain in the interval defined by λmax and λmin at any temperature. Even at a constant temperature, a variation of laser operating wavelengths is expected, due to tolerances in the laser device parameters found in production. The requirements of uncooled operation for CWDM are later discussed in Section 3.2 of this chapter.

TABLE 3.1
Defining the Permissible CWDM Wavelength Range under Operating Conditions

Channel	Center (nm)	Min (nm)	Max (nm)
1	1271	1264.5	1277.5
2	1291	1284.5	1297.5
3	1311	1304.5	1417.5
4	1331	1324.5	1337.5
5	1351	1344.5	1357.5
6	1371	1364.5	1377.5
7	1391	1384.5	1397.5
8	1411	1404.5	1417.5
9	1431	1424.5	1437.5
10	1451	1444.5	1457.5
11	1471	1464.5	1477.5
12	1491	1484.5	1497.5
13	1511	1504.5	1517.5
14	1531	1524.5	1537.5
15	1551	1544.5	1557.5
16	1571	1564.5	1577.5
17	1591	1584.5	1597.5
18	1611	1604.5	1617.5

Lasers and receivers combined into transceivers are used as a compact and modular solution incorporating both transmit and receiver functions into a single device. The transceiver plays a crucial role in CWDM transmission systems. The performance of a typical 16-channel CWDM system is primarily determined by the properties of the lasers used. In Section 3.2, we list typical lasers used for metro applications and also consider them for CWDM. Since low-cost optical transmission is the key philosophy behind CWDM, this feature comes with price/performance restrictions. Typically, CWDM utilizes directly modulated lasers (DMLs), where no external modulator is used. For long haul (LH) systems, directly modulated signals experience degradation when propagating over the link due to chromatic dispersion caused by spectral broadening, an effect more pronounced in directly modulated lasers. The performance of the optical source types is compared in Section 3.3; our focus is on distributed feedback (DFB) lasers, which are widely used for CWDM transmission. Also in Section 3.3, we complement the laser/source discussion with an overview of detectors and their characteristics, going from single components to higher degree of integration. Transmitters and receivers are presented in Section 3.4 and investigated in detail. Finally, Section 3.5 introduces the different types of transceivers used today and also provides an outlook for future developments relevant to CWDM.

3.2 SOURCES FOR CWDM

In this section, we will investigate laser sources for transmitters as one of the key building blocks for those units. Looking into the properties of these devices yields understanding of the transmission performance as well as limitations. Transmitters use either laser diodes or LEDs as sources; where LEDs are normally reserved for short-haul applications with a reach of just a few hundred meters, lasers make up the majority of sources used in medium and long reach systems. Even among the different available laser types, there are significant differences in performance and care must be taken in their selection, depending on the particular application. In general, FP lasers, vertical surface emitting lasers (VCSELs) and DFB lasers are primarily used in transmitters, with higher performance fiber Bragg grating lasers (FGLs) and electro-absorption-based externally modulated lasers (EA-EMLs) are used less frequently. Table 3.2 summarizes the laser types. The requirements for using lasers in CWDM applications are based on the following criteria:

- Performance requirements for lasers in metro CWDM and access are less demanding than for long haul applications due to shorter transmission distance and less transmission capacity (lower bit-rates and wide channel spacing).
- Main requirement is low cost with low power consumption, small footprint, interoperability, simple design being secondary considerations.
- Uncooled lasers can further reduce costs and equipment space.
- Consequences and challenges for low-cost lasers: low output power, high transient chirp due to direct modulation, no temperature stabilization, integration in transceiver packages necessary, simple packaging with often limited radio frequency (RF) performance, optical coupling with no isolator, and no active control of laser performance.

TABLE 3.2
Overview of Light Sources

Type	Relative Cost	Output Power (dBm)	Wavelength Range (nm)	Modulation	Application
LED	very low	<0	850	155 Mb/s	LAN
Fabry Perot	low	3	850, 1310	2.5 Gb/s	access
VCSEL	low	0	850, 1310, 1550	up to 10 Gb/s	access
DFB	medium	6	1270–1610	direct: 2.5–10 Gb/s	CWDM, metro
FGL	medium	3	1550	2.5 Gb/s	metro
EA-EML	high	0	1310, 1550–1590	2.5–40 Gb/s	metro regional

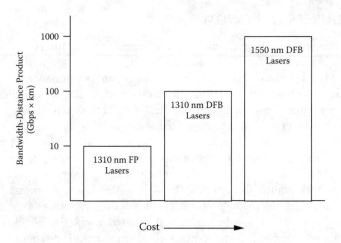

FIGURE 3.2 Comparison of performance and cost of FP and DFB lasers.

3.2.1 Laser Types and Their Properties

There are different measures for assessing the performance of the different lasers used in Table 3.2. The *bit-rate distance product* takes into account modulation performance of the laser and, in addition, the transmission performance. In other words, if a laser modulated at 1 Gb/s allows transmission over 100 km of fiber, without the need to compensate the fiber chromatic dispersion, it has a bit-rate distance product comparable to a directly modulated 10-Gb/s DML with a maximum reach of 10 km. In Figure 3.2, we compare the bit-rate distance product for two of the most common types, the FP and DFB laser. For low-cost applications, FP lasers are the best choice. They have a low bit-rate distance product due to their broad spectral width, which makes these devices very sensitive to fiber dispersion. On the other hand, DFB lasers at 1550 nm benefit from the availability of amplifiers boosting the transmission distance.

We focus on the laser sources of Table 3.2 and discuss their properties next. They are presented in the order of lowest to highest device cost.

3.2.1.1 Fabry Perot Laser

FP lasers are based on GaAs material and are some of the simplest lasers for optical communication systems. Within the gain band, several longitudinal modes experience lasing condition simultaneously so that the resulting spectral width is typically 2 to 3 nm. Commonly, 10 mA threshold current for lasing are expected with 5 to 10 mW maximum output power. For increasing currents, the number of modes also increases, resulting in a broadened spectrum. Due to a simple design, without an optical isolator at the output, the FP lasers are less sensitive to reflections. Laser without temperature stabilization experiences a wavelength drift with temperature. The center wavelength shifts by 0.4 nm toward longer wavelengths when the operating temperature increases by 1°C. The application of FPs to CWDM is rather limited due to the broad linewidth and hence high sensitivity to chromatic dispersion, coupled with a significant wavelength temperature dependence. Practically, the FP sources find their way into single wavelength links in short reach applications; exact span is dependent on the bandwidth of the link. The low dispersion tolerance confines the operation close to the dispersion minimum of the optical fibers, for example, 1310 nm for standard singlemode fiber (SSMF). Therefore, these lasers might be useful for the client interface of a CWDM system or any other forms of low-cost data interconnection.

Future devices with a lower driving voltage and lasing threshold might allow simplified control electronics, and designing a reduced spectral width <2 nm could also overcome the severe dispersion limitation of current devices.

3.2.1.2 Vertical Cavity Surface Emitting Laser

The vertical cavity surface emitting laser (VCSEL) is a lasing device where the light is emitted perpendicular to the layer structure [2]. One-step epitaxy manufacturing of the VCSELs allows a cost advantage for these devices. In contrast to the FP lasers, the short cavity only supports a single lasing mode. The lasers have a low threshold <1 mA and are available for direct modulation at bit-rates from 55 Mb/s to 2.5 Gb/s. Experiments at 10 Gb/s have been performed where the signals have been transmitted over 8.8 km SSMF at 1550 nm using an InP-based VCSEL. Without temperature control, the VCSEL has been shown to operate up to 90°C, making it suitable as a low cost source with a very small footprint. Despite the recent progress, the use of VCSELs in full-spectrum CWDM systems remains challenging since the lasers are available only around 850, 1310, and 1550 nm. The devices are uncolored with a wide wavelength specification. This coupled with a wavelength drift of 0.1 nm/°C make the spectral overlap with CWDM multiplexer passbands an issue. Another challenge is the relatively low output power of a VCSEL, typically less than 0 dBm.

Apart from current short-reach transceivers for LAN/SAN/datacom applications at 850 nm, VCSELs offer a replacement option for existing FP lasers with comparable cost but improved performance. They can be used wherever the low output power is acceptable. Future developments might address the output power issue or tunability to cover a wider spectral bandwidth, thus adapting it closer to the requirements needed for CWDM sources [3].

3.2.1.3 Distributed Feedback Laser

DFB lasers are the workhorse in WDM systems, both as cooled and uncooled devices. They are used for all applications ranging from metro to long-haul. This laser type is based on a Multiple Quantum Well (MQW) structure as the gain medium with a superimposed grating structure as a wavelength selective element to accurately filter out a single longitudinal mode at which the device is emitting its signal. More information on the fabrication process and the structure of the DFB laser can be found in references [2,4]. Due to the more complicated structure, the DFB is more expensive than FP or VCSEL lasers. In the case of high performance, the laser is externally modulated with a separate modulator, a technique mostly used for ULH applications. In metro or CWDM systems where reach is shorter, the technique of direct modulation is employed where the data signal modulates the optical output power via the driving current. The clear advantage is in the simplicity since no additional modulator is needed at the laser output. This aspect is particularly important when no amplifier is available and the link budget is solely determined by the laser output power and the receiver sensitivity. DFB lasers are commonly used for wavelengths within the C-band (1530 to 1565 nm) and also around 1310 nm for legacy systems. Their application is often extended into the L-band and, in some cases, any other wavelengths from 1310 to 1610 nm. The characteristic data are summarized in Table 3.3. DFB lasers are the lasers of choice for most CWDM applications and also used in all examples presented in the subsequent chapters of this book. More information on the challenges of DFB sources at 10 Gb/s or CWDM upgrades is also discussed in Section 7.3 of Chapter 7.

3.2.1.4 Fiber Gating Laser

The use of directly modulated DFB lasers in CWDM systems is restricted by the low dispersion tolerance caused by laser chirp. Other types of directly modulated lasers can overcome this limitation, such as external F(B)GLs. These lasers are built by antireflection coating the fiber-pigtailed facets of a FP laser and forming an external cavity by writing a Bragg grating into the laser's fiber pigtail. While the FP laser has a broad bandwidth, due to the lasing condition

TABLE 3.3
DFB Laser Device Data

Parameter	Typical Value
Wavelength	1300–1600 nm
Output power	up to 40 mW
Linewidth (CW)	<10 MHz
Threshold	10–20 mA
Sidemode suppression ratio	better than 30 dB
Temperature drift	0.1 nm/°C
Package	TO can, butterfly, mini-DIL
Modulation	2.5–10 Gb/s, 40 Gb/s possible
RF driving voltage	2–3 V
Dispersion tolerance	2000–3000 ps/nm at 2.5 Gb/s
	150–200 ps/nm at 10 Gb/s
Coupling	connectorized or pigtailed, either with optical isolator

for several modes, the external cavity of the FBGL acts as a wavelength selective element. Unlike the DFB, this grating is separated from the actual laser chip. The lasers show their advantage under modulation. The laser is directly modulated via the driving current where the maximum achievable bit-rate is determined by the round-trip time of the external cavity. The main properties of these devices are summarized next:

- The laser wavelength is determined by the external cavity grating alone
- Less than 1 nm line-width under modulation, low laser chirp
- Only 0.01 nm/°C temperature drift in uncooled operation
- Commercially available up to 2.5 Gb/s, laboratory samples at 10 Gb/s

Particularly the low wavelength drift could be advantageous to utilize these sources for applications related to CWDM. Since the drift is smaller by an order or magnitude than for DFBs, the channel spacing could be reduced accordingly. However, the operation of the FBGLs without temperature control is prevented by mode hopping during operation [5].

3.2.1.5 Externally Modulated Laser

There are two different approaches to imprint data on the optical carrier. The lasers discussed earlier are directly modulated with the advantage of simplicity, low cost, and potentially higher output power. However, many lasers such as the DFB could also operate in continuous wave mode and the data would be encoded with a separate device, an external modulator either based on the electro absorption (EA) effect modulator or with a Mach Zehnder modulator [6]. We are focusing here on the EA-modulated DFB where a modulator and DFB laser are integrated into a single device. Although strictly speaking two components are used, we still classify it here as a single device type, the externally modulated laser (EML). Unlike the DML, the EA-EML allows the independent control of laser and data modulation. As a result, this source generally achieves better transmission performance than a DML at the same bit rate. Therefore, the EA-EML can be considered in some cases for upgrading existing CWDM transmission systems at wavelengths where better transmission performance, that is, higher tolerance to fiber dispersion is needed. This is generally the case at longer wavelengths of the C- and L-band where the accumulated dispersion is highest and also these devices are commercially available. In Chapter 7, the EA-EML will be used as a 10-Gb/s upgrade solution at longer wavelengths without adding dispersion compensation to the CWDM system. We summarize some of the properties of the EA-EML next:

- Combination of DFB followed by an EA-modulator
- Transmission performance mainly determined by an EA-modulator
- Laser chirp controlled via modulator bias voltage and therefore it can be adjusted for optimized transmission

- Better dispersion tolerance than DMLs and therefore better performance for a small cost premium
- EMLs available at 1310 nm and for C- and L-band
- Device data: 10 Gb/s with 10 to 80 km reach over SSMF
- 0 dBm modulated output power, for increased output power either higher power DFB or integrated SOA/LOA for postamplification

3.2.2 Application to CWDM

Among the discussed laser types, the directly modulated DFB is widely used for the CWDM wavelengths between 1310 and 1610 nm. The lower eight channels between 1310 and 1470 nm can only be used in transmission fibers, such as low water-peak single-mode fiber (LWPF) discussed in Chapter 2. This fiber type, which is increasingly becoming an alternative to SSMF, does not exhibit the typical OH⁻ absorption peak, which for legacy SSMF is centered at 1385 nm and inhibits transmission of up to three CWDM channels. Otherwise, standard LWPF has the same properties as SSMF, in particular the same chromatic dispersion characteristics. In conventional G.652 fibers, only the upper eight channels from 1490 to 1610 nm are used where sometimes a four-channel granularity is assumed for upgrade purposes. Although the average modulated output power of DMLs can reach 6 dBm or more at room temperature, CWDM lasers are normally rated for typically 1 mW (0 dBm) output power. This is because the commonly used uncooled lasers have a lower output power than the cooled lasers. The concept of uncooled lasers is discussed in more detail in Section 3.4 of this chapter.

3.2.2.1 Dispersion Tolerance

The direct modulation of the laser current leads to a modulation of the carrier density and chirp. The chirped pulses of the laser source propagate over the optical fiber where the fiber dispersion determines the performance of the channels, that is, whether the signal degrades due to pulse distortion. The fiber dispersion is described by Equation 3.1 and commonly expressed by the dispersion constant D, which gives the dispersive delay in picoseconds for a wavelength change of $\Delta\lambda = 1$ nm after propagation over $L = 1$ km of fiber. For silica fiber at $\lambda_0 \approx 1.3$ μm, the dispersion is zero ($\beta_2 = 0$) and pulses do not broaden due to zero dispersion.

$$D = \frac{d\beta_1}{d\lambda} = -\frac{2\pi c}{\lambda^2}\beta_2 \qquad (3.1)$$

Generally, the dispersion D has a positive slope and therefore increases at longer wavelength with $D < 0$ for $\lambda < \lambda_0$ and $D > 0$ for $\lambda > \lambda_0$. For SSMF, the dispersion for the first CWDM channel at 1310 nm is close to zero with increased positive accumulated dispersion toward longer wavelengths reaching a maximum at 1610 nm.

Chirped pulses of the modulated lasers broaden, for $D > 0$, when propagating over dispersive fiber, as shown in Figure 3.3. Note that for $D < 0$ (normal dispersion),

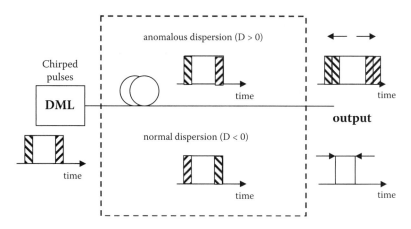

FIGURE 3.3 Interaction of laser chirp and dispersion for DML. (*Left*) Prechirped DML pulses; (*top*) dispersive pulse broadening for $D > 0$; (*bottom*) pulse broadening for $D < 0$; (*right*) resulting output signal which is further broadened ($D > 0$) or recompressed ($D < 0$).

the blue higher-frequency, shorter-wavelength components of the pulse spectrum propagate slower than the red lower-frequency. With longer-wavelength shifted components propagating faster, time-dependent pulses of frequency $\omega(t)$ are generated. In contrast, for longer wavelength $\lambda > \lambda_0$, $D > 0$ (anomalous dispersion), the leading edge of the pulse contains the blue signal components, whereas the trailing edge contains the red signal components, resulting in the opposite chirp. This chirp is due to the fiber dispersion and in addition to any chirp of the laser source itself.

The direct modulation of the DFB results in a blue-shift of the wavelength at the leading edge of an optical pulse. At turn-off, the carrier density decreases and leads to a red-shift of the wavelength at the trailing edge of a pulse. Therefore, the effect of the DML chirp on the pulse has the same effect as the propagation of an unchirped pulse in an anomalous dispersion regime shown in Figure 3.3. With both effects combined, the positive DML chirp interacts with the dispersion of the transmission fiber:

- $D > 0$: The modulated signal experiences further broadening due to the combined effects of chirp and dispersion. This detrimental effect is stronger at longer CWDM wavelengths where the dispersion is higher or when the laser chirp is higher, as observed for a high extinction ratio (ER) directly modulated sources.
- $D < 0$: Here, the pulse broadening of the chirped laser is (partially) compensated by the fiber dispersion and pulses are recompressed. While SSMF has positive dispersion for most CWDM wavelengths, the use of other fiber such as LS or MetroCore with $D < 0$ could help in some cases to improve the transmission performance of DMLs.

TABLE 3.4
Worst-Case Reach (in km) for Different Laser Types (10 Gb/s) and Transmission Fibers

	SSMF G.652, G.652C	NZDSF G.655	MDF G.656
1310 nm			
DFB/VCSEL	55 km	32 km	10 km
FP	14 km	8 km	2 km
1550 nm			
DFB/VCSEL	21 km	50 km	37 km
FP	4 km	12 km	9 km

The performance of such lasers is expressed by the dispersion tolerance parameter indicating the amount of dispersion leading to a certain penalty, where 2 dB are commonly used. Commercial DMLs at 2.5 Gb/s with 1600 to 1800 ps/nm dispersion tolerance are available. Transmission of prototype 2.5-Gb/s DMLs has been reported over 200 km SSMF at dispersion penalties of less than 2 dB. The effects of chirp are discussed further in Chapter 7, where DMLs are considered for operation at higher bit-rate.

In addition to the broadening of optical spectra under direct modulations, there are differences among the intrinsic linewidth of the laser types discussed here. DFB lasers have a single linewidth as compared with a FP laser, which emits many wavelengths over a spectral width of between 5 and 8 nm. This is an important feature when transmitting high bit-rate signals over long distances, because the wider the linewidth of the laser, greater are the effects of chromatic dispersion and ultimately the link performance. Another problem with FP lasers is that they contain multiple wavelengths, which generate mode hopping and give rise to mode partitioning noise. Table 3.4 compares the calculated reach of sources based on FP or DFB lasers.

3.3 DETECTORS FOR CWDM

The basic function of the optical receiver is to detect light and convert it to usable electrical signals. P-intrinsic-n photodiodes (PINs) and avalanche photodiodes (APDs) are the two most commonly used detectors in optical communication systems (see Table 3.5 for comparison of detector properties). Photodetectors can be considered current sources when modeling the behavior of the devices, with PIN photodetectors having a linear relationship between the amount of light input and the output current. This parameter is defined as the responsivity. APDs on the other hand are slightly different in that they have a nonlinear relationship between the input light and the current output. The implication is that APD receivers have higher sensitivity than PIN receivers in high bit-rate telecommunication links.

TABLE 3.5
Comparing the Two Principal Types of Detectors

	Sensitivity	Speed	Cost
PIN	+	++	$
APD	++	+	$$

Note: + and ++ = better performance $ and $$ = higher cost.

Both types of detector-based receivers can be found in optical transceivers. Here, we only focus on the aspects that are relevant to CWDM systems.

3.3.1 PIN Diodes

A PIN diode is a semiconductor device normally grown with a two-step MOVPE/MBE epitaxy process where for operation an electrical field is applied to the p- and n-doped diode structure with reverse bias to act like a capacitor [7]. The incident light gets absorbed and generates carriers proportional to the intensity. The two main parameters characterizing the performance of a PIN diode are responsivity and capacitance. The responsivity describes the conversion efficiency of the diode, that is, the amount photocurrent produced as a function incident optical power. This value can be in the order of 0.5 to 0.9 A/W, Amperes per Watt. Figure 3.4 shows a typical responsivity across the fiber bandwidth.

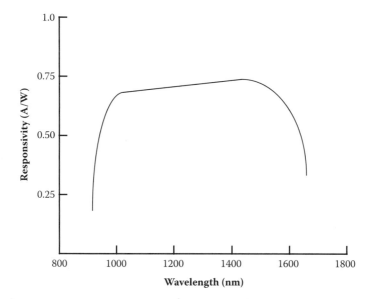

FIGURE 3.4 Responsivity of an InGaAs/InP PIN photodetector. (From Lee, T., Burrus, C.A., and Dentai, A.G., *J. Quantum Electronics*, 15, 1979. With permission.)

GaAs/InP photodetectors that are most commonly used in CWDM system have a broad spectral response from 900 to 1650 nm. The ability of optical receivers to have similar performance over a wide range of input wavelengths allows system designers to use any receiver with any transmitter (with wavelength in the CWDM band). The importance of this will be demonstrated in Section 3.5.1 regarding bidirectional transmission over a single fiber.

Capacitance influences the bit rate in which the PIN is capable of operating. The larger the active area, the greater the capacitance; with larger capacitance comes a lower operating bit rate. PIN diodes have been demonstrated at bit-rates up to 100 Gb/s, but are also available at other lower speeds [9]. The PIN diodes are typically integrated with transimpedance amplifiers (TIAs) and are used in compact receiver designs, as discussed in Section 3.4.

3.3.2 APD Diodes

APD diodes utilize the avalanche effect inside a high electrical field where the incident light generates free carriers, which generate more carriers within the electrical field, thus resulting in a higher sensitivity than comparable PIN diodes. The APDs are an ideal choice for nonamplified CWDM due to high sensitivity and therefore the increased link budget. The maximum bit-rate of commercially available APDs is 10 Gb/s and, although lower than for PIN diodes, sufficient for all CWDM applications. Some practical issues are the low maximum input optical power of around 0 dBm and the relatively high reverse bias DC voltage of approximately 20 to 100 V needed to operate the APD.

3.4 TRANSMITTERS AND RECEIVERS

In the previous section, we have presented the different types of lasers and detectors and also discussed their performance relevant for CWDM applications. In the following, we move from those basic devices to a higher level of complexity and integration and will focus on the application of these devices in transmitters and receivers. We are investigating the use of transmitters and receivers in CWDM systems and review their performance, particularly the challenges under uncooled operation. Combining transmitter and receiver into a single device leads to the concept of the *transceiver*. The transceiver revolutionized many areas of optical communications, and the standards involved as well as recent developments are presented later in Section 3.5.

3.4.1 Evolution of The Optical Transceiver

The transceiver, a combination of transmitter and receiver in a single device, is one of the key building blocks in optical transmission systems. Historically, optical transceivers grew out of system line cards where the sources and detectors were combined with the appropriate electronics to launch signals into fiber in the case of a transmitter, and detect and process signals after fiber transmission in the case of the receiver. An example for early transmitters and receivers is

CWDM Transceivers

(a)

FIGURE 3.5 (a) Early transmitters and receivers.

shown in Figure 3.5a. The presence of transmitter and receiver in pairs facilitates the design of bidirectional fiber links. In its simplest case, the bidirectional transmission is achieved with two separate fibers; each fiber carries a signal in one direction. In this case, the two directions are totally independent and due to the absence of crosstalk, the wavelength allocation for the two directions can be done separately. The concept of bidirectional transmission also works for a *single* fiber. Here, the crosstalk between channels can be minimized to operate the bidirectional link at different wavelengths. Figure 3.5b shows the basic duplex fiber optic transmission system based on transmitter and receiver pairs.

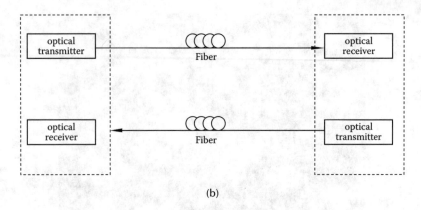

(b)

FIGURE 3.5 (b) Principle of a basic bidirectional link using two separate fibers and pairs of transmitter and receiver.

The system in Figure 3.5b consists of the transmitter, optical fiber, and a receiver. In order to achieve duplex transmission, there needs to be a transmitter and a receiver at each end; thereby enabling data to be sent to and from each location. The earliest implementation of transmitters and receivers were realized on circuit boards. These devices were built from discrete components and placed on a circuit board along with other circuitry to process the signals, provide power, etc. As the volume and speed of transmission increased, there was a push to integrate the functionality of the transmitters and receivers onto a single module. The basic building blocks for an optical transmitter are the source and the driver circuitry for modulation. The digital receiver consists of a detector, amplifier, and comparator.

3.4.2 Laser Transmitters

The most common type of digital laser transmitter consists of driver circuitry to modulate the laser, a monitor photodiode and automatic power control (APC) circuitry, to provide feedback to the driver circuitry in order to adjust modulation and threshold current. In DWDM transmitters, unlike CWDM, the laser wavelength needs to be accurately controlled due to the narrow channel spacing and the narrow passbands of the filters used in multiplexers and demultiplexers of the transmission links. This temperature control is needed because the wavelength of the commonly used DFB laser changes with temperature by approximately 0.1 nm/°C, with the shift going to longer wavelengths as temperature increases and lower when it decreases. Therefore, DWDM laser transmitters need to have not only the APC circuitry found in uncompensated designs, but also temperature monitoring and adjustment capability. This adds a separate control loop to the transmitter, as seen in Figure 3.6.

CWDM Transceivers

Cooled Laser Transmitter

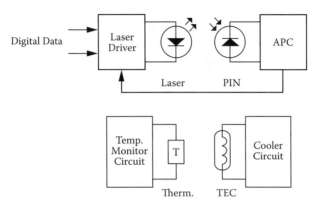

FIGURE 3.6 Laser transmitter using a temperature control circuit for wavelength stabilization.

The temperature is monitored and adjusted, typically using a thermistor in combination with a Peltier cooler [thermoelectrical cooler (TEC)]. The temperature compensation circuitry has an initial set point that is adjusted so that the wavelength resides in the center of the optical passband. As the laser chip temperature fluctuates due to varying ambient conditions, the thermistor detects the changing temperature with respect to the set point and a current is sent to TEC that according to the actual polarity either heats or cools the laser to reach the desired wavelength. In some cases, where wavelengths need to be more tightly controlled, wavelength lockers are added.

3.4.3 Uncooled Transmitters for CWDM

The main difference between DWDM systems and CWDM systems with regard to the electro-optics is the use of cooled DFB lasers as sources in DWDM systems and the use of uncooled DFB lasers as sources in CWDM systems. As a result, the CWDM transmitter is simplified, as shown in Figure 3.7.

Uncooled Laser Transmitter

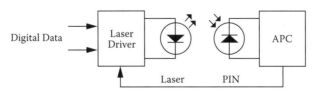

FIGURE 3.7 Transmitter without temperature control as commonly used for CWDM.

Uncooled DMLs are commonly used in CWDM transmission links. The main advantages of these devices associated with uncooled operation are as following:

- No integration of TEC and cooler required
- Less complexity for control electronics
- Reduced power consumption, only laser diode current required
- Smaller footprint
- Lower device cost

At the same time, the following challenges exist:

- Direct modulation of simple transmitters results in broadened linewidth
- Laser wavelength drift needs to be accommodated
- Temperature-induced output power variation needs to be small

3.4.3.1 Wavelength Drift

Figure 3.8 shows the wavelength drift of an uncooled single frequency DFB laser within the passband of a standard 13-nm wide CWDM multiplexer filter. Typically, the operating temperature can vary from 0 to 70°C, thus resulting in a wavelength variation up to 7 nm. Therefore, it is crucial that the CWDM systems design can accommodate the wavelength shift without any loss in performance. As a consequence, the filters have to be sufficiently wide, with a flat top characteristic across the entire passband. More discussion of the filter-related issues can be found in Chapter 4.

As shown in Table 1.1 of Chapter 1, the ITU-specified center wavelengths of the CWDM filters are positioned at 1511, 1531, 1551, 1571 nm, etc. However, the nominal wavelengths of the CWDM lasers are 1510, 1530, 1550, 1570 nm, etc. The reason the filters are offset by 1 nm is to accommodate the typical temperature rise associated with electronic circuit power dissipation, approximately 10°C. This is a typical number, the variation between ambient outside temperature and the ambient temperature found inside the enclosure can be large or small, depending on design and operating conditions. Figure 3.9 summarizes the relationship between the center of the passband and the edges of the filter.

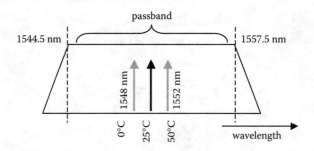

FIGURE 3.8 Wavelength drift of an uncooled CWDM laser within the 1550-nm CWDM passband, the actual wavelength is a function of temperature.

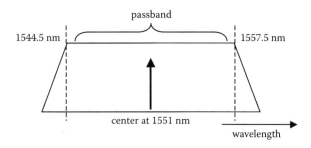

FIGURE 3.9 Definition of CWDM passband center for uncooled lasers, shown for channel at 1550 nm.

The nominal wavelengths are measured at 25°C. The distributions of wavelengths around the nominal DFB value are typically ±2 or ±3 nm, depending on the cost performance trade-off. The reason for this wavelength distribution is due to production variations. If we were to look at the maximum and minimum value of the lasers wavelength due to production variations, the drift within the passband would look like it does in Figure 3.10.

To correctly specify the performance over temperature, one needs to account for the variations in both temperature and production variations. The temperature range can be specified by looking at the minimum and maximum wavelength drift permissible that keep the wavelength in the passband of the filter (Figure 3.11.)

The basic assumption is that the wavelengths are confined to the filter passband. The combined influences of the nominal distributions of laser wavelengths and the thermal performance give the operating temperature range for the laser transmitter. With a 13-nm optical filter passband, this range is from −10 to +80°C. With more tightly controlled laser wavelengths and/or wider filter passbands, the operating temperature range can be increased. A specification of lasers to ±1 nm around the nominal wavelength would give an operating temperature range of −20 to +90°C. Additionally, one could widen the passband from 13 to 14 nm and gain an additional 10° of operating temperature.

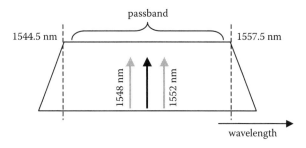

FIGURE 3.10 Statistical distribution of laser wavelengths around 1550 nm at room temperature. (*Center*) Nominal laser wavelength at $T = 25°C$; gray lines: production yield, ±2 nm, also at $T = 25°C$.

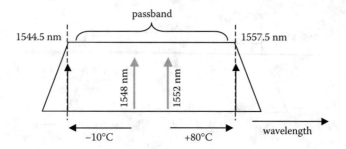

FIGURE 3.11 Temperature variations and production yield with lasers distributed around the passband center.

In order to achieve the industrial temperature operation of –40 to +85°C, heating can be employed as a way to pull-up the operating temperature of the laser and thereby keep the operating wavelength in the filter passband.

3.4.3.2 Power Variation

The second major consequence of uncooled operation is the variation of the laser output power. Not only does the lasing threshold move to higher currents with increasing temperature but also the slope efficiency of the laser decreases. To achieve the same output power and waveform performance obtained at room temperature, the bias current and modulation current need to be higher at elevated temperatures to compensate for the change in threshold current and quantum efficiency. To some degree, this can be accomplished by a simple current control, where the output power of the laser is monitored with a photodiode and the drive current is adjusted accordingly. As the current is increased, a self-heating occurs where the heat dissipation of the laser further increases the chip temperature and therefore an even higher current is needed. Even when no active TEC-assisted cooling is used, a passive heat sink connected to the laser diode can already bring some improvement. While the wavelength drift of the laser can move the signal out of the passband of the passive filters, the power drop at high operating temperatures will deteriorate the link budget; due to the decreased launch power.

Normally, both effects occur simultaneously. When the temperature increases, the laser power decreases and, at the same time, the laser wavelength drifts towards longer wavelengths, showing a red shift. A proper systems design for CWDM needs to account for these two effects.

3.4.4 Optical Receivers

Basic optical receivers consist of two units: detection and amplification. The functional blocks of a receiver are shown in Figure 3.12 where the photodetector as the optical front-end converts incoming light into electrical data signals. Depending on the type of detector, we can classify two types of receivers: either

CWDM Transceivers

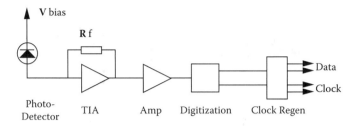

FIGURE 3.12 Building blocks of a typical receiver used in transceiver applications.

PIN-based receivers or APD-based receivers. Following the detection, the amplification part of the receiver is designed to restore the original data, often together with a clock signal derived from the data. The receivers we consider here use direct detection and are working with the nonreturn to zero (NRZ) modulation format. Other, often more complicated receiver concepts are used, which are further described in the literature.

The advantages of receiver integration from a technical perspective are most prevalent in the integration of the photodetector with the TIA, PIN-TIAs or APD-TIAs. Receivers employing the integrated detector/amplifier have better sensitivity due to improved noise immunity and a lower impedance connection between the two devices. In digital systems, the next step in the evolution was a fully integrated receiver with the detector and TIA combined with digitization circuitry, which provides the logic level signals [10]. This construction is often referred to as a 2R-type receiver, where the optical signal is converted to current via the photodetector and then to voltage with the TIA. The electrical signals can then either be amplified or put directly into a comparator circuit, where they are quantized into the logic levels "1" or "0." In some cases, the integration goes even one step further and also re-times the received signal by recovering the clock—this is called a 3R receiver. With the current trend towards standardization of optical transport protocols, the optical receiver could be designed to meet specific requirements. With the constraints of larger volumes and lower costs, the flexibility of having separate elements was not necessary. This was the main driver in the development of the multi-source agreement (MSA) for optical transceivers discussed in the next section. Since the receiver performance is strongly dependent on the optical front end, next we discuss receivers based on the two commonly used photodetectors.

APD receiver circuits
- Allow an increased link budget and therefore enable unamplified transmission with a large fiber and component loss budget and thus ideal for CWDM
- Ideal for detection of wavelengths where no optical amplifiers are available or when transmission line costs have to be low
- Sensitivity better than −30 dBm for 2.5 Gb/s, −24 dBm for 10 Gb/s receivers

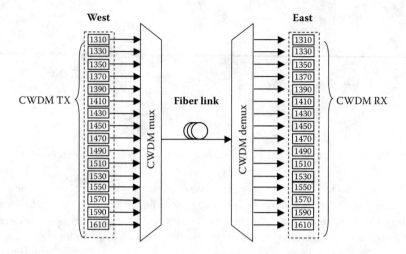

FIGURE 3.13 A 16-channel standard CWDM transmission link using 16 pairs of transmitters (TX) and receivers (RX).

PIN receiver circuits
- The PIN diode for detection at most bit-rates, including >10 Gb/s; however, most common is 1.25 to 2.5-Gb/s operation for CWDM
- Mostly used for low-cost receivers where less sensitivity than APD can be tolerated, for example, −16 dBm at 10 Gb/s, −22 dBm at 2.5 Gb/s
- Compact receivers with low-noise electrical amplifiers, limiting amplifier often integrated

3.4.5 CWDM Transmission Link

In this example, transmitters and receivers are used to build the standard CWDM transmission link comprising of 16 channels between 1310 and 1610 nm. Each of the channels requires a transmitter receiver pair as shown in Figure 3.13 for a conventional, unidirectional link, that is, where all the signals are transmitted parallel from west to east. All channels typically operate at either 1.25 Gb/s or 2.5 Gb/s, although different bit-rates might occur. The DFB lasers are directly modulated with a 1.25/2.5-Gb/s NRZ sequence with a typical ER of 8 dB. The lasers are uncooled and their wavelengths fall within the passbands of the multiplexer/demultiplexer components. After transmission over the fiber link, the channels are demultiplexed and individually detected.

3.5 TRANSCEIVERS

Instead of using discrete lasers, driver circuity, PIN/APDs, and receiver circuitry mounted on boards, these functions are becoming more integrated into modules. This higher degree of integration gave rise to the concept of the transceiver.

CWDM Transceivers 79

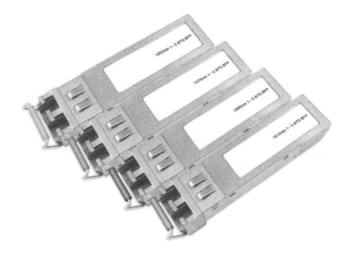

FIGURE 3.14 Set transceivers for use in CWDM; each device has an optical plug for the receiver and transmitter.

These devices are particularly useful when operating bidirectional links, since each site comprises a transmitter as well as a receiver. Laser, receiver diode, and relevant electronics for driving the laser and shaping the received signal are integrated into a module with a standardized interface, as shown in Figure 3.14. Another powerful feature of transceivers is their modularity, that is, the ability to plug the transceivers into electronic circuit boards and to interchange them. This modularity allows separation of the optics from the electronics part of the systems and therefore support a cost-effective system design. One of the requirements for this modularity is the adoption of common standards for transceivers. After highlighting the use of transceivers for CWDM, we address the main standards for 2.5-Gb/s and 10-Gb/s transceivers, which boosted the wide acceptance of these devices in most access and metro systems. Finally, we conclude with a review of new trends and future developments in the field of transceivers.

3.5.1 Building Blocks

Transceivers are highly integrated devices and consist of several elements: laser diode (DFB, VCSEL, or FP), receiver (PIN or APD), and the electronic circuits discussed in the previous section. CWDM transceivers typically use DFB lasers for performance and PIN diodes for simplicity and cost. The transceivers are board mounted so that they electrically directly interface the board electronics. At the optical interface, optical connectors are used that conveniently snap in and allow a simple yet reliable connection to the transmission fiber or other optical components. The connectors are in general standard LC or SC-type connectors.

Next, we want to highlight the use of transceivers as building blocks in CWDM systems. In Figure 3.13, we have shown that for a 16-channel unidirectional CWDM system we need 16 individual transmitters at one side and 16 corresponding receivers at the other side of the fiber link. When using transceivers, three different cases are possible for configuring a bidirectional CWDM link:

- *The 16 transmitters and receivers are replaced by transceivers, resulting in 16 transceivers at each side and two separate fibers.* Consequently, the unidirectional 16-channel reference system of Figure 3.13 is doubled with now a total of 32 channels transmitted over two fibers. Here, the use of transceivers doubles capacity by adding another unidirectional sub-link transmitting into the opposite direction. For example, channel "1" at 1310 nm exists in both links, but the spatial separation by the two fibers eliminates any crosstalk. Typically, optical cables consist of multiple fibers, for example, 48 or 96, so that selecting individual fibers per direction can be easily achieved. Figure 3.15a illustrates the setup of these two unidirectional links. The two optical connectors per transceiver are connected in such a way that one launches the transmitter signal into the first link and the received signal of the second link is coupled into the same transceiver via the other connector. However, this approach requires doubling the number of optical components, leading to increased complexity and cost.
- *At each side, eight transceivers are used with only a single fiber.* In this case, there are only small changes compared to the 16-channel reference link. Only one fiber carries bidirectional traffic and no additional multiplexers are required. At each side, eight transceivers are used where half of the 16 wavelengths propagate west–east, for example, 1310, 1350, 1390 nm, etc., while the remaining wavelength, 1330, 1370 nm, etc., propagate in the other direction. Other channel plans are also possible where the lower eight channels up to 1450 nm propagate in one direction and the other direction uses the longer wavelength channel up to 1610 nm accordingly. Although a single fiber is used, the cross talk among channels is negligible. Unlike the case discussed earlier, only half of the capacity is reached but, on the other hand, the amount of new equipment is considerably reduced to just replacing the discrete transmitters and receiver with the transceiver modules. Figure 3.15b shows an implementation of the discussed scheme with a total of 16 transceivers.
- *Similar to the first case, 16 transceivers for each side are used but only with a single fiber.* Starting from the CWDM link shown in Figure 3.15b, the transmission capacity is doubled by 16 channels propagating in each direction. The challenge here is that in each of the 16 multiplexer passbands there is bidirectional transmission. For example, the channel at 1310 nm is launched into a multiplexer port and, at the same time, the signal from the other end of the transmission link is detected from the same port. As a result, this system is much more compact than the solution depicted in Figure 3.15a, but the requirements on components are higher. In-band crosstalk occurs from reflections in the signal

CWDM Transceivers

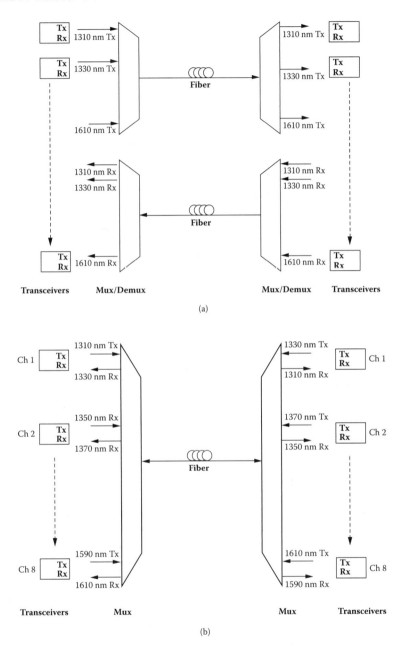

FIGURE 3.15 (a) Bidirectional CWDM link consisting of two separate unidirectional links and 2 × 16 transceivers leading to 16-channel CWDM transmission in each direction; (b) bidirectional CWDM link over a single fiber with 2 × 8 transceivers and a total of 16 wavelengths; and (c) bidirectional link with 16 transceivers at both sides. Each CWDM wavelength is used twice in this set-up and combined into the same multiplexer port.

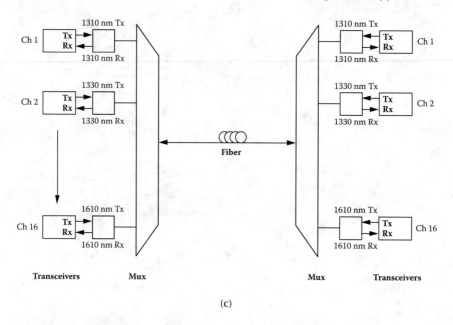

(c)

FIGURE 3.15 (Continued).

path—the reflected light travels in the opposite direction and thus acts as an interference to the detected channel. The performance of a bidirectional CWDM transmission system using the same wavelengths for both directions is investigated in Chapter 7.

3.5.2 STANDARDS

The integrated transmitter and receiver evolved out of the need to provide lower cost, better performance and easier to implement connectivity in optical transport. This trend started in the 1980s with applications in datacom, where computers were just starting to be connected via fiber optic cable. In parallel, discrete boards designs on line cards used in telecom applications were migrated to separate transmitters and receivers. As volumes increased toward the end of the 1980s, for integrated optoelectronics in both datacom and telecom applications, the transmitters and receivers were placed in the same module for further cost and space savings. One of the first standardized transceivers were the OC-3 and Fast Ethernet 1×9 pin modules. These simple transceivers used a receptacle design for the fiber interface rather than a fiber pigtail. The device could be mounted directly on a printed circuit board and have the interface extending through the enclosure for easy access. This saved assembly time and space since the fiber pigtail always needed to have a service loop and bend radiuses need to be kept to a certain diameter in order to maintain reliability as well as performance. The 1×9 modules later had the clock recovery function introduced into the units and had a second row of pins added (Figure 3.16).

CWDM Transceivers

FIGURE 3.16 A 1 × 9 transceiver.

3.5.2.1 Small Form Factor (SFF)

With the increased need for a higher density of interconnects, smaller foot print transceivers were developed to accommodate these requirements. Smaller connectors such as the LC, MT-RJ, and MU enabled transceiver vendors to manufacture smaller devices than were found when they used ST-, SC-, and FC-type connectors. The SFF was one of the first commercially available small form transceivers that used only half the space of the popular conventional SC types (Figure 3.17). These SFF devices found their way into applications ranging from 100 Mb/s all the way up to 2.5 Gb/s. SFF was one of the first high volume transceivers to be manufactured

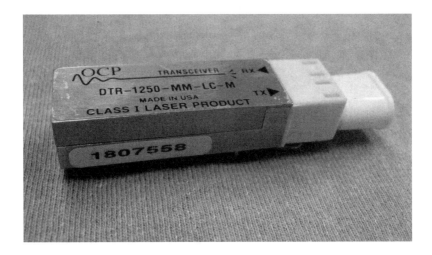

FIGURE 3.17 An SFF transceiver.

under a multi-source agreement (MSA). This enabled multiple vendors to produce parts that have the same form, fit, and function, both electrically and mechanically; however, the MSA did not define performance or reliability standards. In this section, we are focusing on the transceiver standards up to 2.5 Gb/s, while this is complemented by the 10-Gb/s standards presented in the following section.

3.5.2.2 Gigabit Interface Converter (GBIC)

A major step in the drive to make optical transceivers more like connectors rather than networking elements was made with the development of the pluggable transceiver. This allowed line card and blade vendors greater flexibility to reduce their inventory of card types based on their optical interface and instead inventory a standard board and have socket on the assembly to accommodate a pluggable transceiver. It would enable a standard board to be sold and have a variety of different transceiver types available as interfaces. This approach was adopted early on by the Ethernet community, with the GBIC pluggable part. This was the first pluggable MSA transceiver with applications primarily in switch and router blades for Gigabit Ethernet. There were three basic types of Ethernet transceivers: the short reach, an 800-nm VCSEL-laser-based part called the SX with a span of 500 m over multimode fiber; the medium reach, 1310-nm FP laser part called the LX with a span of 2 km over singlemode fiber and a 1550-nm DFB part used for long haul called a LH with a span of 80 km, also over singlemode fiber. An easy migration from the standard LH part was to have specific wavelength DFB lasers; this enabled the development of CWDM GBICs and eventually DWDM GBICs. The GBICs transceivers have been used primarily in Gigabit Ethernet applications, but in some cases both lower speed, multi-rate parts and higher speeds of around 2.5 Gb/s have been developed. The GBIC was one of the first transceivers to have an APD receiver rather than a PIN type for higher sensitivity (Figure 3.18).

FIGURE 3.18 A GBIC transceiver for 1610-nm CWDM wavelength.

CWDM Transceivers

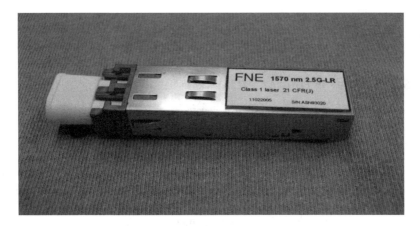

FIGURE 3.19 An SFP transceiver for 1570-nm CWDM wavelength.

3.5.2.3 Small Form Pluggable (SFP)

Combining the space savings of the SFF and the flexibility of the GBIC, the next transceiver to be developed was the small form pluggable (SFP). This device has approximately the same size of the SFF but with the functionality of a GBIC. Besides the size advantages over the GBIC, the SFP had lower power dissipation, since it operates at 3.3 V while the GBICs worked with 5V supply voltage. In addition, the transceivers provide monitoring capabilities for power, temperature, and voltage levels. With technology being pushed to meet the demands of both function and size, CWDM and DWDM SFP transceivers have been developed for use in Gigabit Ethernet as well as 2.5 Gb/s SONET applications. Moreover, SFPs have also been developed for use in 1, 2, and 4 Gb/s fiber channel applications (Figure 3.19).

3.5.3 Transceivers for 10 Gb/s

The next milestones for transceiver performance are found in the 10-Gb/s development occurring at the time of publication of this book. There are multiple MSAs for the 10-Gb/s transceivers, typically denoted by an X in the name, X standing for the Roman numeral 10. XFP, Xenpak, X2, and LX-4 are the names of some of these devices. To enable longer spans at 10 Gb/s, the institute of electronic and electrical engineering (IEEE) has developed a standard for use of CWDM in the design of these high bit-rate devices, IEEE 802.3 ae-2002. Instead of having a serial link running at 10 Gb/s line rate, the LX-4 concept optically transmits and receives four signals at different wavelengths spaced at 1275, 1300, 1325, and 1350 nm. The 25-nm grid is slightly different than the ITU 20-nm grid. The operating bit-rate per channel is 3.125 Gb/s, the aggregate bandwidth of all four channels is 12.5 Gb/s, which includes the associated protocol overhead. The advantage of the LX-4's parallel approach is that each of the four tributary channels has a lower bit-rate and better dispersion tolerance. At the same time, lower-cost lasers and detectors can be used. An overview on the development in the field of 10-Gb/s transceivers is also provided by www.x2msa.org. The technical data of common transceivers are summarized in Table 3.6.

3.5.4 Trends for Future CWDM Transceivers

There are several trends for transceiver development with the majority of activities occurring at 10 Gb/s, where DWDM XFPs are now being deployed. In addition, there is another new MSA standard called quad small form pluggable (QSFP), which is pushing the bit-rate to a 16-Gb/s limit, thus promising a 3× density improvement over current solutions. Work is being done in high channel count DWDM systems for short reach ultra high bandwidth interconnection applications. The next challenge would be 40-Gb/s transceivers and current DWDM technologies would play a more significant role, as bit rates increase and the impairments from chromatic dispersion as well as polarization mode dispersion become harder to manage.

SFP transceivers are currently used for CWDM applications providing a full 16/18-channel support for bit-rates up to 2.5 /s with integrated monitoring functions and hot-plug capability. They represent the building blocks, which allow a simple and cost-effective design of standard CWDM transmission links. With these existing transceivers, variations in the CWDM link design are limited. For new requirements for the CWDM systems, novel transceiver concepts would have to find their way into CWDM.

The trends here will be capacity/performance, modularity, and integration, and are briefly discussed next.

3.5.4.1 Capacity/Performance

When pushing the line rates to 10 Gb/s and beyond, the aforementioned LX-4 or QSFP approach could help to distribute the capacity among four separate channels and therefore revert to the low-cost sources. However, when all 16 channels are allocated and no extra spectral bandwidth is available, this concept no longer works and the existing channels have to be migrated toward higher bit-rate, depending on the individual demand. Here, the new developments in the field of 10-Gb/s transceivers could find their application.

3.5.4.2 Modularity

Another trend for CWDM could be wavelength agnostic transceivers where the operating wavelength can be controlled externally. The idea behind this is to reduce inventory costs, instead of 16 individual transceivers; only one module type could be deployed that covers all wavelengths. This also facilitates the upgrading of existing channels with new wavelengths or the keeping spares for existing channels. Tunable laser devices that are currently used for DWDM are not a viable option for CWDM. This is due to the relatively high cost, the complexity, and the inability of the current technology to extended tuning wavelength range of the current C-band to cover a reasonable number of CWDM channels.

CWDM Transceivers

TABLE 3.6
Overview of 10-Gb/s Transceiver Standards

MSA	300 pin	XENPAK	XPAK	XFP
Application Target	Enterprise switch	Enterprise switch	Server NIC Storage Enterprise switch	Server NIC Storage Enterprise switch
Electrical Interface	16-Bit parallel SFI-4 (OIF)	4-Bit parallel XAUI	4 Bit parallel XAUI	10 Gb/s serial XFI
Optical Reach	10 GbE to 40 km 80 km (future)	10 GbE to 40 km 80 km (future)	up to 10 km 40 km (future)	10 km up to 40 km (future)
Dimensions (L × W × H)	1st Gen: 4.0 × 3.5 × 0.53 2nd Gen: 3.0 × 2.2 × 0.53	4.8 × 1.4 × 0.7	2.7 × 1.4 × 0.4	2.7 × 0.7 × 0.4
Power Consumption	up to 12W	up to 11W	up to 4W	2.5W + external SerDes
Main Features	Protocol independent	Hot pluggable XAUI	Shorter, thinner Single side of PCB Common interface for low integration risk	protocol independent small size requires 10Gbps CDR

3.5.4.3 Integration

The trend toward bandwidth links has also led to the development of multi-channel transmitters and receivers. These devices typically have multiple lasers in bar configuration and a multi-channel driver IC in the transmitter case and a photodetector array and 2R regeneration IC in the receiver implementation. The current SNAP 12 MSA allows for 12 separate transmit channels in a single module with a corresponding 12-channel receiver unit (see Figure 3.20). The SNAP 12 currently operates at 3.125 Gb/s per channel over multimode fiber, with lab results that have shown performance above 8 Gb/s. Without any optical multiplexing element included in the channel, this design would be limited to the number of channels available in a ribbon fiber. Currently, the practical limitation for parallel optics modules is 12 channels, this is due to the ribbon fiber count and number of connections possible in MT-type connectors, (the standard connector type for parallel optics modules today). The integration of multiple wavelength transmitters with optical multiplexers and long width demultiplexers and receivers will be challenging from the point of heat dissipation, crosstalk, and footprint. However, the ultimate goal would be a 16-channel transceiver where all the required optics, that is, transmitter, receiver, and multiplexers along with demultiplexers would be integrated in a single module.

The approach of highly integrated transmitters and receivers has already been shown for 10-Gb/s DWDM systems [11,12]. Following the same path, a single channel fiber link could be easily enabled to provide the capacity of CWDM by adding these modules, as shown in Figure 3.21a and 3.21b. Even extended reach or networking functionality would be possible by multiple optical-electrical-optical sites where the receiver–transmitter arrays operate as repeaters.

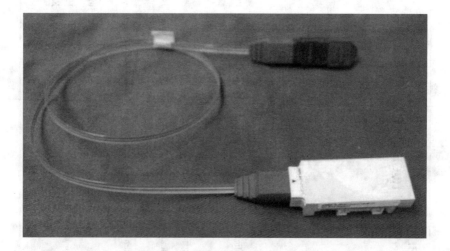

FIGURE 3.20 Integrated transceiver units using parallel optics and ribbon fiber.

CWDM Transceivers

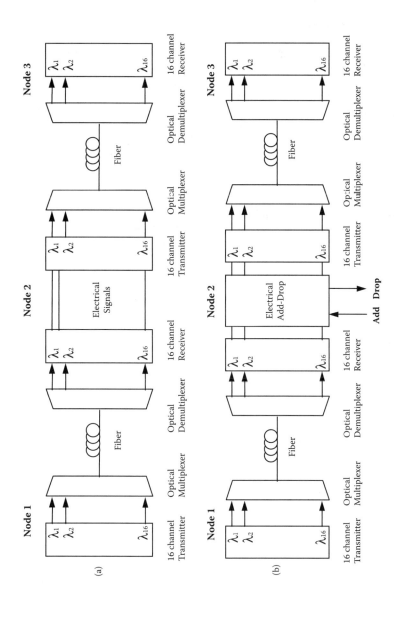

FIGURE 3.21 (a) Concept of highly integrated 16-channel CWDM transceiver as repeater for extended reach CWDM and networking functions; (b) concept of highly integrated 16-channel CWDM transceiver as add/drop multiplexer for extended reach CWDM and networking functions.

REFERENCES

1. Ishio H., Minowa, J., and Nosu, K., Review and status of wavelength-division-multiplexing technology and its application, *J. Lightwave Technol.*, 2(4), 448–463, 1984.
2. Carroll, J., Whiteway, J., and Plumb, D., *Distributed Feedback Semiconductor Lasers*, SPIE Press, 1998.
3. Ghatak, A.K. and Thyagarajan, K., *Optical Electronics*, Cambridge University Press, 1989.
4. Ghafouri-Shiraz, H., *Distributed Feedback Laser Diodes and Optical Tunable Filters*, Wiley Publishers, Optical Communication Series, 2003.
5. Ramaswami, R. and Sivarajan, K., *Optical Networks*, Morgan Kaufman, 1998.
6. Hecht, J., *Understanding Fiber Optics*, Prentice Hall, 1999.
7. Daly, J., *Fiber Optics*, CRC Press, 1984.
8. Lee, T., Burrus, C.A., and Dentai, A.G., Photodetector response, *J. Quantum Electronics*, 15(30), 1979.
9. Behling, A., Bach, H.G., Mekonnen, G.G., Kunkel, R., and Schmidt, D., Miniaturized waveguide integrated PIN photodetector with 120 GHz bandwidth and high responsivity, *Photon. Technol. Lett.*, 17(10), 2152–2154, 2005.
10. Hecht, J., *Understanding Fiber Optics*, Prentice Hall, 1999, p. 225.
11. Nagarajan, R. et al., Wide temperature/coolerless operation at 100 Gb/s DWDM photonic integrated circuit, *Elec. Lett.*, 41(10), 2005.
12. Nagarajan, R. et al., 400 Gb/s DWDM photonic integrated circuit, *Elec. Lett.*, 41(6), 2005.

4 WDM Filters for CWDM

Ralf Lohrmann

CONTENTS

4.1 Introduction 91
    4.1.1 Basic Network Concepts 92
    4.1.2 Requirements for CWDM Filters 93
4.2 Technical Options for Wavelength Division Multiplexing 94
    4.2.1 Gratings, Arrayed Waveguides, and Thin-Film Filters 94
        4.2.1.1 Fiber Bragg Gratings 94
        4.2.1.2 Arrayed Waveguides 99
        4.2.1.3 Thin-Film Filters 101
    4.2.2 Trade-Offs and Performance 102
4.3 Properties of Thin-Film Filters 104
    4.3.1 Thin-Film Multi-Cavity Structures 104
    4.3.2 Manufacturing Process Steps 111
4.4 Thin-Film Filter Packaging Solutions 113
    4.4.1 Cascaded 3-Port Packages, Multi-Ports, and Glass Package 113
        4.4.1.1 3-Ports 113
        4.4.1.2 Multi-Ports 117
        4.4.1.3 Glass Packages 119
        4.4.1.4 CWDM Modules 119
    4.4.2 Multiple Bounce Concept 120
4.5 Future Trends and Requirements 122
References 123

4.1 INTRODUCTION

Fiber-based optical communications can be as simple as a point-to-point link where a single data channel is transmitted between two sites. On the other hand, the demand for capacity and flexibility has led to an evolution of the linear links into more complex networks, including star architectures, rings, and meshed designs. As the complexity increases, so will the demand for additional devices required to support the networking functionality. The most important features for using coarse wavelength division multiplexing (CWDM) in the network include multiplexing and demultiplexing of all optical channels or sub-bands, also the selective add and drop of one or more channels at network nodes. While Chapter 8 will address the role of networking for CWDM in more detail, this chapter clearly defines itself by focusing on optical filters as key components for multiwavelength optical networking in CWDM systems. After Section 4.1 provides a general overview on the benefits for wavelength multiplexing in the metro and access networks, Section 4.2 reviews the technical options for optical filters, including fiber Bragg gratings (FBGs), arrayed waveguide gratings (AWGs), and thin-film filters (TFFs). The properties of the TFF are discussed in more detail, covering manufacturing steps as well as a performance analysis. Section 4.4 addresses the more technical aspects of TFFs, such as 3-ports, multi-ports, modules, and other solutions aiming at a further reduction of packaging size. Finally, the chapter concludes with the investigation of trends toward higher integration and other challenges in the design of CWDM filters.

4.1.1 Basic Network Concepts

The demand for increased transmission capacity is the main driver behind the concepts of wavelength division multiplexing (WDM), time-division multiplexing (TDM), and space division multiplexing (SDM). These three approaches are compared in Figure 4.1. When the information of four different channels A...D is

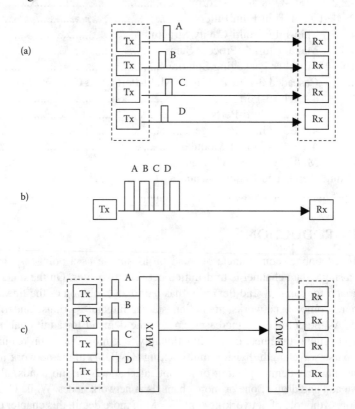

FIGURE 4.1 Basic concepts of multiplexing. Combining four different channels A, B, C, and D in (a) SDM, (b) TDM, and (c) WDM.

WDM Filters for CWDM

transmitted between the transmitter and receiver sites, approach (a) uses four distinct fibers each terminated with a transmitter receiver pair. This solution has the highest demand in total equipment and especially optical fibers, but at the same time offers four totally independent parallel systems that can be all configured individually. In contrast, the technique of TDM in Figure 4.1b combines the information of all four tributaries either electrically (ETDM) or optically (OTDM) onto a single wavelength. Here, we observe a dramatic reduction in the number of fibers, transmitters, and receivers over SDM. However, the bit rate of the resulting single optical channel increases fourfold and is therefore also increasing the requirements for transmitter and receiver and possibly also a more sophisticated dispersion management of the fiber link. Other than the serial transmission in TDM, the technique of WDM in Figure 4.1c uses a parallel transmission of four channels between the sites, but unlike SDM, each channel is assigned a unique wavelength. Therefore, transmitters at the same lower bit rate as for SDM can be used, but colored lasers are required [1].

4.1.2 Requirements For CWDM Filters

In general, WDM allows the combination of multiple channels of different wavelengths into one single fiber where wavelength selective components such as multiplexers and filters play a vital role for combining and separating channels. In CWDM networks, the filters described next are commonly used. Depending on the network topology and capacity demand, the 16 CWDM channels can also be divided into sub-bands of 4 or 8 channels.

- CWDM multiplexer (MUX) to combine up to 16 different wavelengths into a single fiber. Requirements are low cost and a low uniform insertion loss for all channels. As the requirement for crosstalk isolation is low in a multiplexer, this device can be realized with a single stage TFF, while requiring multiple stages for the DEMUX. A wide passband of 12 to 13 nm for each channel accounts for the temperature drift of the uncooled lasers. Further details are discussed in Section 4.2 of this chapter.
- CWDM demultiplexer (DEMUX) is used as a pair with the multiplexer for separating the CWDM channels after transmission. Typically, the isolation of the MUX unit is much smaller than the DEMUX. Example: MUX: 15 dB, DEMUX: 50 dB.
- Add/drop MUX: This device will be covered in detail in Chapter 8. In principle, the add/drop function can be realized by demultiplexing the channels first and then dropping selected channels. All the express channels, that is, wavelengths that are not being changed, are directly connected to a subsequent multiplexer combining all discrete channels into the common fiber again. At the MUX, new channels also can be added at those ports where the corresponding DEMUX side channels have been terminated. Usually, in a practical application, add/drop filters are built from single-wavelength TFFs. The express channels are not demultiplexed and remultiplexed, but stay together and experience relatively low insertion loss.

4.2 TECHNICAL OPTIONS FOR WAVELENGTH DIVISION MULTIPLEXING

Although in today's CWDM networks the most commonly deployed multiplexer technology is based on TFFs, there are several different technologies available to multiplex or demultiplex the wavelengths on a WDM network. The following section intends to provide an overview and compare the different filter concepts and their applicability to CWDM.

4.2.1 GRATINGS, ARRAYED WAVEGUIDES, AND THIN-FILM FILTERS

4.2.1.1 Fiber Bragg Gratings

Figure 4.2 shows an artistic view of a grating that has been written into the core of an optical fiber. The grating is a periodic structure of low and high refractive index segments that reflects light of a specific wavelength and passes others. The grating period determines which wavelengths are being reflected. It is possible to write gratings with changing periodicity (chirped grating) and varying strength (apodized grating) of the refractive index change into the fiber core to obtain optimized response functions. The chirped grating structure in Figure 4.2, for example, can be used for dispersion compensation where longer wavelengths are reflected first and shorter wavelengths later. However, as dispersion compensation is not the topic of this chapter, we are going to concentrate on un-chirped gratings with equal spacings where all wavelengths within the grating bandwidth are reflected equally.

Fabrication: The sensitivity of optical fiber to UV light is the enabling factor to produce FBGs since it allows creation of areas of higher refractive index in the fiber core. The preparation of the fiber with pressurized hydrogen further enhances its receptivity for UV light.

Figure 4.3 shows two methods to obtain FBGs: (i) via a holographic setup or (ii) via phase mask. In both cases, the fiber needs to be stripped before and re-coated after the writing process to allow the UV light to access the fiber core without perturbation or absorption from the cladding. The reproducibility of the phase mask process is much better than the holographic setup, which is why phase masks are commonly used in production environments.

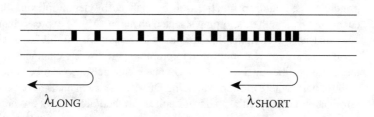

FIGURE 4.2 Principle of chirped fiber Bragg grating reflecting different wavelengths. (From Ref. 7. With permission.)

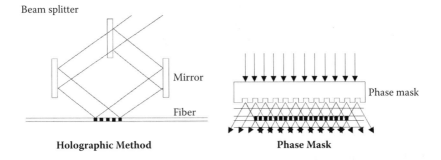

FIGURE 4.3 Two methods to write fiber Bragg gratings. (From Ref. 7. With permission.)

Figure 4.2 reveals another key property of FBGs: the wavelength of interest is reflected by the periodic structure and travels back into the direction where it originated. Since a grating is a device with only two ports (IN and OUT), this makes it necessary to combine the grating with another optical device to access the wavelength of interest. Figure 4.4 shows a configuration of an optical add and drop multiplexer (OADM) for a single wavelength. The device uses one grating and two optical circulators.

A multichannel WDM signal at the IN port passes through the 1st circulator and reaches the Bragg grating (FBG). With the FBG being designed for λ_1, that wavelength is reflected, whereas all other wavelengths pass the device, reach the 2nd circulator and the OUT port. Channel λ_1, which has been reflected by the FBG, travels back to the 1st circulator and appears on the DROP port where it can be further processed by O/E conversion etc.

The ADD port brings in the new traffic, that is, a different voice, data, or video stream that is modulated onto the same wavelength λ_1. Since the FBG is reflecting that wavelength in both directions, the grating can be used to multiplex the new traffic onto the fiber optic link. λ_1-ADD reaches the 2nd circulator and is passed to the FBG port. The FBG reflects that channel so that it is now traveling in the same direction as the other remaining channels on the WDM network that passed through the FGB to the OUT port.

To optimize the grating's performance for OADM applications, a few further technical issues need to be addressed. There are two effects in gratings that have

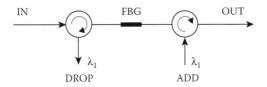

FIGURE 4.4 Optical add drop multiplexer using one grating and two circulators. (From Ref. 7. With permission.)

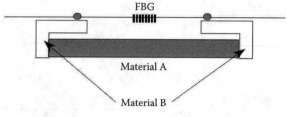

Linear expansion coefficient: A < B

FIGURE 4.5 Fiber Bragg grating with athermalized package design. (From Ref. 7. With permission.)

an impact on the center wavelength of an FBG filter function: (i) temperature and (ii) strain. Both effects need to be considered when designing an FBG package for communication networks where a temperature-dependent behavior is not desired. On the other hand, the response of the FBG's transfer function to temperature and strain makes the device very interesting for sensor applications.

A generic packaging solution is suggested in Figure 4.5. During the packaging process, the grating will be prestrained before being mounted to a structure of different materials and specific thermal expansion coefficients. During a temperature increase of the package, the strain on the grating would be reduced and vice versa; during a temperature reduction, the strain on the grating would be increased. If the changes in the strain are engineered to balance the temperature response of the grating, the shift of the center wavelength can be neglected.

Figure 4.6 shows the results of this design effort. Without compensation the center wavelength shifts at 0.01 nm/°C and has a very linear dependence on

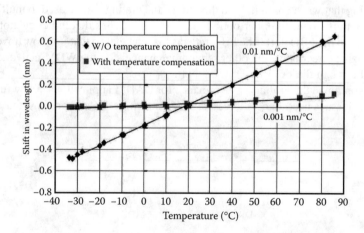

FIGURE 4.6 Fiber Bragg gratings of the center wavelength with temperature and the results of athermalization. (From Ref. 7. With permission.)

WDM Filters for CWDM

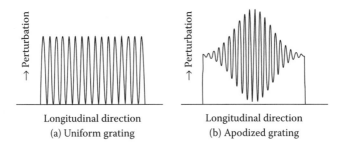

FIGURE 4.7 Relative strength of the refractive index modulation in (a) a uniform grating and (b) an apodized grating. (From Ref. 7. With permission.)

temperature. With an athermalized package design, the temperature response is reduced by one order of magnitude to 0.001 nm/°C. Today's ultra dense WDM networks with 50-GHz or 25-GHz proposed channel spacing would require even tighter control of the center wavelength.

Optimization: The performance of the FBG needs to be further optimized to provide optimum bandpass characteristics for communication networks. The apodization of a grating and the positive effects of cladding mode suppression will be discussed next. Although this FBG optimization is often necessary for DWDM, it could still be used in DWDM over CWDM applications (see Chapter 7) where closer channel spacing puts higher requirements on the crosstalk.

Figure 4.7 shows two different refractive index profiles in an FBG. In Figure 4.7a, all the high-index and low-index elements of the grating have a uniform structure, whereas in Figure 4.7b, the amplitude of the profile variation is a function of the grating length. In the case of a uniform structure, the response function of the grating is not very useful for communication links.

Figure 4.8a shows the calculated response of a uniform grating centered at the 1550 nm. With the gradual increase in isolation, this type of grating cannot

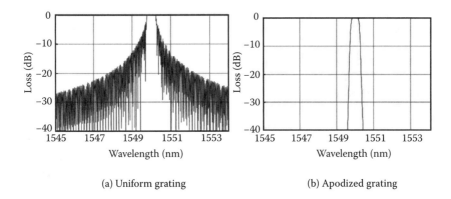

FIGURE 4.8 Calculated response function of an (a) uniform grating and (b) an apodized grating. (From Ref. 7. With permission.)

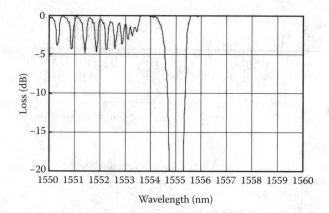

FIGURE 4.9 Measured spectral loss for a fiber Bragg grating written in single mode fiber. Shorter wavelengths experience additional loss. (From Ref. 7. With permission.)

be used in communications links with narrow channel spacing. The crosstalk from adjacent and nonadjacent channels would make it impossible to recover the signal in the desired quality. Figure 4.8b demonstrates the effectiveness of apodized gratings. The gradual increase and decrease of the refractive index perturbation has a very positive influence on the adjacent and nonadjacent channel isolation. Side lobes of the grating's spectral response function are suppressed beyond –40 dB, thus allowing an excellent suppression of the neighbor channels.

A further issue is shown in Figure 4.9: shorter wavelengths compared to the center wavelength of the grating are exposed to an additional periodic loss structure if the grating is written in standard single mode fiber (blue-side loss). This effect can be explained by the coupling of the forward-traveling waveguide mode with the backward-traveling cladding mode. In an OADM configuration (see Figure 4.4), this will impact the pass-channels or express channels in a negative way. The index profile of the grating fiber can be changed in such a way that the mode coupling is suppressed and the spectral region of the periodic pattern is moved out of the transmission window.

Figure 4.10 shows the spectral loss of an optimized FBG for narrow channel spacing. All previously discussed artifacts of Figure 4.9 have been corrected via apodization and use of high-NA optical fiber.

FBGs have found a wide range of applications in DWDM networks, as gain-flattening filters (GFFs) for optical amplifiers and dispersion compensators. However, they are not commonly used in CWDM networks because of two issues: (i) the spectral range of CWDM networks is very wide with only a few channels. Channel 1 (1271 nm) starts right above the cut-off wavelength of single mode fiber at 1260 nm. The spectral range of all 18 CWDM channels ends at 1611 nm, sometimes even including a monitor path at 1625 nm. (ii) Because of cost: the

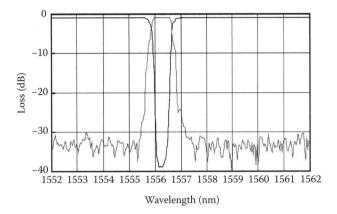

FIGURE 4.10 Measured spectral loss for an apodized fiber Bragg grating in nonstandard fiber with high numerical aperture, transmitted and reflected signal. (From Ref. 7. With permission.)

combination of FBGs with circulators does not meet the targets for low-cost capacity upgrades in the metro/edge infrastructure. Outside of communication, GFFs are used in sensing applications for temperature and stress.

4.2.1.2 Arrayed Waveguides

With arrayed waveguides, a new class of production methods is brought into the telecom/datacom world that are derived from silicon chip manufacturing [2]. The manufacturing processes rely on CAPEX intense high precision machinery and are designed to bring down prices at high volumes with the intention that the chip (AWG) design is similar to an ASIC design. The question arises whether the telecom/datacom market is big enough in offering comparable volumes to make effective use of this equipment, and it looks like the answer has been "no" at least until today. Some long-haul networks deploy high performance AWGs with narrow DWDM channel spacing in the range of 100 to 50 GHz or even 25 GHz. But real volume applications with quantities expected in consumer-level FTTH networks that play toward the strength of the manufacturing processes have not been seen yet. Figure 4.11 shows the different structural elements of an arrayed waveguide in more detail. A waveguide is an area with high refractive index that is embedded into an environment with lower refractive index. Single mode fiber is using a similar principle. AWGs are formed of arrays of individual waveguides that meet certain well-defined length and distance requirements. The planar carrier materials are typically silicon as used by the semiconductor industry. The waveguide structures can be made of silicon, silica glass, or even polymers. The most commonly used process for arrayed waveguides is silica-on-silicon, that is, glass on semiconductors. The other techniques are more or less in a research stage.

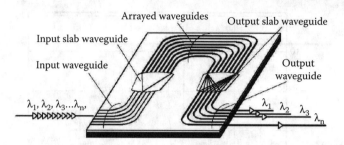

FIGURE 4.11 Structural elements of an arrayed waveguide demultiplexer. (From Ref. 8. With permission.)

As shown in Figure 4.11, the input fiber with channels $\lambda_1, \lambda_2, \ldots, \lambda_n$ is coupled into a single input waveguide [3]. This waveguide leads the light to the slab waveguide in which free-space propagation distributes the optical power evenly across the arrayed waveguides. The waveguides in the array section have well-defined optical retardation. This results in interference of the different optical signals in the output slab waveguide. The phase delay in the array is designed in such a way that through interference the center wavelength of each communication channel is focused on a different output waveguide. Each channel appears on a separate fiber.

Typically, the response function of an arrayed waveguide has a Gaussian profile, as shown in Figure 4.12.

Another important issue is the temperature stability of the center wavelength. The thermal expansion of the materials and the temperature dependence of the refractive index causes the center wavelength to drift at about 0.011 nm/°C in the 1550-nm band. Considering an operating temperature range of -5 to 70°C, this drift

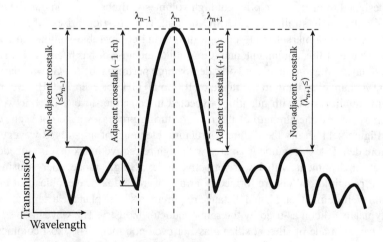

FIGURE 4.12 Response function of an arrayed waveguide. The spectrum shows the typical Gaussian profile. (From Ref. 8. With permission.)

results in a center wavelength shift of about 0.8 nm or 100 GHz. This corresponds to the full channel spacing in a 100-GHz DWDM link, which is unacceptable for such communications systems. For CWDM, the typically narrow passband of the AWG and the CWDM laser wavelength have different temperature coefficients so that the spectral overlap of filter center wavelength and the uncooled laser wavelength cannot be ensured over the entire temperature operating range.

As solutions for performance degradation due to temperature drift, two approaches have been implemented for AWGs: the first employs compensation for temperature effects in athermal AWGs [4], whereas the second uses active temperature stabilization through a heating element where a constant temperature is maintained via a control loop. The drawback of controlling the temperature is that it adds another active network element in the optical link that requires a power supply, alarm management, and the support of the network management software.

For deployment in CWDM metro or access networks, AWGs do not play an important role. To work properly with uncooled DFB lasers, the CWDM multiplexers and demultiplexers require a passband of >13 nm, which is not practical to obtain in a cost effective way with arrayed waveguides.

4.2.1.3 Thin-Film Filters

Compared to FBGs and arrayed waveguides, the TFFs are dominating the market today mainly because of three things:

1. TFFs provide the *reliability* of a passive product. The MUX/DEMUX functionality can be provided in a simple device that is made of very few sub-elements. Given a well-engineered design and stable manufacturing processes, this eliminates the need for a power supply and sophisticated alarm management.
2. TFFs provide *scalability* to accommodate all kinds of channel plans. Simple 2-channel WDM systems just using 1310 and 1550 nm can be upgraded into full 192-channel systems with combinations of band splitters and narrowband filters. The scalability also supports the common pay-per-growth scenario for operators to keep first installed cost at a manageable size.
3. Compared to the other technologies, TFFs offer a very attractive *price* per channel. Gratings demand a higher price since they need a circulator to drop or add the wavelengths. Arrayed waveguides have high cost at low volumes to amortize a wafer fab. AWGs have cost advantages of TFFs with higher channel count systems, typically 16 and above.

TFFs successfully address the need of the market for customization. This includes rapidly changing channel plans, increased performance requirements such as insertion loss, return loss, and isolation and physical size (Table 4.1).

The following sections will provide a more detailed discussion of the properties of TFFs and the different packaging solution.

TABLE 4.1
Comparison of Different Filter Type-Fused Coupler, FBG, AWG, and TFF

	Number of Channels	Channel Spacing	Insertion Loss	Isolation	Cost per Channel	Size
Fused biconic taper devices	2	wide	low	medium	low	small
Gratings & circulators	low	wide or narrow	medium	high	high	large
Arrayed waveguides	high	narrow	medium	medium	high	large
Thin-film filters	low to medium	wide or narrow	medium	medium to high	medium	small to medium

4.2.2 Trade-Offs and Performance

Passive CWDM devices are characterized by a set of optical parameters, which are defined in Figure 4.13. The following list intends to provide an overview of some basic parameters that are used to specify performance expectations of devices for CWDM networks.

- Insertion loss [dB]: Loss of light within the passband of the optical filter. This parameter is used for power budget calculations and network planning.
- Passband width [nm]: wavelength range around the ITU center wavelength of the channel. The device typically exceeds this to compensate for temperature and polarization effects.
- Passband ripple [dB]: difference of min. to max. insertion loss within the passband. The number is an indication for output power variations in case a laser changes wavelength with operating temperature.
- Isolation [dB]: This value is an indication for the amount of light that is leaking from adjacent and nonadjacent channels into the channel under test.
- Polarization-dependent loss [dB]: Since the transmitters typically launch highly polarized light into the fiber, the polarization sensitivity of the optical devices in the network needs to be as low as possible.
- Center wavelength thermal stability [nm/°C]: The wavelength stability describes the amount of thermal drift of the channel center wavelength; this number is not critical for CWDM systems.
- Return loss [dB]: The amount of light that is reflected from the device back into the launch fiber. A low return loss (= strong back reflection) could lead to instabilities for the laser source.
- Directivity [dB]: The amount of light that appears on the wrong port when the device is used as a multiplexer. Like return loss, a low

WDM Filters for CWDM

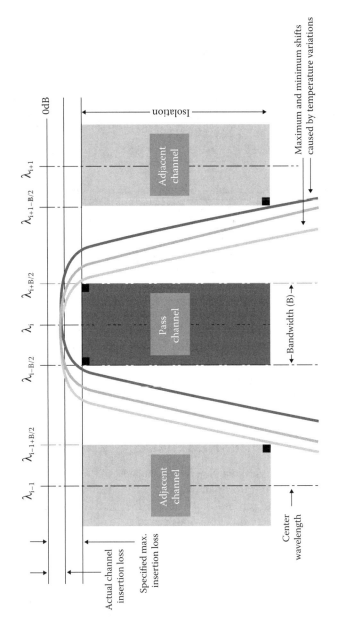

FIGURE 4.13 Definition of some basic performance parameters.

TABLE 4.2
Performance Data of TF Filters for CWDM

Parameter	Units	Min.	Typical	Max.
Insertion loss	[dB]	1.5	2.0	2.2
Passband width	[nm]	±6.5	±7.0	±7.5
Passband ripple	[dB]	0.1	0.15	0.2
Isolation — adjacent channel	[dB]	25	30	35
Isolation — nonadjacent channel	[dB]	40	45	50
Reflect isolation	[dB]	12	13	15
Polarization-dependent loss	[dB]	0.1	0.15	0.2
Center wavelength stability	[nm/°C]	0.001	0.002	0.005
Return loss	[dB]	45	50	55
Directivity	[dB]	50	55	60
Operating temperature range	[°C]	−5/65	−40/+85	−40/120

directivity could have a negative impact on the transmitter performance of the adjacent and nonadjacent channels.
- Operating temperature range [°C]: The temperature range over which the device is expected to be fully functional and within all other specification requirements. There is a strong trend toward outside plant deployment of CWDM components, therefore making an operating temperature range of −40 to +85°C a basic requirement. Some surveillance or security applications even require operation up to 120°C.

Table 4.2 provides a performance summary of a typical CWDM 8-channel device. The parameters Min., Typical, and Max. are intended as an overview over the range of performance requirements that are requested by the market. As an example, most customers request an insertion loss of around 2.0 dB for the 8-channel device without connectors, some really demanding applications need 1.5 dB, and some relaxed link budgets can tolerate up to 2.2 dB.

4.3 PROPERTIES OF THIN-FILM FILTERS

In this section, we will review the underlying concepts of dielectric TFFs in more detail. The intention is to provide an overview of the technology and to outline a generic manufacturing process that enables the performance requirements, which have been discussed in the previous section.

4.3.1 THIN-FILM MULTI-CAVITY STRUCTURES

TFFs are made of materials with different refractive index in an alternating layer structure. These layers are called dielectric coatings in contrast to metallic coatings such as gold, silver, aluminum, etc. Simple coatings are made out of a few layers of dielectrics such as Tantalpentoxide (Ta_2O_5) as a high index material and Silicondioxide (SiO_2) as the low index material. Figure 4.14 shows a side view photo of a CWDM optical filter element, including the dielectric coating layer.

WDM Filters for CWDM

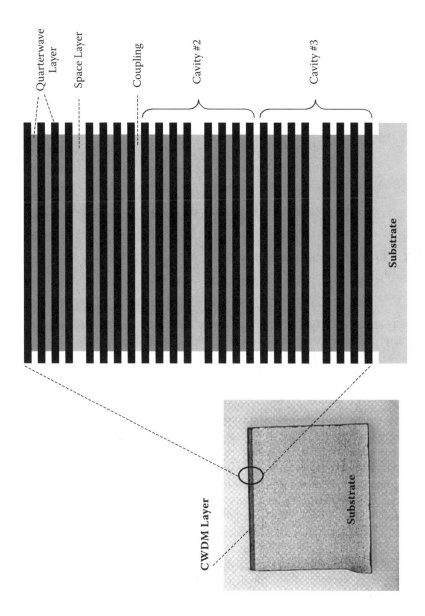

FIGURE 4.14 Coating layer structure of a 3-cavity thin-film filter.

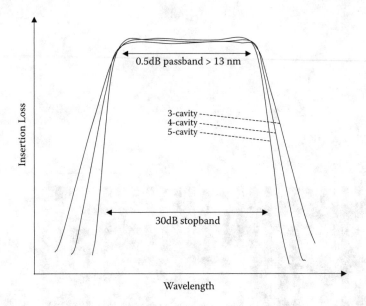

FIGURE 4.15 An increasing number of cavities results in a smaller stop-band or improved adjacent channel isolation.

The coating layer typically consists of a number of Fabry-Perot cavities, which determine the performance of the optical filter. Figure 4.14 shows an example with three cavities. Each cavity has a "spacer layer" to establish a fixed and stable spacing between both reflectors. The reflectors are made of a sequence of low-index and high-index quarterwave layers. The coupling between the different cavities is accomplished by a coupling layer [5].

The bandpass characteristics of this type of TFF can be modified by changing the number of cavities. Figure 4.15 shows a performance comparison of a 3-, 4-, and 5-cavity filter. With the increased number of layers, it is possible to obtain steeper filter slopes at the same width of the passband. This results in an improved adjacent channel isolation of the demultiplexer.

For typical CWDM filters, the dielectric coating structure has a thickness of about 30 to 40 μm and can be clearly identified under a microscope (see Figure 4.14). The number of optical layers ranges between 130 and 160, and can be as high as 200 layers for high performance applications.

The backside of a dielectric filter element typically has a wideband antireflection (AR) coating to reduce the losses otherwise caused by internal reflection. For CWDM filters, such AR coatings need to perform over a wide spectral range, typically 1260 to 1620 nm and have 60 to 80 quarterwave layers.

Figure 4.16 shows a simple configuration of optical beams going through a single filter element. A collimated beam (IN) enters the filter substrate under a certain

WDM Filters for CWDM

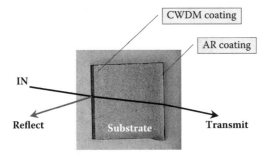

FIGURE 4.16 Definitions for input and output beams, IN, Transmit and Reflect.

angle of incidence (AOI). Depending on the packaging concept, AOIs can range from <2° to well beyond 10°. The transmit beam experiences a slight offset due to the higher refractive index of the substrate.

There are two basic types of dielectric filters that are used in TFFs devices, which are named by the characteristic spectral performance: (i) edge pass filters and (ii) bandpass filters.

The *edge filters* have a single slope or "edge" that separates the spectral range of high transmission from the spectral range of high blocking or reflection. Edge filters split into two subgroups: (i) short pass filters and (ii) long pass filters. The short passes transmit the light of the shorter wavelengths and reflect the longer wavelengths. Its counterpart is the long pass filter that transmits the longer wavelengths. A typical long pass filter for 1310/1550 nm is shown in Figure 4.17. The reflect path provides access to the 1310-nm channel, whereas the transmit path provides access to the 1550-nm band, including all eight CWDM channels from 1471 to 1611 nm.

The *bandpass filters* are typically used to provide access to an individual CWDM channel. Figure 4.18 shows a typical spectral scan for a CWDM bandpass filter centered at 1551 nm. The light that belongs to the CWDM channel at 1551 nm will pass through the filter element (transmit) while all other wavelengths will be reflected.

Other more complex types of bandpass filter elements are also deployed in modern CWDM networks. Figure 4.19 shows an optical filter element called "4-skip-0." Its bandpass characteristics are designed in such a way that four CWDM channels (1511 to 1571 nm) are transmitted by the filter element, whereas the two channels left (1471 and 1491 nm) and right (1591 and 1611 nm) appear on the reflect path. This device is useful for pay-per-growth and in-service upgrade concepts, as well as DWDM over CWDM. It provides access to the C-band (1530 to 1565 nm), which enables system designers to leverage DWDM combined with erbium-doped fiber amplifiers (EDFAs) to further boost the number of transmission channels.

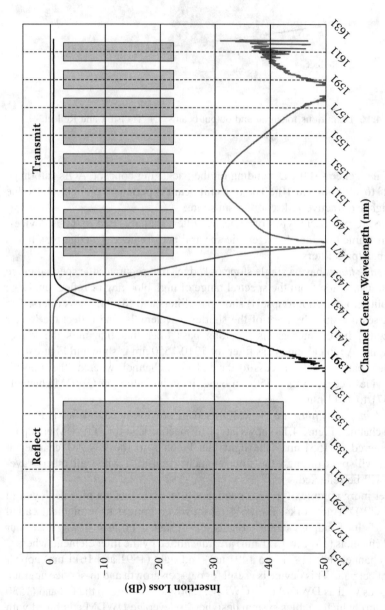

FIGURE 4.17 Edge pass filter (long pass) splitting the 1310-nm window from 1550-nm window.

WDM Filters for CWDM

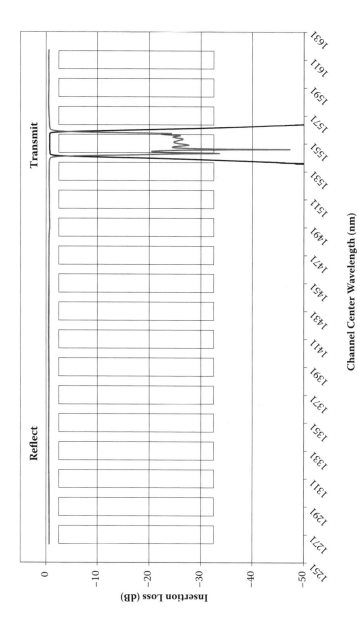

FIGURE 4.18 Example for a CWDM bandpass filter. (Reference in the table was changed to min of 25 dB.) The filter element transmits the light corresponding to the 1551-nm channel while all other wavelengths appear on the reflect port.

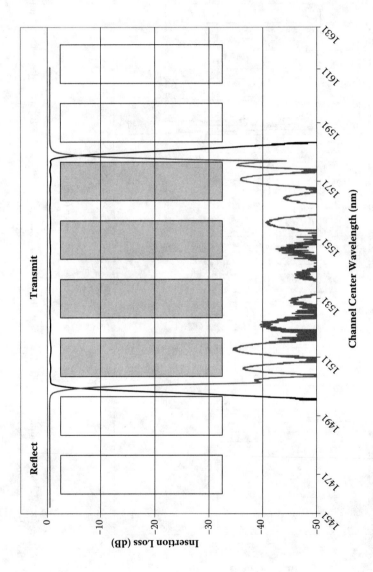

FIGURE 4.19 Example for a special type of bandpass filter called "4-skip-0."

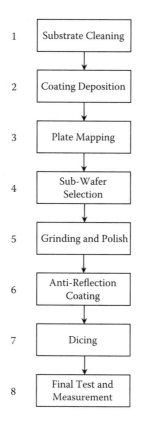

FIGURE 4.20 Process flowchart for thin-film dielectric coatings.

4.3.2 MANUFACTURING PROCESS STEPS

The manufacturing process for thin-film coatings can be best described in eight basic process steps. The flowchart in Figure 4.20 shows the process steps from initial substrate cleaning to the final inspection.

The glass wafers typically have a 6 to 8 inch diameter and a thickness of about 6 to 10 mm. These glass substrates need to be extremely clean before they can be coated. Very complex cleaning and rinsing processes have been developed to support these requirements. The processes involve using automated wafer handling, highly purified water, and ultrasound cleaning. However, the final step of the cleaning process is usually manual inspection by an experienced operator. Substrates are then moved into the coating chamber where the first of two coating processes starts.

To monitor the process during the earlier days of CWDM and DWDM filter production, the chambers typically used a glass substrate with a resonant frequency that was tested/interrogated by ultrasound excitation. As the deposition process progressed, the resonance frequency of the substrate changed, giving the operator

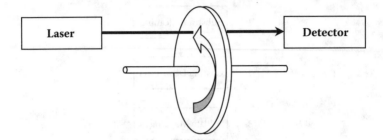

FIGURE 4.21 *In situ* monitoring capability of modern coating chambers. A laser beam is directed through the spinning substrate to monitor the progress of the deposition process.

an indication of how far along they were in the deposition process. However, the optical properties of the wafer remained unclear until the deposition process was completed and the wafer was removed from the chamber for characterization.

Today's modern chambers are equipped with optical *in situ* examination capability, which enables monitoring of the deposition process, thereby yielding much better process control. The underlying principle is simple (see Figure 4.21): an optical beam ideally from a laser source outside the chamber is directed through the windows of the chamber onto the wafer. Typically, the laser is set to a wavelength where the coating that is processed in the chamber would have high transmission and low reflection. The transmitted light after passing through the wafer is then directed onto a photo detector [6]. The switch point between high-index and low-index material can be controlled by a decision circuit that compares the actual absorption of the layer structure with the theoretical curve.

Depending on the complexity of the coating, not all areas of the wafer meet the targeted specification requirements. For very demanding filter designs, only a small ring around the monitor beam might yield into usable product. Therefore, most wafers are subjected to a plate-mapping process where individual spots on the wafer are characterized for optical performance. This approach provides a landscape overview of the product performance throughout the wafer and determines which parts of the wafer are useful for further production. A limitation of this process is that the optical characteristics of the coating can only be determined in transmission, while for certain applications also, the reflection characteristics, for example, reflection isolation, are of interest.

A quick assessment of the reflection characteristics from a transmission spectrum can be obtained by assuming no absorption in the coating

$$100\% = T\% + R\% \Rightarrow \mathrm{Iso\_R\ [dB]} = 10 \cdot \log\left(1 - 10^{\frac{\mathrm{Ripple\_T[dB]}}{10}}\right) \quad (4.1)$$

where T = transmission, R = reflectance, and Ripple_T = passband ripple in transmission (<0).

WDM Filters for CWDM   113

With this approach in Equation 4.1, the Rx spectrum can be calculated from a Tx spectrum of a coating, giving a first indication of the Rx performance. Figure 4.22 shows a comparison of the calculated Rx spectrum with the actual Rx spectrum for a fully assembled device. This simple concept clearly underestimates the reflectance isolation but still gives a first performance indication.

After plate mapping is completed, the wafer is typically cut into smaller subsections that are expected to yield good product. These sub-wafers are then polished to the desired thickness of the target specification. This step also determines the wedge angle between the dielectric coating and the back surface. Some packaging technologies require about 0° of wedge, whereas others are designed for a nominal wedge of 0.7° to avoid multipath interference within the glass substrate. With high performance anti reflection coatings (AR-coatings) on the back surface, these interference effects can be avoided.

The sub-wafers have coatings on both sides, typically a WDM coating on one side and a more simple AR coating on the opposite side. The final handling step is the dicing process where the sub-wafers are cut to the final size. Dicing is an art and clearly goes beyond simple "cutting" of the wafer. A simple CWDM coating has a thickness of about 30 to 40 µm; this generates stress on one side of the sub-wafer. The speed of the blade, depth, etc. very much determines the appearance of the filter. Inappropriate settings can lead to chipped coating surfaces that further limit the performance of the filter. The dicing process is followed by an intense cleaning process to remove residues from the previous step. The individual filters are placed into gel packs or similar arrayed packages in preparation for the final step.

Final inspection and packaging is typically done with automated test and measurement equipment. A pick and place machine positions a filter element on a pivot mount filter holder. The filter needs to be angle tuned to ensure that the CWDM coating is tested at 0° AOI. Other angles would lead to an error in the center wavelength, insertion loss, and passband width. After adjusting the AOI, the spectral performance of the filter is recorded and stored in a database for further processing.

4.4 THIN-FILM FILTER PACKAGING SOLUTIONS

Over the past years, many different packaging concepts for TFFs have been developed and introduced to the markets. This section will provide an overview of the different packaging concepts, starting with the basic approach to package individual filter elements.

4.4.1 CASCADED 3-PORT PACKAGES, MULTI-PORTS, AND GLASS PACKAGE

4.4.1.1 3-Ports

Figure 4.23 shows a typical 3-port package. The name "3-port" is derived from the number of ports or pigtails that these devices have. In most cases, these

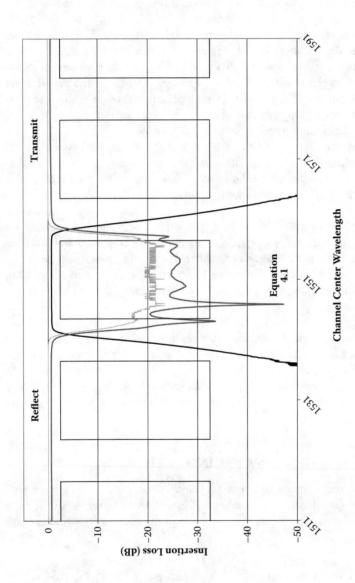

FIGURE 4.22 Transmittance and reflectance spectrum of a CWDM 1551-nm filter; the light trace shows the calculated reflectance.

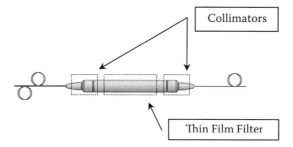

FIGURE 4.23 A typical 3-port package.

devices are between 35 and 50 mm long, depending on the requirements for strain relieves to protect the pigtails. Package diameters are between 5 and 6 mm. In any case, the 3-port devices will have fibers on both sides of the package.

As shown in Figure 4.24, light from the input port or "common" port fiber pigtail enters into a dual fiber ferrule, which is a structural element that holds two optical fibers at a certain distance from each other. The light from the common port enters a gradient index lens (GRIN) which collimates the beam. This sub-assembly of the dual fiber ferule and a GRIN lens is called "dual fiber collimator."

The beam emitting from the common port is directed onto the dielectric filter element that is closely mounted in front of the dual fiber collimator. All wavelengths reflected by the dielectric coating travel back into the dual fiber collimator, while the GRIN lens focuses the light into the 2nd fiber in the ferrule assembly. Because this port collects all the reflected light, it is sometimes called the "reflect port." One of the key advantages of this assembly is that the design allows very low AOI, typically something between 1.8 to 2.8°. At these low angles, the loss variation between s- and p-polarized light that is implied by the dielectric coating is very low. As a consequence, the 3-port device has low polarization-dependent loss (PDL) and shows no sensitivity to polarized light.

The light that passes the dielectric filter element is collected by a 2nd collimator that is aligned opposite to the common port of the dual fiber collimator. This 2nd collimator consists of a GRIN lens and a single fiber and is therefore referred to as "single fiber collimator." Since the light is transmitted through the filter, this port is called the "transmit port."

The typical performance specifications that can be achieved with this type of assembly are transmittance insertion loss IL-T < 1.0 dB and a reflectance insertion loss IL-R < 0.3 dB over operating temperature from 0 to +70°C. A further extension of the operating temperature range from −40 to +85°C is possible, but approaches the technical limitations of this design. Typically, the package is assembled into two major alignment steps: (i) dual fiber collimator to filter and (ii) filter to the single fiber collimator. In both steps, the optimum position is found in an active alignment process, that is, light is traveling through the assembly while an operator or computer algorithm makes small position corrections until

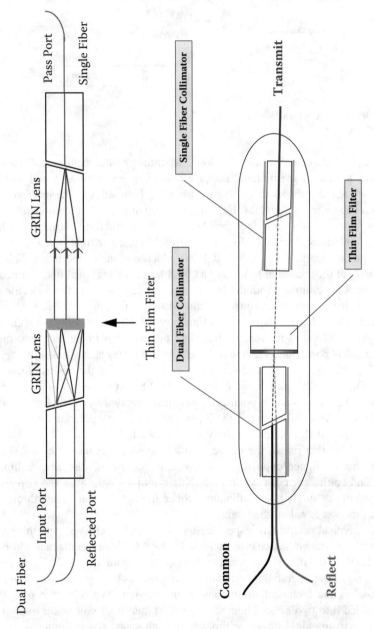

FIGURE 4.24 Key optical elements of a 3-port package; details are discussed in the text.

WDM Filters for CWDM

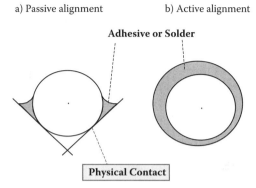

FIGURE 4.25 Two basic concepts to align optical fiber: (a) through physical contact to a passive alignment structure and (b) via active alignment.

the maximum throughput is reached. This position is then fixed either through adhesives or solder joints.

The implications of fixing two parts after a multidimensional alignment means that the joint structure is very asymmetric and becomes part of the alignment structure, as shown in Figure 4.25. With increased stability requirements over a wider temperature range, the asymmetric joint structure is at a disadvantage over the passive alignment structure. The thermal expansion of the adhesive or solder is unevenly distributed in the joint, leading to undesired changes of the position of the single fiber collimator. Such changes in the position lead to coupling losses with temperature, sometimes also called temperature-dependent loss (TDL).

The basic concept of the 3-port package has been developed further into two directions: (i) the number of ports has been increased to create "multi-ports" around a single filter element and (ii) the original material combination (brass– solder) has been changed to glass–adhesive to create "all-glass" packages or "green 3-ports."

Especially, the green products are increasingly interesting for the market since as of July 1, 2006 the EU directive 2002/95/EC for the Restriction of Hazardous Substances (RoHS) requires a limitation of lead-based solder in electronics assemblies. However, there is an exception for telecom infrastructure equipment in place that allows a certain weight% of lead in the assembly.

4.4.1.2 Multi-Ports

Figure 4.26 shows the concept behind the idea of using the same filter several times, for example, as OADM, for a single wavelength in a network node. The dual fiber collimator has been replaced by a quad fiber collimator and the single fiber collimator has been changed into a dual fiber collimator. With this concept, two independent optical paths have been overlaid to use the same TFF. The packaging density doubled, that is, one device replaced the functionality of two original 3-ports.

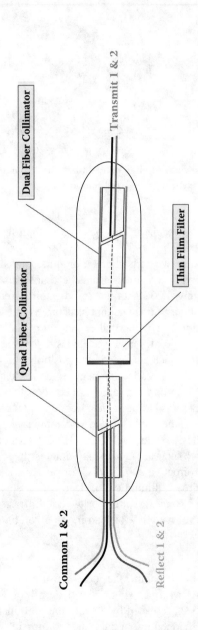

FIGURE 4.26 Conceptual drawing of a 6-port device; the same filter is used twice in two independent 3-ports sharing the same package.

This is an excellent design strategy for increasing the channel density and lowering the device cost through shared filter use. This is possible where multiplexer and demultiplexer are in close proximity of each other. However, a careful design of the package is required to take full advantage of the concept.

Some considerations need to be given to scattered light traveling inside the package. One example of a source could be a high-powered transmitter that is added to the network using the add-function of the OADM while a highly sensitive receiver is connected to the drop port of the device. Since the drop function and the add function share the same physical package in this 6-port design, especially the directivity becomes an important parameter.

4.4.1.3 Glass Packages

The concept of an all glass package follows the same design approach as a classical 3-port package, while all materials have been changed to glass. The all glass concept addresses a fundamental issue of the earlier 3-port packages: the mismatch of the thermal expansion coefficient of the different materials (brass, solder, glass, adhesive, etc.), which has been seen as the root cause of the temperature sensitivity of these devices.

In addition to an improved performance over temperature, the all glass concept also addresses the issue with new regulations such as the RoHS directive 2002/95/EC of the European Community for Restriction of Hazardous Substances. The glass packages obsolete the use of lead-based solder with low melting temperatures.

4.4.1.4 CWDM Modules

In order to build multi-channel devices like 8-channel multiplexer or demultiplexer, several 3-port devices need to be cascaded and connected with optical splices. This makes it necessary to provide a secondary housing that holds all the required 3-ports, splices, and sometimes also band splitters.

The design of such a secondary package is expected to pass module level endurance and reliability testing per Telcordia GR-1221. The three critical elements are: (i) the technique to mount the 3-ports and the splice protectors to withstand mechanical shock and vibration testing, (ii) ensure that the minimum bend radius requirements of the fibers are met, and (iii) the fiber feed through is designed to meet retention and side pull requirements.

Depending on the complexity of the system and the number of channels, the size of an 8-channel CWDM module that uses concatenated 3-ports is approximately $100 \times 80 \times 10$ mm^3.

Another interesting aspect is that the insertion loss increases for each channel in the channel sequence is giving very different launch power conditions for the individual channel. This insertion loss sequence has been calculated for MUX and DEMUX in Figure 4.27. At the demultiplexer in the receiver sites, this effect can be compensated by reversing the cascade of channels relative to the multiplexer. This approach to optimize the power budget of an optical link is possible by pairing

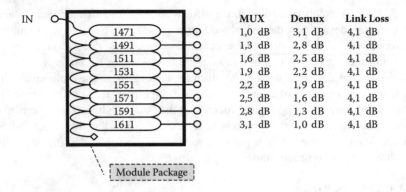

FIGURE 4.27 Cascaded 3-port devices in a CWDM 8-channel module. A secondary housing needs to protect the fiber management.

a multiplexer with a demultiplexer. The technical improvement makes it necessary to clearly distinguish between multiplexer and demultiplexer and to identify them via different part numbers. A channel assignment taking into account the varying losses for the particular CWDM channels is discussed in Chapter 5.

4.4.2 Multiple Bounce Concept

The multiple bounce concept is a different approach to cascade optical filter elements with different center wavelengths and to build a device that combines or separates optical channels. In this concept (Figure 4.28), a collimated beam is reflected by each filter element without re-launching the light back into a fiber

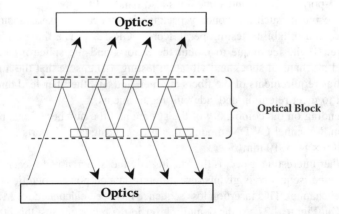

FIGURE 4.28 Conceptual drawing of a free-space multiple bounce architecture. Individual thin-film filter elements are attached to an optical block to form the core of the device.

WDM Filters for CWDM   121

in between. The concept itself is not new. Compared to 3-ports it has several advantages but also some drawbacks, which will be discussed in the following.

The multiple bounce concept offers an interesting potential for further size reduction and allows cost saving on the bill of material. For an 8-channel device, the number of collimators is reduced to nine pieces, that is, one input + 8 output collimators, all of them being single fiber collimators. For comparison, an 8-channel device made of concatenated 3-ports needs 16 collimators, half of it being the more complex dual fiber collimators.

However, the manufacturing process presents some intrinsic challenges that are a lot easier to overcome in 3-port designs. In a 3-port assembly process it is possible to build many of the same parts, except maybe for the center wavelength of the optical filter inside. Each channel (3-port) individually can be optimized for performance using the full set of degrees of freedom. For comparison, after the input collimator of a multiple bounce unit has been fixed, it is fixed for all channels, which makes this an extremely critical and sensitive alignment step.

The real breakthrough of the multiple bounce concept came with the advent of CWDM systems. The relaxed system requirements allowed higher tolerances in the mechanical design of the passive optical multiplexer.

The current design concepts can be further differentiated into two groups: The first group is deploying the standard collimator technology with gradient index or C-lenses in the assembly. The devices typically have fibers on both sides of the package. The second group is using arrayed optical elements to generate collimated beams and combines these optics arrays with the optical filter block. This unique combination opens a path to further size reduction, as illustrated in Figure 4.29.

FIGURE 4.29 Size comparison of a standard module based on 3-port technology versus a new miniature design concept.

All of the multiple bounce designs use a higher AOI compared to 3-ports. Typical AOI numbers are between 10° to 14°. Assuming a fixed size of the optical filter elements between 1.0 and 1.4 mm^2, this enables a very compact design. On the other hand, the high AOI causes stronger s- and p-polarization splitting in the CWDM coating, leading to an increased PDL. For CWDM systems, this is still acceptable (passband and bounce angle are related) because of the relaxed filter tolerances and the large channel spacing.

Typically, CWDM device manufacturers would specify wider optical passbands at the filter level to be able to meet the PDL requirements at the device level. The tighter channel spacing for DWDM 200 GHz and 100 GHz can be best supported by low AOI packaging concepts because of even more demanding PDL requirements within the channel passband and tighter center wavelength tolerances. Multiple bounce concepts with high AOI do not seem suitable for this application.

4.5 FUTURE TRENDS AND REQUIREMENTS

With continued deployment of bandwidth hungry applications in the metro/edge and access markets, some key attributes can be identified that will drive future product requirements.

First, more and more equipment is installed in outside plant environments such as street cabinets, splice closures, on aerials or in pedestals. This makes it necessary to adopt the operating temperature range of CWDM devices and modules to these harsher conditions, calling for outside plant reliability and endurance testing. These new conditions include operating temperatures starting as low as −40°C and going up to +85°C and beyond while moisture is present at 95% relative humidity levels. These environmental conditions will further drive packaging and sealing technology, as the market requires more and more cost-effective solutions.

The second trend seems to be a tendency to retro-fit existing optical links and upgrade the capacity of installed network connections rather than writing off the "old" equipment and installing brand new off-the-shelf solutions. In most cases, this "older" equipment is probably not that old and has seen less than 5 years of field life. Future passive components need to be flexible enough to address this situation. An example is the upgrade of an existing 1310-nm SDH or SONET link with a CWDM overlay. Ideally, the multiplexer and demultiplexer have an embedded 1310-nm port that can handle the existing traffic, while adding four or eight more CWDM channels to the optical link.

The third trend, which is probably further out on the roadmap, is the need for further integration of optics with electronics. There is a wide range of proposals to bring the world of optics and the world of electronics closer together by integrating lasers and detectors into smaller packages or onto a single chip. The tremendous success of pluggable optical transceivers clearly demonstrates the market acceptance of integrated products as the transceivers solve a major issue at the optoelectronic interface. Other optoelectronic interface solutions are designed today to access four and more channels without fiber management

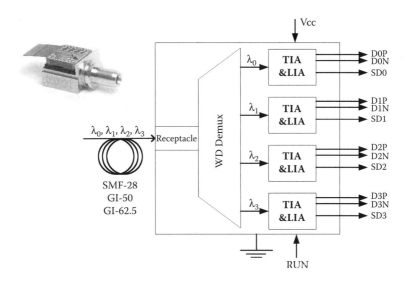

FIGURE 4.30 A new generation of integrated devices for optoelectronic interfaces will enable more compact and cost-effective system solutions.

between the demultiplexer and the receiver. Figure 4.30 shows an example of a 4-lane receiver optical subassembly that integrates a 4-channel demultiplexer (WD demux) with four independent PIN photodiodes as well as transimpedance (TIA) and limiting amplifiers (LIAs).

REFERENCES

1. Agraval, G.P., *Fiber-Optic Communication Systems*, Wiley Series in Microwave and Optical Engineering, 2nd edition, 1997.
2. Kaneko, A., Recent progress on arrayed waveguide gratings for WDM applications, Proceedings LEOS Summer Topical Meetings, 1999, pp.29–30.
3. Dutton, H., *Understanding Optical Communications*, Prentice Hall, 1998.
4. Hasegawa, J. and Nara, K., Ultra-wide temperature range (−30–70°C) operation of athermal AWG module using pure aluminum plate, Proceedings OFC 2006.
5. Ramaswami, R. and Sivarajan, K., *Optical Networks — A Practical Perspective*, Morgan Kaufman, 1998.
6. Beißwenger, S., Götzelmann, R., Matl, K., and Zöller, A., Low Temperature Optical Coatings with High Packing Density Produced with Plasma Ion-Assisted Deposition, Society of Vacuum Coaters, Proceedings SVC Conference, 1994.
7. Ota, I. et al., Development of optical fiber gratings for WDM systems, *Furukawa Review* 19; 35–40, 2000.
8. Saito, T. et al., 16-Ch arrayed waveguide grating module with 100-GHz spacing, *Furukawa Review* 19; 47–52, 2000.

5 Optimizing CWDM for Nonamplified Networks

Charles Ufongene

CONTENTS

5.1 Introduction 125
5.2 Optimized Design of CWDM Networks 128
    5.2.1 Attenuation Slope Compensating Wavelength Assignment 128
        5.2.1.1 Discussion 130
    5.2.2 Ring Perimeter 131
    5.2.3 Channel Power 133
    5.2.4 Impact of Loss 133
5.3 Generalized Case of CWDM Design 135
    5.3.1 Discussion 136
5.4 Application: Wavelength Assignment in 4-Node Ring 138
    5.4.1 Wavelength Assignment 138
    5.4.2 Calculation of Filter Losses 143
5.5 Analysis of Results for 4-Node Ring 152
    5.5.1 Nonattenuation Slope Compensating Wavelength Assignment 152
    5.5.2 Attenuation Slope Compensating Wavelength Assignment 152
    5.5.3 Impact of Ring Perimeter 154
    5.5.4 Extension to 8-Node Ring 154
5.6 N-Node CWDM Meshed Network 156
    5.6.1 Wavelength Assignment 156
    5.6.2 Optimization Techniques 162
5.7 Conclusions and Future Work 168
References 169

5.1 INTRODUCTION

Coarse wavelength division multiplexing (CWDM) is set to introduce multiwavelength optical systems cheaply into the metro network. Compared to dense wavelength division multiplexing (DWDM), CWDM achieves cost reduction via the use of cheaper wide channel spacing filters, which in turn allow the use of cheaper uncooled lasers in CWDM systems. However, because CWDM systems are nonamplified, the attainable system reach is severely limited by filter and fiber attenuation losses. While the use of amplifiers is discussed in Chapter 6, we are going to focus here on per-channel techniques to increase the link budget and thus enhance the reach of nonamplified CWDM transmission links.

This chapter presents a CWDM network design approach that can be employed to maximize the perimeter of a 16-channel CWDM ring network based on a G.652D-class fiber [zero water peak fiber (ZWPF)], such as AllWave. All 16 channels of the full spectrum (FS) CWDM channel plan extending from 1310* nm in the O-band to 1610 nm in the L-band can be supported on a single fiber, as already described in Chapter 2. A wavelength assignment algorithm assigns wavelength bands to the nodes in such a way that the accumulated filter losses incurred by the O-band wavelengths around the ring are minimized and increase to its maximum in the L-band. In contrast, ZWPF attenuation as a function of wavelength reaches a maximum in the O-band and monotonically decreases to a minimum value in the L-band. Hence, the increasing attenuation slope of the filters toward longer wavelengths is partially offset by a falling attenuation slope of the fiber loss curve, yielding a ring (combined fiber and filter) loss that is minimized across the entire CWDM channel spectrum, thereby maximizing the link budget for the optical path. This technique is further applied to a logical star network topology in which four local exchanges nodes (central offices) transmit and receive signals, each on a set of four wavelengths reconfigured as a 4-node CWDM hubbed ring network. The derived wavelength assignment algorithm is employed to maximize the perimeter of this ring to 54 km, including fiber and component losses that are within the order of typical metro ring perimeters**. Therefore, the main purpose of this approach is to increase the application range of nonamplified CWDM for metro rings, to demonstrate the competitiveness of CWDM for new segments of the metro market that are dominated by short-reach DWDM, time division multiplexing (TDM), and space division multiplexing (SDM) for metro deployment. This will further enhance the competitiveness of G.652D fiber since these fibers offer a seamless support of all 16 (18) CWDM channels on a single fiber compared to standard G.652 (SSMF), which only supports a maximum of 12 CWDM channels.

The FS-CWDM channel plan being standardized by the ITU-T [1] comprises 16 nonamplified channels, with center wavelengths starting at 1310 nm with 20-nm channel spacing as shown in Table 5.1 and discussed in more detail in Chapter 1. The channel plan extending from O-band to the L-band is made possible by the development of G.652C/D fiber with very low or ZWPF attenuation at 1400 nm [2,3]. The CWDM system achieves cost reduction in comparison to DWDM through the use of multiplexing and demultiplexing filters with wide channel spacing and passband. Filters used in DWDM systems have very narrow Gaussian passbands and are typically spaced at 50, 100, or 200 GHz, (0.4, 0.8, and 1.6 nm) [4]. To prevent inter-channel crosstalk, DWDM systems must therefore

---

* In the following, the channels are denoted 1310, 1330 nm, etc, although the corresponding nominal center wavelengths are 1311, 1331 nm. See also Chapter 1 on CWDM standards.
** Over 90% of metro rings in the U.S. have perimeters of less than 100 km.

TABLE 5.1
ITU-Based CWDM Channel Plan for 12 CWDM Channels

Fiber Att. [dB/km]	Laser Wavelength (nm)	Channel #	
0.360	1310 nm	1	
0.335	1330 nm	2	O-Band
0.322	1350 nm	3	
0.311	1370 nm	4	
0.333	1390 nm	5	
0.291	1410 nm	6	E-band
0.281	1430 nm	7	
0.272	1450 nm	8	
0.266	1470 nm	9	
0.260	1490 nm	10	
0.254	1510 nm	11	
0.252	1530 nm	12	S-, C-, L-band
0.250	1550 nm	13	
0.250	1570 nm	14	
0.257	1590 nm	15	
0.266	1610 nm	16	

use wavelength-stabilized optical sources with a sufficiently narrow spectral width demanding the use of externally modulated lasers. CWDM systems, on the other hand, employ filters with 20-nm channel spacing and a 13- to 14-nm passband that allow the filters to be more cheaply produced due to the less restrictive constraints. Further, the wider filter passband also allows us to use cheaper directly modulated uncooled lasers in CWDM systems. However, since the CWDM systems are nonamplified, the attainable link reach is limited by the filter, connector, splice, and fiber losses. As component losses can only be minimized but not eliminated, the task here is to develop a suitable wavelength assignment scheme that, when applied to a 16-channel CWDM n-node ring, minimizes those total losses and increases the reach. Figure 5.1 shows the ZWPF attenuation as a function of wavelength in the 16-channel CWDM channel plan. The intrinsic fiber attenuation decreases from 0.36 dB/km at 1310 nm to 0.25 dB/km at 1570 nm and thereafter increases (due to micro bending losses) to 0.266 dB/km at 1610 nm. Across the CWDM channel spectrum, fiber attenuation reaches its maximum in the O-band and approximately decreases to a minimum in the L-band. Therefore, if the filter loss* incurred by the wavelengths in an n-node ring hubbed network is minimum in the O-band and approximately increases to a maximum in the L-band, then the filter loss will have an inverse slope to the fiber loss slope across the CWDM channel spectrum. The filter loss then uniformly compensates the fiber loss, resulting in a ring loss (the combined filter and fiber losses) minimization

---

* This includes connector losses.

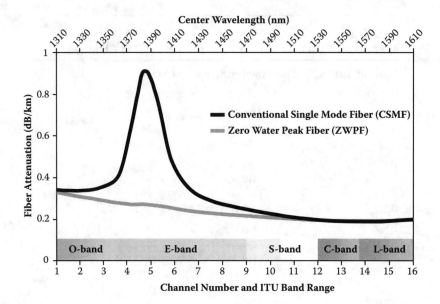

FIGURE 5.1 Measured AllWave fiber (ZWPF) spectral attenuation as a function of wavelength for all considered 16 CWDM channels compared to conventional single mode fiber.

across the CWDM channel spectrum. A wavelength assignment algorithm that achieves this objective is presented next [5].

Terms used in this chapter:

Express/through loss: incurred by wavelengths that pass through a node without being added or dropped.
Drop loss: incurred by wavelengths that are dropped at a node.
Add loss: incurred by wavelengths or band that are added at a node.
Demux loss: incurred by wavelengths when demultiplexed into individual wavelengths.
Mux loss: incurred by wavelengths when multiplexed together into a single band.

5.2 OPTIMIZED DESIGN OF CWDM NETWORKS

5.2.1 Attenuation Slope Compensating Wavelength Assignment

Figure 5.2 shows a 16-channel CWDM n-node hubbed ring equivalent network used in this assignment scheme. The n nodes are numbered clockwise sequentially from the hub: N_1, N_2, N_k, N_n. The 16 CWDM channels $\{\lambda_1, \lambda_2,...,\lambda_{16}\}$ are divided into n ($n = 4$, 8, etc.) sets or bands: B(1), B(2), B(j),...,B(n), ($j = 1...n$) satisfying the condition:

Fiber attenuation of B(1) ≥ fiber attenuation of B(2)

≥...≥ fiber attenuation of B(n).

Optimizing CWDM for Nonamplified Networks

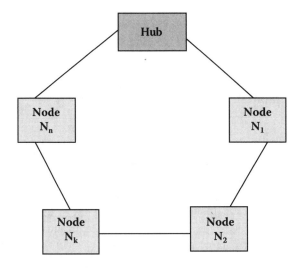

FIGURE 5.2 A 16-channel CWDM n-node hubbed ring network.

In other words, the intrinsic fiber attenuation loss is maximum in band $B(1)$ and decreases to a minimum in band $B(n)$. The wavelength assignment scheme that minimizes the ring loss (combined fiber, connector, splicing, and filter losses) for a wavelength channel λ_i assigned to node-N_k is as follows and will be deduced later in this section:

$$N_k \xrightarrow{\text{is assigned}} B \begin{cases} (n+1-2k) & \text{if } n+1 > 2k \\ (2k-n) & \text{if } 2k \geq n+1 \end{cases} \quad (5.1)$$

where $B(j)$ is the wavelength band assigned to node-N_k, ($j = n + 1-2k$ or $j = 2k-n$), n is the number of nodes in the ring, k is the node number and $k = 1...n$. This wavelength assignment is shown to be attenuation slope compensating, yielding a minimum filter loss in band $B(1)$ and increases to a maximum in band $B(n)$; in other words:

Filter loss of $B(1) \leq$ filter loss of $B(2) \leq ... \leq$ filter loss of $B(n)$.

The attenuation slope compensating wavelength assignment of Equation 5.1 yields a maximized ring perimeter R given by:

$$R = \min\left(\left\{\left(\frac{n+1}{n}\right)\left(\frac{P - S - F(\lambda_i, k) - C(\lambda_i, k)}{A(\lambda_i)}\right)\bigg| \text{ for } i = 1...16\right\}\right) \quad (5.2)$$

where

P = source power
S = receiver sensitivity
$C(\lambda_i, k)$ = connector loss incurred by wavelength assigned to node N_k
$F(\lambda_i, k)$ = filter insertion loss incurred by wavelength assigned to node N_k
$A(\lambda_i)$ = fiber attenuation coefficient at wavelength λ_i

5.2.1.1 Discussion

Since the loss incurred by a wavelength channel between the hub and a node on the hub-to-node path is generally different from the loss the wavelength incurs on the node-to-hub path, the ring loss budget must be based on the path with the higher loss. We make a distinction between two types of losses in the ring: (i) fiber loss, which is dependent on the channel wavelength and (ii) filter, connector, and splice losses, which are nonfiber attenuation-dependent. The nonfiber attenuation-dependent losses incurred by a wavelength are independent of the fiber and thus the span lengths. Rather, they are a function of the number of mux/demux filter stages and the number of spans that a wavelength traverses on its path between transmission and reception. To simplify the analysis, we initially assume that the nodes in Figure 5.2 are equally spaced around the ring, with the ring spans having equal lengths. The generalized case where the ring spans are of different lengths is addressed at the end of this section.

When the ring spans are of equal lengths, to determine the higher nonfiber attenuation-dependent loss path between hub-to-node and node-to-hub paths, it is sufficient to determine simply the path with more spans. Referring again to Figure 5.2 and proceeding clockwise on the node-N_k-to-hub path, the number of spans traversed by a wavelength λ_i assigned to node-N_k is given by:

$$S_{NkH} = (n+1-k) \tag{5.3}$$

Similarly, proceeding clockwise on the hub-to-node-N_k path, the number of spans traversed by a wavelength λ_i assigned to node-N_k is given by:

$$S_{HNk} = k \tag{5.4}$$

Therefore, the number of spans on the higher-loss-path is:

$$S_{HLP} = \max([n+1-k], k) \tag{5.5}$$

From Equation 5.3 and Equation 5.5, if the node-N_k-to-hub path is the higher-loss-path, then

$$(n+1-k) > k \Rightarrow (n+1) > 2k \tag{5.6a}$$

Optimizing CWDM for Nonamplified Networks

From Equation 5.4 and Equation 5.5, if the hub-to-node- N_k path is the higher-loss-path, then

$$k > (n+1-k) \Rightarrow 2k > (n+1) \tag{5.6b}$$

The algorithms for assigning the wavelength bands to the nodes are developed for the two cases represented in Equation 5.6a and Equation 5.6b.

Case-1: If the node- N_k -to-hub path is the higher-loss-path, we obtain Equation 5.6a.
Number of bands at node $N_k = (n - k)$.
Number of nodes already assigned with bands $= (k - 1)$.
Hence, available bands at node- $N_k = [(n - k) - (k - 1)] = (n + 1 - 2k)$ for $(k = 1, 2.. < (n + 1)/2$.
Therefore, node- N_k is assigned the $[(n + 1) - 2k]$th band as follows:

$$N_k \xrightarrow{\text{is assigned}} B(n+1-2k) \quad \text{if } (n+1) > 2k \tag{5.7}$$

Case-2: If the hub-to-node- N_k path is the higher-loss-path, we obtain Equation 5.6b.
Available bands at node $N_k = [k - (n - k)] = (2k - n)$ for $(2k > n + 1)$
Therefore, node- N_k is assigned the $(2k - n)$th band as follows:

$$N_k \xrightarrow{\text{is assigned}} B(2k-n) \quad \text{if } (2k > n+1) \tag{5.8}$$

Hence, combining Equation 5.7 and Equation 5.8 yields the expressions of Equation 5.1. This assignment is symmetric, that is, the nodes could have been numbered in a counterclockwise order with the same result. Also, we will show in this section that the assignment is equally valid for rings with irregularly spaced spans.

5.2.2 Ring Perimeter

The total ring loss incurred by a channel consists of filter and fiber losses and is a function of wavelength λ_i and the node N_k to which λ_i is assigned. If the total ring loss is denoted $L(\lambda_i, k)$, then

$$L(\lambda_i, k) = F_i(\lambda_i, k) + W(\lambda_i, k) + C(\lambda_i, k) \tag{5.9}$$

where
$F(\lambda_i, k)$ = filter insertion loss incurred by wavelength λ_i assigned to node N_k
$W(\lambda_i, k)$ = fiber attenuation loss incurred by wavelength λ_i
$C(\lambda_i, k)$ = connector loss + system margin incurred by wavelength λ_i

The ring perimeter is constrained to the smallest circumference $R(\lambda_i)$ attained by the wavelength λ_i with the highest loss. The maximum number of spans traversed by a signal around the n-node ring is between the hub and its adjacent nodes, that is, nodes N_n and N_1. In both cases, a signal traverses n spans. Therefore, since the nodes are assumed to be equally spaced around the ring then, the fiber loss is:

$$W(\lambda_i,k) = nDA(\lambda_i) \tag{5.10}$$

where D is the ring span length of Figure 5.2. If $R(\lambda_i, k)$ denotes the ring perimeter attained by wavelength λ_i, then:

$$R(\lambda_i,k) = (n+1)D \tag{5.11}$$

Eliminating D between Equation 5.11 in Equation 5.10 yields

$$W(\lambda_i.k) = \left(\frac{n}{n+1}\right)A(\lambda_i)R(\lambda_i,k) \tag{5.12}$$

From Equation 5.9 and Equation 5.12, we obtain

$$L(\lambda_i,k) = F(\lambda_i,k) + \left(\frac{n}{n+1}\right)A(\lambda_i)R(\lambda_i,k) + C(\lambda_i,k) \tag{5.13}$$

If P is the power in each channel and S is the receiver sensitivity, then:

$$P = L(\lambda_i,k) + S = F(\lambda_i,k) + \left(\frac{n}{n+1}\right)A(\lambda_i)R(\lambda_i,k) + C(\lambda_i,k) + S \tag{5.14}$$

Hence, it follows from expression (5.14) that for source power P and detector sensitivity S, the ring perimeter attainable by wavelength λ_i is:

$$R(\lambda_i) = \left(\frac{n+1}{n}\right)\left(\frac{P - S - F(\lambda_i,k) - C(\lambda_i,k)}{A(\lambda_i)}\right) \tag{5.15}$$

Thus, for same source power P at the transmitter, the 16 channels will yield different ring perimeters due to differences in fiber and filter losses incurred by the wavelengths. Thus, as i varies over the set of 16 CWDM wavelengths, a set of 16 ring circumferences is generated. If R is the minimum value of the set $\{R(\lambda_i)|_{\text{for } i = 1...16}\}$, then

$$R = \min(\{R(\lambda_i)|_{\text{for } i=1...16}\}) \tag{5.16}$$

or, substituting Equation 5.15 in Equation 5.16, we obtain as already presented in Equation 5.2:

$$R = \min\left(\left\{\left(\frac{n+1}{n}\right)\left(\frac{P - S - F(\lambda_i,k) - C(\lambda_i,k)}{A(\lambda_i)}\right)\bigg| \text{ for } i = 1...16\right\}\right) \quad (5.17)$$

5.2.3 Channel Power

For the same receiver sensitivity S, the other lower loss wavelength channels require power $P(\lambda_i) < P$ to yield the same ring perimeter R. Hence, the power required in wavelength channel λ_i assigned to node N_k to attain the ring perimeter R is given by:

$$P(\lambda_i) = F(\lambda_i,k) + \left(\frac{n}{n+1}\right) A(\lambda_i,k) R + C(\lambda_i,k) + S \quad (5.18)$$

5.2.4 Impact of Loss

Here, we summarize the impact of loss originating from various sources. First, there is the spectrally dependent loss of the optical transmission fiber and the different channels of the CWDM multiplexer with a significant variation between the CWDM channels due to the wide transmission bandwidth of 300 nm. Secondly, there are other sources of loss such as connector loss that are generally independent of wavelengths or particular channel numbers.

a) **Fiber loss $W(\lambda_i, k)$**
From Equation 5.12 and Equation 5.15, fiber loss in the ring is:

$$W(\lambda_i,k) = P - S - F(\lambda_i,k) - C(\lambda_i,k) \quad (5.19)$$

b) **Filter loss $F(\lambda_i, k)$**
Referring to Figure 5.3, a typical thin-film filter (TFF) is shown as a model for the add/drop functionality at a node [4]. Details about TFF-based multiplexers and demultiplexers can be found in Chapter 4 while here we concentrate on the wavelength (de-)multiplexing alone. The advantage of parallel filtering as illustrated in the set-up of Figure 5.3 is the mux/demux losses are the same for all wavelengths.
In Figure 5.3 (left), the ring fiber carries 16 CWDM channels into the node on the demultiplexing side as shown. If it is desired to add/drop a wavelength band $B(1) = \{\lambda_1, \lambda_2, \lambda_3, \lambda_4\}$ from the ring, then using the edge filtering technique [6] at point A_d, the TFF transmits band $B(1)$ and reflects bands $B(2)$ to $B(4) = \{\lambda_5, \lambda_6...\lambda_{16}\}$. Further, at point B_d, through a series of transmissions and reflections, TFFs demultiplex band $B(1)$ into its individual constituent wavelengths.

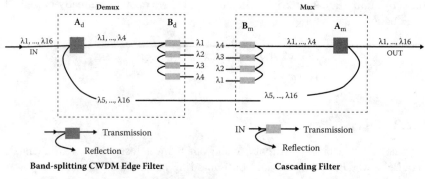

FIGURE 5.3 Thin film filter implementation of OADM to add/drop wavelength band B(1) = $\{\lambda_1, \lambda_2, \lambda_3, \lambda_4\}$ and node loss model.

In Figure 5.3 (right), the reverse process takes place on the multiplex side in which the individual wavelengths $\{\lambda_1, \lambda_2, \lambda_3, \lambda_4\}$ are first multiplexed together as to obtain band B(1) at point B_m. Band B(1) then combines with bands B(2) to B(4) at point A_m. The 16 CWDM channels are again available for transmission on the next fiber section.

Bands B(2) to B(4) = $\{\lambda_5, \lambda_6 ... \lambda_{16}\}$ in Figure 5.3 incur express/through loss at points A_d and A_m. Band B(1) = $\{\lambda_1, \lambda_2, \lambda_3, \lambda_4\}$ incurs a drop loss at point A_d. The wavelengths $\{\lambda_1, \lambda_2, \lambda_3, \lambda_4\}$ incur demux losses at point B_d. The wavelengths $\{\lambda_1, \lambda_2, \lambda_3, \lambda_4\}$ incur mux losses at point B_m as they are multiplexed into B(1) and subsequently band B(1) = $\{\lambda_1, \lambda_2, \lambda_3, \lambda_4\}$ incurs loss at point A_m. The ring filter loss is evaluated for the two cases satisfying Equation 5.6: that is, (i) the node-N_k-to-hub path is higher-loss-path and (ii) the hub-to-node-N_k path is higher-loss-path:

Case-1: *node-N_k-to-hub* path is the higher-loss-path:

From Figure 5.2, on the path node-N_k-to-hub, the filter loss $F(\lambda_i, k)$ incurred by a wavelength λ_i is calculated as follows:

$$F(\lambda_i, k) = f_{mux} + f_{add} + (n-k)f_{exp} + f_{drop} + f_{demux} \quad \text{if } n+1 > 2k \quad (5.20)$$

where

f_{mux} = mux loss at node N_k
f_{add} = add loss at node N_k
f_{exp} = express losses at nodes N_{k+1} to N_n
f_{drop} = drop loss at the hub
f_{demux} = demux loss at the hub.

Case-2: *hub*-to-*node*- N_k path is the higher loss path:
On the path from the hub-to-node- N_k, the filter loss incurred by a wavelength λ_i is:

$$F(\lambda_i, k) = f_{mux} + f_{add} + (k-1)f_{exp} + f_{drop} + f_{demux} \quad \text{if } 2k > n+1 \quad (5.21)$$

where

f_{mux} = mux loss at the hub
f_{add} = add loss at the hub
f_{exp} = express losses at nodes N_1 to N_{k-1}
f_{drop} = drop loss at node N_k
f_{demux} = demux loss at node N_k.

Therefore, combining Equation 5.20 and Equation 5.21, the filter loss incurred by a wavelength λ_i assigned to node- N_k is given by:

$$F(\lambda_i, k) = \begin{cases} f_{mux} + f_{add} + (n-k)f_{exp} + f_{drop} + f_{demux} & \text{if } n+1 > 2k \\ f_{mux} + f_{add} + (k-1)f_{exp} + f_{drop} + f_{demux} & \text{if } 2k \geq n+1 \end{cases} \quad (5.22)$$

c) Connector loss $C(\lambda_i, k)$

Following the same procedure as for the filter loss derivations, the connector loss incurred by a wavelength λ_i assigned to node- N_k is given by:

$$C(\lambda_i, k) = \begin{cases} (n+1-k)c_{sp} & \text{if } n+1 > 2k \\ kc_{sp} & \text{if } 2k \geq n+1 \end{cases} \quad (5.23)$$

where C_{sp} is the connector loss per span, which is the same for each node and independent of wavelength.

5.3 GENERALIZED CASE OF CWDM DESIGN

In Section 5.2, we have made the assumption that the nodes are equally spaced around the ring, which simplified the analysis. However, in practice, the lengths of the ring spans will be generally different due to the individual network topologies. In this section, we present a generalized analysis for the case of unequal span lengths.

Figure 5.4 shows a 16-channel CWDM n-node hubbed ring network. As before, the n nodes are numbered clockwise, or counterclockwise sequentially from the hub: N_1, N_2, N_3, N_n, with spans $d_1, d_2, d_k, d_n, d_{n+1}$ as shown. The 16 CWDM channels $\{\lambda_1, \lambda_2, ..., \lambda_{16}\}$ are divided into n (n = 4, 8, etc.) sets or bands: B(1), B(2), B(j),...,B(n), (j = 1...n), satisfying the condition:

Fiber attenuation of B(1) ≥ fiber attenuation of B(2)

≥...≥ fiber attenuation of B(n).

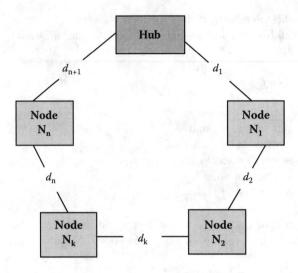

FIGURE 5.4 A 16-channel CWDM n-node hubbed ring network with variable span lengths.

In other words, the intrinsic fiber attenuation loss is maximum in band B(1) and decreases to a minimum in band B(n). Then, a wavelength assignment scheme that minimizes the ring loss (combined fiber attenuation loss) and nonfiber attenuation losses (filter, connector, and splice losses) for a wavelength channel λ_i assigned to node- N_k is as follows:

$$N_k \text{ is assigned B} \begin{cases} (n+1-2k) & \text{if } A(\lambda_i)\sum_{i=k}^{n} d_{i+1} + (n+1)f_{\exp} > A(\lambda_i)\sum_{i=1}^{k} d_i + (2k)f_{\exp} \\ (2k-n) & \text{if } A(\lambda_i)\sum_{i=1}^{k} d_i + (2k)f_{\exp} > A(\lambda_i)\sum_{i=k}^{n} d_{i+1} + (n+1)f_{\exp} \end{cases}$$

(5.24)

For the case of equal spans, this is shown to reduce to the previously discussed case.

5.3.1 Discussion

To evaluate the nonfiber attenuation losses, recall Equation 5.22. Only the through/express components of the fiber attenuation losses are required to determine the higher-loss-path. Therefore, in Figure 5.4, the filter loss on the higher-loss-path is:

$$F_{\text{HLP}} = \max([n-k]f_{\exp}, [k-1]f_{\exp})$$

(5.25)

On the node- N_k -to-hub path, the number of spans traversed by a wavelength λ_i assigned to node- N_k is given by Equation 5.3. Therefore, the fiber attenuation

Optimizing CWDM for Nonamplified Networks

loss incurred by the wavelength λ_i on the node-N_k-to-hub path is:

$$W_{NkH} = A(\lambda_i) \sum_{i=k}^{n} d_{i+1} \qquad (5.26)$$

Similarly, on the hub-to-node-N_k path, the number of spans traversed by a wavelength λ_i assigned to node-N_k is given by Equation 5.4. Therefore, the fiber attenuation loss incurred by the wavelength λ_i on the hub-to-node-N_k path is:

$$W_{HNk} = A(\lambda_i) \sum_{i=1}^{k} d_i \qquad (5.27)$$

Thus, the fiber attenuation loss incurred by the wavelength λ_i on the higher-loss-path between the hub and the node-N_k is:

$$W_{HLP} = \max\left(A(\lambda_i) \sum_{i=k}^{n} d_{i+1},\ A(\lambda_i) \sum_{i=1}^{k} d_i\right) \qquad (5.28)$$

From Equation 5.25 and Equation 5.28, the ring loss, that is, the combined fiber attenuation and nonfiber attenuation losses on the higher-loss-path is:

$$R_{HLP} = \max\left[\left(A(\lambda_i) \sum_{i=k}^{n} d_{i+1} + (n+1)f_{\exp}\right),\ \left(A(\lambda_i) \sum_{i=1}^{k} d_i + (2k)f_{\exp}\right)\right] \qquad (5.29)$$

Therefore, it follows that if node-N_k-to-hub path is the higher-loss-path, then

$$A(\lambda_i) \sum_{i=k}^{n} d_{i+1} + (n+1)f_{\exp} > A(\lambda_i) \sum_{i=1}^{k} d_i + (2k)f_{\exp} \qquad (5.30a)$$

Similarly, if hub-to-node-N_k path is the higher-loss-path, then

$$A(\lambda_i) \sum_{i=1}^{k} d_i + (2k)f_{\exp} > A(\lambda_i) \sum_{i=k}^{n} d_{i+1} + (n+1)f_{\exp} \qquad (5.30b)$$

Therefore, a wavelength assignment scheme that minimizes the ring loss for a wavelength channel λ_i assigned to node-N_k is given by the expression in Equation 5.24. For the particular case when the ring span lengths are equal, that is, $d_1 = d_2 = d_1 = D$, then, from Equation 5.3 and Equation 5.26,

$$W_{NkH} = A(\lambda_i)\sum_{i=k}^{n} d_{i+1} = D(n+1-k)A(\lambda_i).$$

Similarly, from Equation 5.4 and Equation 5.27, it follows

$$W_{HNk} = A(\lambda_i)\sum_{i=1}^{k} d_i = D(k)A(\lambda_i).$$

After substituting in Equation 5.24, it yields:

$$N_k \xrightarrow{\text{is assigned}} B\begin{cases} (n+1-2k) & \text{if } (n+1)(DA(\lambda_i)+f_{\exp}) > 2k(DA(\lambda_i)+f_{\exp}) \\ (2k-n) & \text{if } 2k(DA(\lambda_i)+f_{\exp}) > (n+1)(DA(\lambda_i)+f_{\exp}) \end{cases},$$
(5.31)

which simplifies into the known expression (5.1) already given in Section 5.2 for the case of equal span lengths.

5.4 APPLICATION: WAVELENGTH ASSIGNMENT IN 4-NODE RING

A practical application for studying the described loss-optimized wavelength assignment scheme to the 16-channel CWDM network is shown in Figure 5.5a in which four local exchanges or central offices (COs), numbered CO-1 to CO-4, are transmitting to and receiving signals from a tandem exchange H, each on a set of four wavelengths such that there are 16 wavelengths at 2.5 Gb/s each in the network. CO-1 transmits/receives from H on sub-band λ_1 to λ_4, CO-2 transmits/receives from H on λ_5 to λ_8, CO-3 transmits/receives from H on λ_9 to λ_{12}, and CO-4 transmits/receives from H on λ_{13} to λ_{16}. This logical star network is implemented as a physical unidirectional hubbed ring network (logical star/physical ring), as shown in Figure 5.5b, with all 16 wavelengths in the network supported on a single G.652D fiber (ZWPF). The hubbed ring configuration also allows the easy implementation of protection schemes. Next, we demonstrate an optimal wavelength assignment to the nodes, using the attenuation slope compensating algorithm of Equation 5.1.

5.4.1 WAVELENGTH ASSIGNMENT

For the band-wise wavelength assignment, the 16 CWDM wavelength channels are divided into four bands, B(1) = {$\lambda_1 - \lambda_4$}, B(2) = {$\lambda_5 - \lambda_8$}, B(3) = {$\lambda_9 - \lambda_{12}$}, and B(4) = {$\lambda_{13} - \lambda_{16}$}, as shown in Table 5.2. All 16 CWDM channels are

Optimizing CWDM for Nonamplified Networks

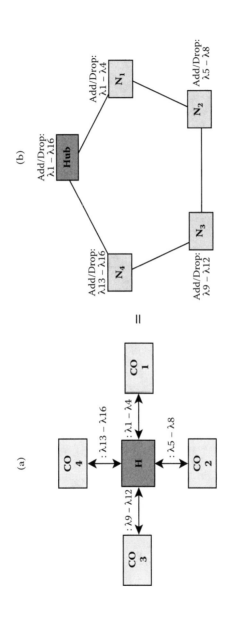

FIGURE 5.5 (a) Logical star network of four local exchanges (COs) and a tandem exchange and (b) equivalent hubbed ring configuration.

TABLE 5.2
A 16-CWDM Channel Grid Divided into Four Wavelength Bands

FS-CWDM Channel Plan	O-Band				E-Band				S +C +L-Band							
Band #	$B(1) = \{\lambda_1 - \lambda_4\}$				$B(2) = \{\lambda_5 - \lambda_8\}$				$B(3) = \{\lambda_9 - \lambda_{12}\}$				$B(4) = \{\lambda_{13} - \lambda_{16}\}$			
Channel #	1	2	3	4	5	6	7	8	9	10	11	12	13	14	15	16
Center wavelength	1310	1330	1350	1370	1390	1410	1430	1450	1470	1490	1510	1530	1550	1570	1590	1610
Attenuation (dB/km)	0.360	0.335	0.322	0.311	0.333	0.291	0.281	0.272	0.266	0.260	0.254	0.252	0.250	0.250	0.257	0.266

Optimizing CWDM for Nonamplified Networks

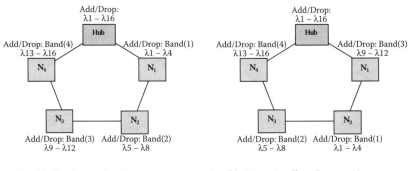

FIGURE 5.6 Application of wavelength assignment in 4-node ring with unidirectional traffic.

with add/drop at the hub. At each node, 4 channels are add/drop while the remaining 12 channels are through/express channels.

Figure 5.6 shows the wavelength assignment procedure in the 16-channel CWDM 4-node unidirectional hubbed ring network. Two principal cases are discussed in the following. In the first case, with nonattenuation slope compensating wavelength assignment (Figure 5.6, case a), the wavelength bands are assigned sequentially to the nodes. The analysis results will demonstrate that this type of wavelength assignment does not maximize the ring perimeter.

In the second case, attenuation slope compensating wavelength assignment (Figure 5.6, case b), the attenuation slope compensating wavelength assignment algorithm of Equation 5.1, is employed to optimally assign wavelength bands to the nodes such that the ring fiber, filter, connector, and splice losses are minimized, consequently maximizing the ring perimeter.

Analysis assumptions

Optical source power	−1 dBm
APD receiver sensitivity	−29 dBm at 2.5 Gb/s
⇒Allowable loss	28 dB
Connectors per span	2
Connector loss	0.5 dB
System margin (splicing etc.)	1 dB

The wavelength band assignment shown in Table 5.3 is carried out. A wavelength band is assigned to each node- N_k, according to the attenuation slope

TABLE 5.3
Wavelength Band Assignment in 4-Node Unidirectional Hubbed Ring

4-Node Hubbed Ring N_k	Node #	Higher-Loss-Path	Number of Spans on Higher-Loss-Path	Case (a): Nonattenuation-Slope Compensating Wavelength Assignment	Case (b): Attenuation Slope Compensating Wavelength Assignment
Node N_1	1	Node-N_1-to-hub	4	B(1)	B(3)
Node N_2	2	Node-N_2-to-hub	3	B(2)	B(1)
Node N_3	3	Hub-to-node-N_3	3	B(3)	B(2)
Node N_4	4	Hub-to-node-N_4	4	B(4)	B(4)

compensating wavelength assignment algorithm described in Equation 5.1:

$$N_k \xrightarrow{\text{is assigned}} B \begin{cases} (n+1-2k) & \text{if } n+1 > 2k \\ (2k-n) & \text{if } 2k \geq n+1 \end{cases}$$

where $B(j)$ is the wavelength band assigned to node- N_k, ($j = n + 1-2k$ or $j = 2k - n$), $n = 4$ is the number of nodes in the ring, k is the node number, and $k = 1…4$. Note that in this wavelength band assignment, bands $B(1)$ and $B(2)$ traverse 3 spans, whereas bands $B(3)$ and band $B(4)$ traverse 4 spans. In other words, higher fiber attenuation O- and E-bands [bands $B(1)$ and $B(2)$] incur less filter, connector, and splice losses than the lower fiber attenuation S-, C-, and L-bands [bands $B(3)$ and $B(4)$]. Consequently, the ring loss — the combined filter, fiber, connector, and splice losses — is minimized across the CWDM channel spectrum, leading to a maximized ring perimeter as shown in the results of Section 5.4.

5.4.2 CALCULATION OF FILTER LOSSES

Figure 5.7 shows TFF loss models for the hub. All wavelengths are added and dropped at the hub. Therefore, 16 CWDM channels coming in on the ring fiber on the demultiplex side pass through a first stage TFF at point A_d. Using an edge filtering technique, wavelength bands $B(1)$ and $B(2) = \{\lambda_1 - \lambda_8\}$ are transmitted while bands $B(3)$ and $B(4) = \{\lambda_9 - \lambda_{16}\}$ are reflected by the TFF as shown. At point B_d, band $B(1) = \{\lambda_1 - \lambda_4\}$ is transmitted while band $B(2) = \{\lambda_5 - \lambda_8\}$ is reflected. Also at B_d, band $B(3) = \{\lambda_9 - \lambda_{12}\}$ is transmitted while band $B(4) = \{\lambda_{13} - \lambda_{16}\}$ is reflected. All the bands are then available for demultiplexing into individual wavelengths via a series of transmissions and reflections at point C_d, using cascade filters. The process is reversed for multiplexing the individual wavelength channels into bands at point C_m. Bands $B(1)$ and $B(2)$ and bands $B(3)$ and $B(4)$ are combined at point B_m. Finally, all the bands are combined at point A_m to obtain the 16 CWDM channels for transmission on the next fiber section.

Transmission and reflection losses for the TFF are taken as 0.5 dB. Hence, the loss incurred at each point by each band or individual wavelength is as shown in Figure 5.7. In order to achieve a balanced multiplex/demultiplex loss, the bands are demultiplexed into individual wavelengths in reverse order to the order of multiplexing. Figure 5.8 shows TFF loss models for the node where band $B(1) = \{\lambda_1 - \lambda_4\}$ is added and dropped and bands $B(2)$, $B(3)$, and $B(4) = \{\lambda_5 - \lambda_{16}\}$ pass through without dropping or adding. Therefore, they incur express or through channel loss at the node. The express loss incurred by band $B(2) = \{\lambda_5 - \lambda_8\}$ is 2 dB, and 1 dB for bands $B(3)$ and $B(4) = \{\lambda_9 - \lambda_{16}\}$. The TFF losses for all the nodes can be similarly modeled. The express loss is shown to vary from a minimum of 1 dB to a maximum of 2 dB for all bands. However, for computational convenience, a uniform express loss of 2 dB is assumed for all bands.

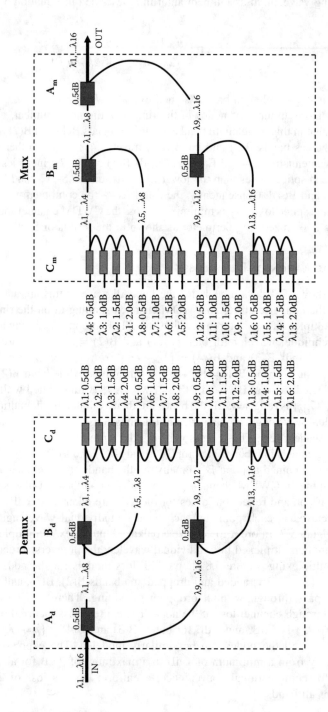

FIGURE 5.7 Hub loss model and TFF implementation of OADM to add/drop wavelength band B(1) to B(4); channels $\{\lambda_1,\ldots,\lambda_{16}\}$.

Optimizing CWDM for Nonamplified Networks 145

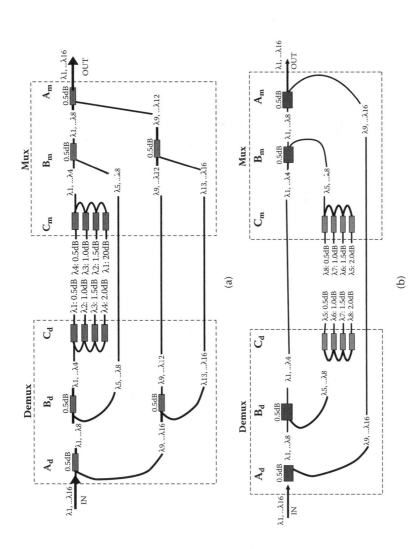

FIGURE 5.8 (a) Node loss model and TFF implementation of OADM to add/drop wavelength band B(1) = $\{\lambda_1, \lambda_2, \lambda_3, \lambda_4\}$. (b) Node loss model and TFF implementation of OADM to add/drop wavelength band B(2); channels $\{\lambda_5...\lambda_8\}$. (c) Hub loss model and TFF implementation of OADM to add/drop wavelength band B(3); channels $\{\lambda_9...\lambda_{12}\}$. (d) Node loss model and TFF implementation of OADM to add/drop wavelength band B(4); channels $\{\lambda_{13}...\lambda_{16}\}$.

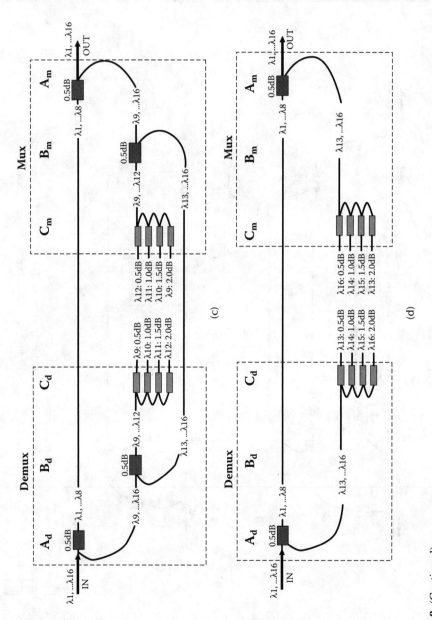

FIGURE 5.8 (Continued).

Optimizing CWDM for Nonamplified Networks

For example, to calculate the filter loss incurred by wavelength λ_1 assigned to node- N_2 in the 4-node ring of Figure 5.6b, then from Equation 5.22:

$$F(\lambda_i,k) = \begin{cases} f_{mux} + f_{add} + (n-k)f_{exp} + f_{drop} + f_{demux} & \text{if } n+1 > 2k \\ f_{mux} + f_{add} + (k-1)f_{exp} + f_{drop} + f_{demux} & \text{if } 2k > n+1 \end{cases}$$

with number of nodes $n = 4$ and node $N_k = N_2 \Rightarrow k = 2$. Therefore, from Equation 5.6, $n + 1 = 5 > 2k = 4 \Rightarrow$ higher-loss-path is from node- N_2 -to-hub. Hence, from Equation 5.22,

$$F(\lambda_1,2) = f_{mux} + f_{add} + 2f_{exp} + f_{drop} + f_{demux}$$

Referring to Figure 5.6b: Attenuation slope compensating wavelength assignment

f_{exp} Express losses at nodes N_3 and $N_4 = 2$ dB/node

Referring to Figure 5.7: TFF filter loss model for the hub,

f_{demux} Demux loss incurred by wavelength λ_1 at the hub (point C_d) = 0.5 dB.

f_{drop} Drop loss at the hub (points B_d and A_d) = 0.5 dB + 0.5 dB = 1 dB

Referring to Figure 5.8a: Node loss model to add/drop wavelength band B(1)

f_{add} Add loss at node N_2 (points A_m and B_m) = 0.5 dB + 0.5 dB = 1 dB

f_{mux} Mux loss at node N_2 (point C_m) = 2.0 dB.

Therefore, the total filter loss incurred by wavelength λ_1 assigned to node- N_2 in the 4-node ring of Figure 5.6b is $F(\lambda_1,2) = 2$ dB + 1 dB + 2 × 2 dB + 1 dB + 0.5 dB = 8.5 dB. Detailed filter losses for all 16 CWDM wavelength channels are shown in Table 5.4, with band B_1 assigned to node N_1 and so on. The fiber and connector losses and ring perimeters are calculated from Equation 5.19, Equation 5.23, and Equation 5.15, respectively, and presented in Table 5.5 for all CWDM wavelengths with a reach from 43 km for λ_1 to 77 km for λ_{12}. Table 5.6 and Table 5.7 show the same calculations for a different wavelength assignment, which minimizes the spread among the channels to distances from 53.8 to 71.3 km.

TABLE 5.4
Filter Loss Calculation for 4-Node Unidirectional Hubbed Ring, Incrementing Node Assignment

Band	Channel #	Node Assignment	Number of Spans on Longer Path	Express Loss (dB)	Mux Loss (dB)	Add Loss (dB)	Drop Loss (dB)	Demux Loss (dB)	Filter Loss (dB)
B(1)	1	N_1	4	6.0	0.5	1.0	1.0	2.0	10.5
	2		4	6.0	1.0	1.0	1.0	1.5	10.5
	3		4	6.0	1.5	1.0	1.0	1.0	10.5
	4		4	6.0	2.0	1.0	1.0	0.5	10.5
B(2)	5	N_2	3	4.0	0.5	1.0	1.0	2.0	8.5
	6		3	4.0	1.0	1.0	1.0	1.5	8.5
	7		3	4.0	1.5	1.0	1.0	1.0	8.5
	8		3	4.0	2.0	1.0	1.0	0.5	8.5
B(3)	9	N_3	3	4.0	0.5	1.0	1.0	2.0	8.5
	10		3	4.0	1.0	1.0	1.0	1.5	8.5
	11		3	4.0	1.5	1.0	1.0	1.0	8.5
	12		3	4.0	2.0	1.0	1.0	0.5	8.5
B(4)	13	N_4	4	6.0	0.5	1.0	1.0	2.0	10.5
	14		4	6.0	1.0	1.0	1.0	1.5	10.5
	15		4	6.0	1.5	1.0	1.0	1.0	10.5
	16		4	6.0	2.0	1.0	1.0	0.5	10.5

Optimizing CWDM for Nonamplified Networks 149

TABLE 5.5
Losses and Ring Perimeter for 4-Node Unidirectional Hubbed Ring Case, Incrementing Node Assignment

Band	Channel #	Node Assignment	Number of Spans on Path	Connector loss + 1 dB	Filter + Connector Loss (dB)	Fiber Att. (dB/km)	Fiber Loss (dB)	Ring Perimeter (km)	Allowable Loss (dB)
B(1)	1	N_1	4	5	15.5	0.360	12.50	43.40	28.00
	2		4	5	15.5	0.335	11.63	46.63	27.13
	3		4	5	15.5	0.322	11.18	48.52	26.68
	4		4	5	15.5	0.311	10.81	50.17	26.31
B(2)	5	N_2	3	4	12.5	0.333	11.58	58.11	24.08
	6		3	4	12.5	0.291	10.09	66.65	22.59
	7		3	4	12.5	0.281	9.75	69.02	22.25
	8		3	4	12.5	0.272	9.44	71.27	21.94
B(3)	9	N_3	3	4	12.5	0.266	9.23	72.87	21.73
	10		3	4	12.5	0.260	9.01	74.66	21.51
	11		3	4	12.5	0.254	8.83	76.22	21.33
	12		3	4	12.5	0.252	8.74	77.00	21.24
B(4)	13	N_4	4	5	15.5	0.250	8.67	62.58	24.17
	14		4	5	15.5	0.250	8.68	62.52	24.18
	15		4	5	15.5	0.257	8.91	60.88	24.41
	16		4	5	15.5	0.266	9.24	58.69	24.74

TABLE 5.6
Filter Loss Calculation for 4-Node Unidirectional Hubbed Ring, Attenuation Slope Compensating Wavelength Assignment

Band	Channel #	Node Assignment	Number of Spans on Higher-Loss-Path	Express Loss (dB)	Mux Loss (dB)	Add Loss (dB)	Drop Loss (dB)	Demux Loss (dB)	Filter Loss (dB)
B(1)	1	N_2	3	4	0.5	1.0	1.0	2.0	8.50
	2		3	4	1.0	1.0	1.0	1.5	8.50
	3		3	4	1.5	1.0	1.0	1.0	8.50
	4		3	4	2.0	1.0	1.0	0.5	8.50
B(2)	5	N_3	3	4	0.5	1.0	1.0	2.0	8.50
	6		3	4	1.0	1.0	1.0	1.5	8.50
	7		3	4	1.5	1.0	1.0	1.0	8.50
	8		3	4	2.0	1.0	1.0	0.5	8.50
B(3)	9	N_1	4	6	0.5	1.0	1.0	2.0	10.50
	10		4	6	1.0	1.0	1.0	1.5	10.50
	11		4	6	1.5	1.0	1.0	1.0	10.50
	12		4	6	2.0	1.0	1.0	0.5	10.50
B(4)	13	N_4	4	6	0.5	1.0	1.0	2.0	10.50
	14		4	6	1.0	1.0	1.0	1.5	10.50
	15		4	6	1.5	1.0	1.0	1.0	10.50
	16		4	6	2.0	1.0	1.0	0.5	10.50

Optimizing CWDM for Nonamplified Networks   151

TABLE 5.7
Losses and Ring Perimeter for 4-Node Unidirectional Hubbed Ring Case, Attenuation Slope Compensating Wavelength Assignment

Bands	Channel #	Assignment	Number of Spans on Higher-Loss-Path	Connector Loss + 1 dB	Filter + Connector Loss (dB)	Fiber Att. (dB/km)	Fiber Loss (dB)	Ring Perimeter Attained at 28 dB Allowable Loss (km)	Allowable Loss for Attained Ring Perimeter of 53.82 km (dB)
B(1)	1	N_2	3	4.00	12.50	0.360	15.50	53.82	28.00
	2		3	4.00	12.50	0.335	14.43	57.82	26.93
	3		3	4.00	12.50	0.322	13.87	60.16	26.37
	4		3	4.00	12.50	0.311	13.41	62.21	25.91
B(2)	5	N_3	3	4.00	12.50	0.333	14.36	58.11	26.86
	6		3	4.00	12.50	0.291	12.52	66.65	25.02
	7		3	4.00	12.50	0.281	12.09	69.02	24.59
	8		3	4.00	12.50	0.272	11.70	71.27	24.20
B(3)	9	N_1	4	5.00	15.50	0.266	11.45	58.76	26.95
	10		4	5.00	15.50	0.260	11.17	60.21	26.67
	11		4	5.00	15.50	0.254	10.94	61.47	26.44
	12		4	5.00	15.50	0.252	10.83	62.10	26.33
B(4)	13	N_4	4	5.00	15.50	0.250	10.75	62.58	26.25
	14		4	5.00	15.50	0.250	10.76	62.52	26.26
	15		4	5.00	15.50	0.257	11.05	60.88	26.55
	16		4	5.00	15.50	0.266	11.46	58.69	26.96

5.5 ANALYSIS OF RESULTS FOR 4-NODE RING

5.5.1 Nonattenuation Slope Compensating Wavelength Assignment

Figure 5.9 shows the spectrally resolved filter, fiber, and ring losses plotted as functions of the nodes N_1 to N_4 and their assigned wavelength bands B(1) to B(4). In this case, the nodes are assigned wavelengths with no preference as follows: Node N_1 is assigned band B(1), node N_2 is assigned band B(2), node N_3 is assigned band B(3), and node N_4 is assigned band B(4), as shown in Table 5.5. The filter loss for bands B(1), B(2), B(3), and B(4) are 15.5, 12.5, 12.5, and 15.5 dB, respectively, that is, filter loss of B(1) > filter loss of B(2) = filter loss of B(3) < filter loss of B(4). For the available link budget of 28 dB, this leaves 12.5, 15.5, 15.5, and 12.5 dB for fiber loss in bands B(1), B(2), B(3), and B(4), respectively. The higher attenuation O-band achieves a ring perimeter of 43 km due to the worst channel of the sub-band B(1), as shown in Figure 5.10. In contrast, the lower attenuation S + C + L-bands achieve much higher ring perimeters, however, the ring size is constrained here to only 43 km due to the O-band limit.

5.5.2 Attenuation Slope Compensating Wavelength Assignment

Figure 5.11 shows the spectral loss of the filter, fiber, and ring plotted as functions of the nodes N_1 to N_4 and their assigned wavelength bands B(1) to B(4). The wavelength

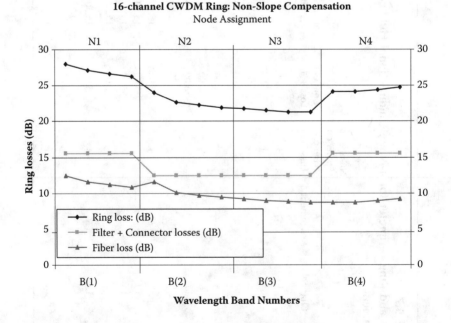

FIGURE 5.9 Filter (including connectors and splices), fiber, and ring losses plotted as functions of the nodes N_1 to N_4 and their assigned wavelength bands B(1) to B(4) for nonattenuation slope compensating wavelength assignment.

Optimizing CWDM for Nonamplified Networks 153

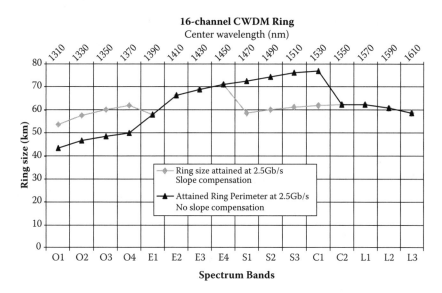

FIGURE 5.10 Calculated ring perimeter for both wavelength assignment approaches across the CWDM band, 4-node hubbed ring.

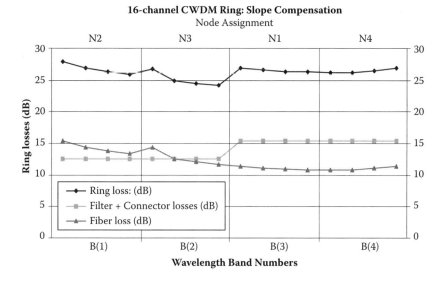

FIGURE 5.11 Filter, fiber, and ring losses plotted as functions of the nodes N_1 to N_4 and their assigned wavelength bands B(1) to B(4) for attenuation slope compensating wavelength assignment.

assignment algorithm of Equation 5.1 is used with $n = 4$:

$$N_k \xrightarrow{\text{is assigned}} B \begin{cases} (n+1-2k) & \text{if } n+1 > 2k \\ (2k-n) & \text{if } 2k \geq n+1 \end{cases}$$

Node N_1 is assigned to band B(3), N_2 to B(1), N_3 to B(2), and N_4 to band B(4), with B(1) = $\{\lambda_1 - \lambda_4\}$, B(2) = $\{\lambda_5 - \lambda_8\}$, B(3) = $\{\lambda_9 - \lambda_{12}\}$, and B(4) = $\{\lambda_{13} - \lambda_{16}\}$, and channel number, center wavelength, and fiber attenuation according to Table 5.2. The filter loss for bands B(1), B(2), B(3), and B(4) are 12.5, 12.5, 15.5, and 15.5 dB, respectively, that is, filter loss of B(1) $\leq$ filter loss of B(2) $\leq$ filter loss of B(3) $\leq$ filter loss of B(n). For the available budget of 28 dB, this leaves 28 dB $-$ 12.5 dB = 15.5 dB for fiber loss in the higher attenuation O- and E-bands — bands B(1) and B(2), respectively. Therefore, this wavelength assignment achieves an extra 15.5 $-$ 12.5 = 3 dB in the O- and E-bands over the lower attenuation S + C + L-bands. This 3-dB extra power margin compensates the higher fiber attenuation in the O- and E-bands, consequently yielding an enhanced ring perimeter of 54 km as shown in Figure 5.10.

5.5.3 Impact of Ring Perimeter

In Figure 5.10, the attainable ring perimeters for the two different scenarios of Section 5.5.1 and Section 5.5.2 are shown as a function of the CWDM center wavelengths across the O-, E-, S-, C-, and L-bands. It can be seen from the graph that in the case of no attenuation slope compensation, the ring perimeter varies from a minimum of 43 km for the 1310-nm channel in the O-band to a maximum of 77 km for the 1530-nm channel of the C-band. Hence, the maximum attainable ring perimeter is constrained to only 43 km. By employing the wavelength assignment algorithm of Section 5.5.2, the ring perimeter now varies from a minimum of 54 km for the 1310-nm channel in the O-Band to a maximum of 71 km for the 1450-nm channel of the E-band. The maximum perimeter in this case is 54 km, a 25% increase.

5.5.4 Extension to 8-Node Ring

In the following, we extend the ring structure to 8 nodes as shown in Figure 5.12 and apply the attenuation slope compensating wavelength assignment algorithm to this new network. In Table 5.8, a wavelength band is assigned to each node-N_k, as described in Equation 5.1. Note that in this example the wavelength bands B(1) to B(8) contain two wavelengths each, and the assignment of particular CWDM wavelengths is shown in Table 5.9 and Table 5.10.

First, the ring losses are calculated as a function of wavelength bands for the 8-node ring with incremental wavelength assignment. The losses are based on an allowable loss budget of 28 dB. We find that nonfiber (filter plus connector) losses are highest at the wavelengths $(\lambda_1-\lambda_2)$ and $(\lambda_{15}-\lambda_{16})$ assigned to nodes N_1 and

Optimizing CWDM for Nonamplified Networks

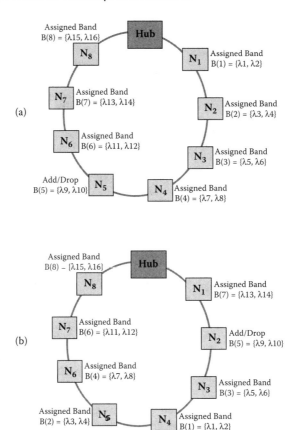

FIGURE 5.12 Wavelength assignment in 8-node unidirectional ring. (a) configuration with bands assigned to nodes with no preference and (b) configuration with attenuation slope compensating wavelength assignment.

N_8, respectively, since the wavelengths traverse seven nodes between nodes N_1 and N_8 and the hub. Again, the ring size is constrained by the higher fiber attenuation O-band (λ_1–λ_2) to a length of approximately 38 km. Next, the ring losses are determined for the wavelength bands assigned to the eight nodes in the case of attenuation slope compensating wavelength assignment. As mentioned earlier, losses are based on an allowable loss budget of 28 dB. Nonfiber (filter plus connector) losses are now highest at wavelengths (λ_{13}–λ_{14}) and (λ_{15}–λ_{16}) assigned to nodes N_1 and N_8, respectively, since the wavelengths traverse seven nodes. Nonfiber losses are now minimum in the O-band for wavelengths (λ_1–λ_2) and (λ_3–λ_4) assigned to nodes N_4 and N_5, respectively, since they traverse only four nodes. Therefore, higher available power in the higher attenuation O-bands thereby maximizes the ring size despite the fiber loss. The ring perimeter is consequently extended from 38 to 51 km as shown in Figure 5.13, a 35% increase.

TABLE 5.8
Wavelength Band Assignment in 8-Node Unidirectional Hubbed Ring

8-Node Hubbed Ring N_k	Higher-Loss-Path	Number of Spans on Higher-Loss-Path	Case (a): Attenuation Slope Compensating Wavelength Assignment	Case (b): Nonattenuation Slope Compensating Wavelength Assignment
Node N_1	Node-N_1-to-hub	8	B(7)	B(1)
Node N_2	Node-N_2-to-hub	7	B(5)	B(2)
Node N_3	Node-N_3-to-hub	6	B(3)	B(3)
Node N_4	Node-N_4-to-hub	5	B(1)	B(4)
Node N_5	Hub-to-node-N_5	5	B(2)	B(5)
Node N_6	Hub-to-node-N_6	6	B(4)	B(6)
Node N_7	Hub-to-node-N_7	7	B(6)	B(7)
Node N_8	Hub-to-node-N_8	8	B(8)	B(8)

Compared to the 4-node ring analyzed earlier in this section, the benefit of this assignment algorithm for CWDM wavelengths is even greater for the 8-node ring.

5.6 N-NODE CWDM MESHED NETWORK

In this section, we show how the attenuation slope compensating wavelength assignment that was in previous sections discussed for unidirectional rings can also be applied to an N-node meshed ring to maximize the ring perimeter. Figure 5.14a shows an example for the N-node mesh ring network in which each node has a transparent path to all the other nodes. Therefore, the required number of wavelengths is $N(N-1)$. The logical mesh network can be configured as a single hop, that is, a network in which optical signals carried between a pair of transmitting and receiving nodes are not converted into electrical signals at intermediate nodes but remain in the optical domain until the destination node, unidirectional physical ring network, as shown in Figure 5.14b. In a unidirectional ring, a single wavelength can be used between a pair of nodes, thereby reducing the required number of wavelengths in the network to $N(N-1)/2$. For example, a 6-node mesh network requires a minimum of 15 wavelengths. Therefore, the 16-channel CWDM system can support a maximum of 6-nodes mesh. Consequently, the 6-node ring is used as the case study. A sufficient number of wavelengths must be assigned to enable each node in the network to transmit and receive signals to and from all other nodes. The wavelength assignment must be nonblocking.

5.6.1 Wavelength Assignment

In the unidirectional ring under consideration, each wavelength can be used twice in different portions of the network. If node j transmits to node k on the wavelength

TABLE 5.9
An 8-Node Ring Channel to Node Assignment for Figure 5.12a

Node Number	N_1		N_2		N_3		N_4		N_5		N_6		N_7		N_8	
Channel assignment	1	2	3	4	5	6	7	8	9	10	11	12	13	14	15	16
Center wavelength (nm)	1310	1330	1350	1370	1390	1410	1430	1450	1470	1490	1510	1530	1550	1570	1590	1610
Attenuation (dB/km)	0.360	0.335	0.322	0.311	0.333	0.291	0.281	0.272	0.266	0.260	0.254	0.252	0.250	0.250	0.257	0.266

TABLE 5.10
An 8-Node Ring Channel to Node Assignment for Figure 5.12b

Node Number	N_1			N_2			N_3			N_4			N_5			N_6			N_7			N_8	
Channel assignment	13	14	9	10	5	6	1	2	3	4	7	8	11	12	15	16
Center wavelength (nm)	1550	1570	1470	1490	1390	1410	1310	1330	1350	1370	1430	1450	1510	1530	1590	1610
Attenuation (dB/km)	0.250	0.250	0.266	0.260	0.333	0.291	0.360	0.335	0.322	0.311	0.281	0.272	0.254	0.252	0.257	0.266

Optimizing CWDM for Nonamplified Networks

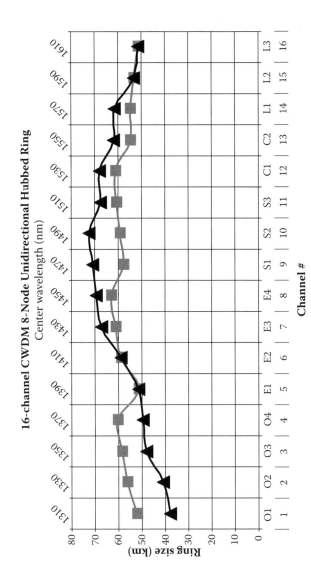

FIGURE 5.13 Calculated ring perimeter for both wavelength assignment approaches across the CWDM band, 8-node hubbed ring. Squares: Slope compensation; triangles; no slope compensation.

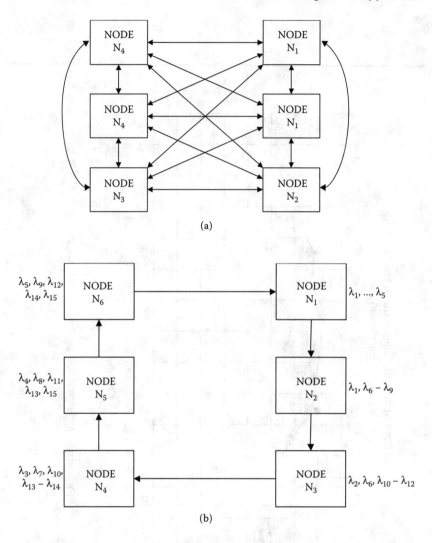

FIGURE 5.14 (a) Topology of a meshed network and (b) equivalent N-node unidirectional ring.

λ_p on a clockwise or counterclockwise direction, the node k can also transmit to node j on same λ_p in the same clockwise or counterclockwise direction. Hence, the minimum number of wavelengths required in the network is $N(N-1)/2$. The simple wavelength assignment scheme mentioned subsequently can be used to assign wavelengths to the nodes in a nonblocking manner and results in a $N \times N$ matrix of transmitting and receiving nodes, as shown in Table 5.11. Next, an algorithm is shown that can be used to assign wavelengths in such a network in a static nonblocking fashion. The $\kappa(j, k)$ obtained from this loop occupy the 6×6 matrix in the upper half of Table 5.11.

TABLE 5.11
Nonattenuation Slope Compensating Wavelength Assignment in 6-Node Meshed Network

	Wavelength Assignment in a 6-Node Ring Receiving Nodes					
	Node-1	Node-2	Node-3	Node-4	Node-5	Node-6
Node-1		1	2	3	4	5
Node-2	1		6	7	8	9
Node-3	2	6		10	11	12
Node-4	3	7	10		13	14
Node-5	4	8	11	13		15
Node-6	5	9	12	14	15	

	15 Wavelengths assigned to the nodes (nm) Receiving Nodes					
	Node-1	Node-2	Node-3	Node-4	Node-5	Node-6
Node-1		1310	1330	1350	1370	1390
Node-2	1310		1410	1430	1450	1470
Node-3	1330	1410		1490	1510	1530
Node-4	1350	1430	1490		1550	1570
Node-5	1370	1450	1510	1550		1590
Node-6	1390	1470	1530	1570	1590	

Note: Assigned CWDM wavelengths cover O + E + S + C + L-bands.

```
For j = 1 To N
      For k = 1 To N
If (j < k) Then
Counter_Lambda = Counter_Lambda + 1
k(j, k) = Counter_Lambda
End If

If (j > k) Then
Counter_Lambda = Counter_Lambda + 1
k(j, k) = N(N - 1)/ 2 - (Counter_Lambda - 1)
k(j, k) = k(k, j)
End If
```

5.6.2 Optimization Techniques

Each node in the ring uses a set of $(N - 1)$ wavelengths to transmit and receive signals from the other nodes in the ring. In other words, a set of $(N - 1)$ wavelengths can be part of add/drop at each node. When a node transmits signals to other nodes, the set of transmitted wavelengths passes through different fiber lengths, number of mux/demux stages, splices and connectors, thereby incurring different losses due to fiber attenuation, filter, splice, and connector losses. Similarly, when a node receives transmitted signals from other nodes, each signal passes through different fiber lengths, number of WDM mux/demux stages, splices and connectors before arriving at a receiving node, again incurring a different amount of losses in their paths. The analytical approach of this chapter models the network link budget here as a function of the path-lengths, number of mux/demux stages, number of splices and number of connectors traversed by a signal between any pair of transmitting and receiving nodes in a ring. This enables the losses associated with each loss mechanism in the network to be determined. For CWDM operation, 16 wavelengths are available from 1310 to 1610 nm between O + E + S + C + L bands. In the analysis, the group of

TABLE 5.12
Wavelength Assignment in 6-Node Meshed Network: (Top) Number of Traversed Ring Spans and (Bottom) Attenuation Slope Compensation Assignment

	Number of Ring Spans Traversed by Each Wavelength					
	Receiving Nodes					
	Node-1	Node-2	Node-3	Node-4	Node-5	Node-6
Node-1	–	1	2	3	4	5
Node-2	5	–	1	2	3	4
Node-3	4	5	–	1	2	3
Node-4	3	4	5	–	1	2
Node-5	2	3	4	5	–	1
Node-6	1	2	3	4	5	–
	Wavelength Assignment					
	Receiving Nodes					
	Node-1	Node-2	Node-3	Node-4	Node-5	Node-6
Node-1	–	12	9	2	10	11
Node-2	12	–	13	8	3	5
Node-3	9	13	–	14	7	4
Node-4	2	8	14	–	15	6
Node-5	10	3	7	15	–	16
Node-6	11	5	4	6	16	–

Optimizing CWDM for Nonamplified Networks

wavelengths in the network is treated as a set and it is shown how the losses incurred by each individual wavelength due to various loss mechanisms — fiber attenuation, WDM insertion, and splice and connector losses — may be computed, for the amplified and nonamplified cases. The results of this case study are presented for $N = 6$ in Table 5.12.

The number of actual spans traversed between transmitting and receiving node-pairs (j, k) by all the wavelengths in the network is the set $\{N_{span}(j, k)\}$ in Table 5.12 (top) and is given by:

$$\{N_{span}(j, k)\} = \begin{cases} (k - j) & \text{if } j \leq k \\ N - (j - k) & \text{if } j > k \end{cases} \quad \text{for} \quad \{j = 1, 2 \ldots N\} \quad \text{and} \quad \{k = 1, 2 \ldots N\}$$

The CWDM channels utilized for this particular wavelength assignment are presented in Table 5.12 (bottom). These data are subsequently used to calculate ring losses in the CWDM 6-node logical mesh/unidirectional ring as a function of wavelengths assigned to the nodes. The results are shown in Figure 5.15 where the ring perimeter or actual maximum transmission length for a particular CWDM wavelength is presented. We compare two cases: For simple assignment as in Table 5.11, we observe a pronounced variation of the ring size with wavelength ranging from 45 km at 1310 nm to more than 90 km at 1530 nm. In this scenario, the strong wavelength-dependent reach requires a careful selection of the appropriate CWDM wavelength when configuring the mesh network. This disadvantage can be avoided with the attenuation slope compensating assignment scheme where the corresponding ring size for all CWDM channels only varies between 58 and 75 km and, therefore, considerably minimizes the dependence of reach on CWDM wavelength numbers.

Finally, the steps for wavelength assignment in the meshed network in Visual Basic is presented. The logical mesh network is implemented as a unidirectional physical ring. In a unidirectional ring, this wavelength assignment will be nonblocking.

Case A: Regular assignment

```
Sub N_NodeRingLambdaAssign()

    Dim j As Integer
    Dim k As Integer
    Dim Path_Length(100, 100) As Variant
    Dim Span_length As Variant
    Dim Counter_Lambda As Integer
    Dim Counter1_Lambda As Integer
    Dim Numb_Nodes As Integer
```

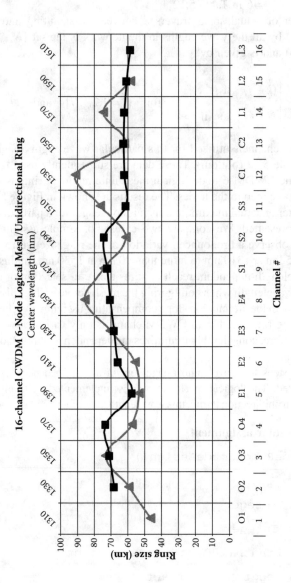

FIGURE 5.15 Attainable ring perimeters calculated as function of the CWDM center wavelengths for the CWDM-based 6-node logical mesh/unidirectional ring; triangles: simple wavelength assignment versus squares: attenuation slope compensating assignment.

```
Application.ScreenUpdating = False
   Sheets("N_NodeRing").Activate
Span_length = 8
     Counter_Lambda = 0
     Counter1_Lambda = 0
     Numb_Nodes = 4


For j = 1 To Numb_Nodes
       For k = 1 To Numb_Nodes


   If (j < k) Then
   Counter_Lambda = Counter_Lambda + 1
   Path_Length(j, k) = Counter_Lambda


End If


    If (j > k) Then


Counter1_Lambda = Counter1_Lambda + 1
   Path_Length(j, k) = Numb_Nodes * (Numb_Nodes - 1) / 2 - (Counter1_Lambda - 1)
    Path_Length(j, k) = Path_Length(k, j)


    End If


    With Sheets("N_NodeRing")


             .Cells(1, 1 + k).Value = "Node-" & k
             .Cells(1 + j, 1).Value = "Node-" & j
          .Cells(1 + j, 1 + k).Value = Path_Length(j, k)
```

```
        .Cells(Numb_Nodes + 4, 1).Value = "Ring Span Length
(km.)"
        .Cells(Numb_Nodes + 4, 2).Value = Span_length
          .Cells(Numb_Nodes + 5, 1).Value = "Number
of Spans"
        .Cells(Numb_Nodes + 5, 2).Value = Numb_Nodes
            .Cells(Numb_Nodes + 6, 1).Value = "Ring
circumf (km.)"
        .Cells(Numb_Nodes + 6, 2).Value = Numb_Nodes
* Span_length


          .Cells(Numb_Nodes + 7, 1).Value = "Number
of Wavelengths"
        .Cells(Numb_Nodes + 7, 2).Value = Numb_Nodes
* (Numb_Nodes - 1) / 2


            .Cells(Numb_Nodes + 8, 1).Value = "Path
Lengths"


        End With

    Next k
        Path_Length(j, k) = Path_Length(j, k)
    Next j
```

Case B: Optimal Wavelength Assignment in the meshed network

```
Sub N_NodeRingSpan_length()


Dim Numb_Nodes As Integer
Dim j As Integer
Dim k As Integer
Dim Path_Length(100, 100) As Variant
Dim Span_length As Variant
```

```
Application.ScreenUpdating = False
   Sheets("N_NodeRing").Activate


      Numb_Nodes = 6
      Span_length = 1


For j = 1 To Numb_Nodes
       For k = 1 To Numb_Nodes


    If (j <= k) Then


   Path_Length(j, k) = (k - j) * Span_length


    End If
If (j > k) Then


    Path_Length(j, k) = (Numb_Nodes - (j - k)) *
Span_length


    End If


    With Sheets("N_NodeRing")


          .Cells(1, 1 + k).Value = "Node-" & k
          .Cells(1 + j, 1).Value = "Node-" & j
       .Cells(1 + j, 1 + k).Value = Path_Length(j, k)
.Cells(Numb_Nodes + 4, 1).Value = "Ring Span Length
(km.)"
       .Cells(Numb_Nodes + 4, 2).Value = Span_length
         .Cells(Numb_Nodes + 5, 1).Value = "Number
of Spans"
```

```
            .Cells(Numb_Nodes + 5, 2).Value = Numb_Nodes

              .Cells(Numb_Nodes + 6, 1).Value = "Ring
    circumf (km.)"

              .Cells(Numb_Nodes + 6, 2).Value = Numb_Nodes
    * Span_length


              .Cells(Numb_Nodes + 7, 1).Value = "Number
    of Wavelengths"

              .Cells(Numb_Nodes + 7, 2).Value = Numb_Nodes
    * (Numb_Nodes - 1) / 2


              .Cells(Numb_Nodes + 8, 1).Value = "Path
    Lengths"


            End With
            Next k
        Next j


     End Sub
```

5.7 CONCLUSIONS AND FUTURE WORK

This chapter has presented a wavelength assignment algorithm for maximizing the perimeter of the CWDM hubbed ring. The algorithm assigns wavelengths to nodes such that the filter, connector, and splice losses incurred by the wavelengths are at a minimum in the high fiber attenuation O-band. The filter loss therefore compensates fiber loss such that the ring loss (combined fiber, filter, connector, and splice losses) is minimized across the CWDM channel spectrum, thereby increasing the attainable ring size or perimeter. By employing this algorithm, the attainable perimeter of a 16-channel CWDM hubbed ring, with each channel operating at 2.5 Gb/s on a 28-dB loss budget, was increased from 43 km (for the case where the algorithm was not applied) to 54 km in the case employing the algorithm in the wavelength assignment process. This represents a 25% increase. To achieve a 43-km perimeter, only a 25-dB loss budget was required. This Wavelength Assignment algorithm therefore leads to a 3-dB saving in power budget. By our estimation, a 3-dB budget approximately equals $50 in cost. We believe that the algorithm can thus achieve up to $800 savings in the cost of optical sources for a 16-channel CWDM system.

As mentioned earlier, reach limitation is the single most important shortcoming for nonamplified CWDM systems. CWDM is cost competitive versus DWDM due to its simplicity and the components used, but since the CWDM system is nonamplified, it is severely constrained by fiber and filter losses, thus limiting a typical CWDM ring to perimeters on the order of 40 km. Under these constraints, the wavelength assignment algorithm presented here was shown to increase reach without the use of amplifiers. The main interest in this technique is that if CWDM system reach is significantly extended, this increases the domain of applicability of CWDM and better positions the technology to compete, versus DWDM, TDM, and SDM for metro network deployment.

The wavelength assignment algorithm presented here minimizes the filter loss incurred around the ring by the high intrinsic fiber loss O-band channels, which compensates the higher fiber budget in the O-band to maximize the ring perimeter. TFFs process light signals serially. It is possible to exploit this to advantage and design multiplexing and demultiplexing filters such that the through loss incurred by the wavelengths as they traverse a filter is minimum in the O-band and maximum in the L-band. Employing this filter design technique could increase the attainable ring perimeter beyond 60 km for a 16-channel 6-node network.

REFERENCES

1. ITU-T Recommendation G.694.2 (2003), Spectral Grids for WDM Applications: CWDM Wavelength Grid.
2. Nebeling, M., *CWDM — lower cost for more capacity in the short haul*, white paper http://www.fn-eng.com/images/CWDMwhitepaper.pdf, Lightwave, August 2001.
3. Keiser, G., *Optical Fiber Communications*, McGraw-Hill Higher Education, 3rd ed., pp. 93–96.
4. Lam, J. and Zhao, L., Design Trade-offs For Arrayed Waveguide Grating DWDM MUX/DEMUX, SPIE Proc. Volume 3949, WDM and Photonic Switching Devices for Network Applications, pp. 90–98, April 2000.
5. Ufongene, C. and Boncek, R., Zero-water-peak fiber extends metro CWDM reach. WDM Solutions, November 2002.
6. Guétré, S., *CWDM Filter Toolbox*, CIG Presentation, OFC Conference, Anaheim, 2002.

6 Amplifiers for CWDM

Leo H. Spiekman

CONTENTS

6.1 Introduction 171
6.2 Principles of Optical Amplification 172
    6.2.1 Optical Amplifier Types 172
        6.2.1.1 EDFA 173
        6.2.1.2 SOA 173
        6.2.1.3 Raman...174
    6.2.2 Gain..174
    6.2.3 Noise..175
    6.2.4 Gain Saturation and Gain Dynamics 176
6.3 Challenges of Amplifying CWDM 177
    6.3.1 Wideband Amplification..177
    6.3.2 Influence of Broadband Noise 178
6.4 Doped Fiber-Based CWDM Amplifiers...180
6.5 Semiconductor-Based CWDM Amplifiers..182
    6.5.1 Device Specifics 182
    6.5.2 Challenges and Solutions: Dynamics, Gain Clamping................184
    6.5.3 Fiber Transmission Using SOAs 187
6.6 Raman and Hybrid Amplifiers 190
6.7 Amplified CWDM Transmission Line...193
6.8 Summary...194
References 194

6.1 INTRODUCTION

A good case can be made that the fast progress made in optical telecommunication systems over the past decade has been mainly due to the introduction of optical amplification, and more specifically due to the erbium-doped fiber amplifier (EDFA). This development allowed for transmission distances to multiply, and for line capacities to increase by orders of magnitude, thanks to dense wavelength division multiplexing (DWDM). These days, it is possible to transmit hundreds of DWDM channels down a fiber link over many thousands of kilometers. Line amplifiers placed at every 80 km or so regenerate the signal level for all channels simultaneously. Transmission distance is no longer limited by signal power, but by amplifier noise and by the accumulation of transmission impairments such as chromatic dispersion and fiber nonlinearities. In this way, optical amplifiers have radically changed the economics of optical fiber transmission. Before, a regenerator (basically a receiver–transmitter pair) was needed for each separate wavelength channel at each regeneration site. Now, a single amplifier can be used by many channels at once, dramatically reducing the cost per channel. Most, if not all, modern DWDM systems use optical amplification or are at least prepared for its use. Not so with CWDM.

CWDM systems have not yet seen the introduction of amplification, at least not in installed systems, although some other techniques have been applied as described in Chapter 5. This results from several factors, such as the relatively short transmission distances usually required or the unavailability of suitable amplifiers; these subjects will be discussed in this chapter. That is not to say that CWDM amplification has not been considered. Initially, this was done with a C-band EDFA and a set of lasers spaced by 600 GHz [1]. Even though coarse, this spacing does not conform to the established ITU-T G.694-2 standard for CWDM. Therefore, amplifiers with wider bandwidth have been developed, as will be described later. It is clear that the desire for more transmission margin lives among CWDM system designers. The use of forward error correction (FEC) has been proposed to provide the extra margin for upgrading the transmission speed from 2.5 to 10 Gb/s [2]. It depends on the details of the economics of the system whether FEC alone, or amplifiers, or maybe a combination of both will be used. But if significantly more margin is needed, amplification is the only way out.

6.2 PRINCIPLES OF OPTICAL AMPLIFICATION

An optical amplifier duplicates incoming photons by a process known as stimulated emission. This is the same process responsible for light generation in lasers. An ideal amplifier would just increase the signal strength without any disadvantageous side effects, and function as a kind of negative loss to offset span and component losses. However, in reality, the process of spontaneous emission, which is inseparably linked with stimulated emission, gives rise to optical noise that degrades the signal. Both the advantageous gain and the disadvantageous noise will be discussed later.

6.2.1 Optical Amplifier Types

The idea of amplifying CWDM systems is fairly young, see reference [3] and the references therein, but amplifiers have been used in optical communications for many years. The most well-known amplifier types are the EDFA, the semiconductor optical amplifier (SOA), and the Raman amplifier. The properties of each of these amplifier types relevant to their use in CWDM systems will be discussed in more detail subsequently. Here, we will only cover the general principles of operation of each type and address the properties they have in common.

Amplifiers for CWDM

6.2.1.1 EDFA

EDFA is the most well-known case of a rare earth doped fiber amplifier. A fiber is doped with rare earth elements, which are pumped optically. The pumping light excites the rare earth atoms, and a passing photon can give rise to stimulated emission when its wavelength corresponds to the energy difference between the appropriate energy levels of the atom [4].

Erbium has become very popular as a doping element, since its energy levels are such that amplification occurs in the low-loss window of standard fiber around 1550 nm. The standard amplification band of an EDFA, ranging approximately from 1530 to 1565 nm, has become known as the C-band. By appropriately choosing doping and pump levels, the wavelength range of 1565 to approximately 1610 nm, known as the L-band, can be covered. Even an S-band amplifier has been demonstrated, providing 20 nm of gain centered around 1500 nm [5].

An EDFA can be pumped at 1480 or 980 nm, wavelengths that are readily accessible using semiconductor lasers. The most simple EDFA consists of a piece of erbium-doped fiber (typically tens of meters long), a 980-nm or 1480-nm pump laser, and a wavelength-dependent coupler to multiplex the signal and pump wavelengths. Other rare earth elements can be used to amplify wavelength bands outside of the C- and L-bands. However, many of these are not easily incorporated in standard silica fiber or suffer from other disadvantages such as more complicated pumping schemes or faster gain dynamics, which makes them impractical for the use with CWDM.

6.2.1.2 SOA

The SOA [6] is a structure that is very similar to a semiconductor laser. Where amplification in the EDFA occurs through stimulated emission by excited erbium atoms, in the SOA recombination of electrons and holes in the conduction and valence bands of the semiconductor gives rise to emission of photons. The emission wavelength corresponds to the band gap energy of the semiconductor material.

A typical SOA, like a semiconductor laser, consists of a gain chip containing an active waveguide, with fibers coupled to it at both ends. One or more lenses adapt the fiber mode to the much smaller waveguide mode. Since the gain per unit length is very high compared to the EDFA gain, the chip can be very short (typically ~1 mm).

The gain chip contains a semiconductor heterostructure that confines both light and injection current in the active area. The injection current excites electrons into the conduction band, which gives rise to light emission. In a laser, the facets of the device provide optical feedback, which leads to lasing. In an SOA, on the other hand, reflection at the facets is suppressed by using antireflection coatings and by having the gain stripe not at a right angle with respect to the facets. This way, the signal injected by the input fiber experiences gain during its single pass through the active waveguide, until it reaches the output fiber.

In an SOA, the gain medium commonly used is indium gallium arsenide phosphide (InGaAsP). This material has a band gap corresponding to an emission wavelength of 0.9 to 1.65 µm, depending on the composition, which can be smoothly varied from binary InP to ternary InGaAs [7]. This means that, in principle, the entire wavelength range used for CWDM is accessible by appropriate choice of the active layer composition.

6.2.1.3 Raman

Raman amplification occurs in an optical fiber, but it does not rely on the presence of rare earth dopants. Instead, it is based on nonlinear interaction between the signal and a high-power optical pump. Stimulated Raman scattering occurs when a signal photon causes inelastic scattering of a higher energy pump photon, producing another signal photon, while the excess energy is dissipated by phonons (heat). In optical fiber, the frequency downshift is approximately 13 THz, meaning that a pump of approximately 1450 nm is required in order to amplify the C-band [8].

Raman amplification has proved beneficial for DWDM transmission since the transmission fiber itself can be used as the gain medium. Since the fiber can be pumped in a counter-propagating fashion, the signal experiences gain many kilometers before it reaches the amplifier site, which can significantly improve the apparent noise figure (NF) referenced to that site. Often, the Raman amplifier is referred to as a distributed amplifier, whereas the EDFA and SOA are more typical "lumped element" amplifiers. Raman amplification can also provide broadband gain by the appropriate choice of a comb of pump wavelengths.

6.2.2 Gain

The key parameter for each of the amplifier types described earlier is the gain. Ideally, amplifier gain is a constant irrespective of signal wavelength or polarization state, and constant in time independent of the (instantaneous or average) signal power. The actual behavior of amplifier gain over time will be discussed in Section 6.2.4 on gain dynamics.

The gain spectrum of an EDFA has a particular shape, which depends on the stimulated emission cross-section of the erbium atoms. In long transmission systems with many cascaded EDFAs, specially designed gain flattening filters are usually employed to limit the overall gain variation over the entire transmission length. Since the gain medium (the erbium-doped fiber) has rotational symmetry, the polarization dependence of the gain (known as polarization-dependent gain, PDG) of the overall amplifier usually only depends on the polarization-dependent loss (PDL) of components such as the pump multiplexer and isolators.

Raman gain depends strongly on the frequency offset between pump and signal. Since many different phonon states are available in the material, the gain spectrum is broad, increasing almost linearly with wavelength offset between pump and signal, peaking at an offset of 100 nm and then dropping rapidly in a characteristic triangular shape. Since the pump wavelength can be freely chosen

Amplifiers for CWDM

and multiple pumps can be applied, the gain spectrum can be readily tailored. The main disadvantage of Raman amplification is that very high pump powers are needed to generate reasonable gain. Since stimulated Raman scattering only occurs when pump and signal are co-polarized, the pumps usually have to be depolarized to obtain amplification with low PDG.

In the SOA case, the gain spectrum is determined by the shape of the conduction and valence bands. The longest wavelength gain occurs at the band gap, and the more the bands are filled with carriers, that is, the harder the gain medium is pumped electrically, the further the gain spectrum extends toward shorter wavelengths. The spectrum is very wide (typically 60 to 80 nm full width half maximum) and has a smooth parabolic shape so that gain flattening filters are not needed for most applications. The only structure that can be found on an SOA gain spectrum is Fabry-Perot ripple due to the standing-wave resonance between the two device facets. This gain ripple can be observed on an optical spectrum analyzer with standard resolution due to the short length of the cavity (typical ripple period is a few tenths of a nanometer), and it is minimized (usually to less than 0.2 dB depth) by applying antireflection coatings and an angled gain stripe. Note that in the EDFA and Raman domains, the term "gain ripple" refers to the maximum excursion of gain versus wavelength as determined by the erbium gain cross-section or spectrum of phonon states.

Because the SOA is a planar device, the waveguide has a rectangular shape. This leads to different confinement of the horizontally and vertically polarized modes and therefore to PDG. Various methods are used to compensate for this PDG, most based on engineering the crystal strain of the active layer, and at present a polarization dependence of <0.5 dB can routinely be achieved.

6.2.3 Noise

In addition to the stimulated emission that produces the desired gain, all optical amplifiers exhibit spontaneous emission of photons. The fraction of these noise photons that couple into the fiber or gain waveguide is subsequently amplified and leads to the amplified spontaneous emission (ASE) spectrum of the amplifier. In the receiver of a telecommunication system, ASE beats with the signal and with itself, resulting in noise that places a lower limit on the error rate with which the signal can be received.

The noise properties of an amplifier can be quantified by its NF, which for optical amplifiers is defined as the deterioration in signal-to-noise ratio (SNR) when the input signal is purely shot noise limited. A more practical definition, in common use among amplifier manufacturers, neglects the shot noise terms in the full NF expression in favor of the usually dominant beat noise terms, and makes the approximation $NF = 2n_{sp}/\eta_i$, in which $n_{sp} = N_2/(N_2 - N_1)$ is the inversion parameter of the amplifier (i.e., the degree of population inversion, with N_1 and N_2 the fractional number of erbium atoms or carriers in the ground and excited states, respectively) and η_i is the sum of all loss the signal experiences from the input of the amplifier to the beginning of the gain medium.

In all amplifier types we discuss here, strong population inversion (i.e., a n_{sp} close to 1) can usually be achieved. In the EDFA, the loss term only comprises the isolator, pump coupler, and splice from input fiber to erbium-doped fiber, which can be very small. For this reason, high-end EDFAs can have an NF as small as 4 dB, although values up to 6 dB are more common for low-cost versions. For an SOA, the main contribution to the loss term is mode mismatch from fiber-chip coupling, leading to a typical NF of 6 to 7 dB for these devices. In a Raman amplifier, since the fiber is the gain medium, the input loss can be close to zero, allowing Raman gain to come close to the theoretical limit of 3 dB. Since Raman gain starts before the pump site in a backward-pumped fiber, the apparent NF referenced to that site can even be negative.

6.2.4 Gain Saturation and Gain Dynamics

When the input signal power into an optical amplifier is increased, more and more of its (electrical or optical) pump power will be consumed to maintain its gain. Eventually, the amplifier will saturate and its gain will decrease because the excited erbium atoms, or the excited carriers, or the optical pump power, are being depleted. All three amplifier types essentially saturate according to the same curve, but the dynamics of the amplifier determine the practical consequences.

In steady-state operation, the saturation of the amplifier determines the maximum output power that it can supply: upon increasing input power, the dropping gain lets the output power approach a maximum value. However, the amplifier may not be able to track fast changes in input power. The excited state of the erbium atom in an EDFA has a lifetime of about 10 ms. Therefore, the modulation of an intensity-modulated signal (typically 155 Mb/s to 40 Gb/s for optical telecommunication systems) is too fast for it to follow. The EDFA will "see" a constant-power signal, and its degree of saturation will correspond to the average signal power. Note that some other rare earth dopants have faster lifetimes so that gain dynamics may become an issue for low-bitrate signals.

When the average power in the EDFA changes, for example, in a packet switched network, or due to provisioning or restoration actions that change the number of active channels, it will readjust to the new average power with a change in gain. Therefore, it is straightforward to amplify modulated signals using an EDFA in transmission links with a fixed number of channels, but more challenging to operate this amplifier in a dynamic network [9].

The carrier lifetime of an SOA, on the other hand, is very short, typically 100 ps. Since this is of the order of the bit period in a 10-Gb/s data stream, the SOA will try to track the instantaneous power of the intensity-modulated data, and when the power in the "1"-bits saturates the amplifier, this will lead to reduced extinction ratio and the appearance of inter-symbol interference (ISI). In multi-channel systems this is compounded by inter-channel crosstalk mediated by the gain changes of the amplifier. These effects limit the maximum usable output power of the SOA to a point well below the maximum power it is capable of delivering in the

TABLE 6.1
Summary of Amplifier Properties

	SOA	EDFA (low cost)	EDFA (high end)	Raman
Gain	Medium	Medium	Medium/high	Low
Wavelength	Any	C- or L-band	C + L-band	Any
Bandwidth	Wide	Small	Medium	Wide
Polarization independence	Sufficient	Good	Good	Good
Noise figure	Sufficient	Sufficient	Low	Very Low
Output power	Medium	Medium	High	Dep. on pump
Dynamics	Fast	Slow	Slow	Averaged out
Cost	Low	Low	High	High

steady-state case. Experimentally, it has been determined that operating the device at a gain saturation of about 1 dB below small-signal gain is an optimal compromise between maximizing the optical signal to noise ratio and minimizing the signal distortion due to saturation [10]. Since the SOA has to be operated away from saturation, it is in a better position to cope with dynamic networks.

The gain dynamics associated with stimulated Raman scattering are very fast, but since the pump and signal travel at different speeds or can even be counter-propagating, pump depletion due to "1"-bits is averaged over many bit periods so that Raman gain essentially behaves like erbium gain as far as gain dynamics are concerned.

Table 6.1 reviews the properties of the amplifier types that have been discussed.

6.3 CHALLENGES OF AMPLIFYING CWDM

One fundamental and one significant engineering challenge are associated with the idea of having amplification in CWDM system. The engineering challenge is to cover a much larger bandwidth than has been necessary for DWDM amplification. Several approaches are available for the amplifier types that are discussed. The fundamental challenge is that of broadband noise into the receiver. In a DWDM system, optical noise into the receiver is largely rejected by a narrow bandpass filter so that the only relevant noise term is signal-spontaneous beat noise [11]. But the very essence of CWDM is the relaxed wavelength registration, which necessitates the use of fairly wide filters. The result is that optical noise in a bandwidth of about 13 nm reaches the receiver, making the spontaneous-spontaneous beat noise term non-negligible.

6.3.1 WIDEBAND AMPLIFICATION

Amplifiers with a bandwidth broad enough to cover the full CWDM spectrum have not been constructed, but work on broadband amplifiers has long been underway.

Most work in this area aimed to increase the capacity of DWDM transmission as for example shown in ref. [12]. The gain bandwidth of EDFAs has been increased by using new host materials such as fluoride and tellurite glass and by using other dopants such as thulium. The Raman bandwidth has been extended by using multiple pumps, and hybrid EDFA/Raman and thulium-doped fiber amplifier (TDFA)/Raman amplifiers have been constructed in which the Raman gain helps flatten the rare earth gain spectrum. 3-dB gain bandwidths of over 100 nm have been realized.

The bandwidth of an SOA is already fairly large (60 to 80 nm of 3-dB bandwidth, depending on its gain). The design of the active layer can be engineered to broaden this. Quantum well active structures can have very broad gain peaks, leading to devices with very wide gain bandwidth [13,14]. Chips with high gain and low PDG over a bandwidth of 95 nm have been demonstrated.

In order to fill the entire CWDM spectrum, a banded approach has been proposed [15]. Since the location of the gain peak of an SOA can be relatively easily tuned, this device is an obvious candidate for this approach. As shown in Figure 6.1, a full CWDM amplifier could consist of a band demultiplexer, four SOAs covering the wavelength ranges 1300 to 1380 nm, 1380 to 1460 nm, 1460 to 1540 nm, and 1540 to 1620 nm, and finally a band multiplexer.

6.3.2 Influence of Broadband Noise

The CWDM standard calls for 20-nm spaced channels, where the wavelength of each channel can vary by at most 13 nm. This configuration calls for filters with a fairly flat passband of 13 nm, and therefore any optical amplifier noise in such a bandwidth will reach the receiver. More details about filters for CWDM can be found in Chapter 4.

In a DWDM system, optical filtering is usually done with a bandpass filter just wide enough to pass the signal without causing ISI. This ensures that beat noise caused by the spontaneous emission alone can be neglected. Only components of the ASE that are so close in frequency to the signal that their beating with the signal falls within the electrical bandwidth of the receiver will degrade reception. Components of the ASE that are within the same delta in frequency of each other have such low power that they do not contribute to the noise appreciably. However, when a wide band of noise hits the receiver, the components of the ASE beating with each other need to be integrated over the full noise bandwidth, and in this case they will constitute a noise contribution that cannot be neglected. Figure 6.2 shows the sensitivity of an optically preamplified receiver in which preamplifier noise is filtered at different optical bandwidths. The preamplifier in this measurement was an SOA with an NF of 8 dB. The smallest filter bandwidth corresponds to the case of an optimally filtered receiver. As the filter bandwidth is increased, the receiver sensitivity is reduced. At a bandwidth of 10 nm, close to the 13-nm bandwidth of a CWDM filter, the reduction is 3.5 dB with respect to the DWDM case. This graph shows the somewhat reduced benefit of optical preamplification in a CWDM transmission link

Amplifiers for CWDM

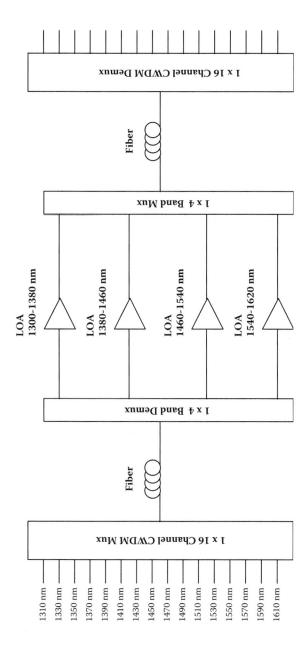

FIGURE 6.1 CWDM architecture with semiconductor-based amplifiers using band (de)multiplexers. (Taken from Thiele, H.J., Nelson, L.E., Thomas, J., Eichenbaum, B., Spiekman, L.H., and van den Hoven, G.N., Proceedings OFC 2003, vol. 1, pp. 23–24, 2003. With permission.)

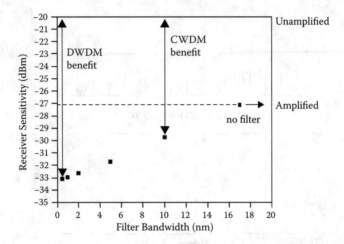

FIGURE 6.2 Measured receiver sensitivity at 10^{-9} BER of an optically preamplified pin-diode receiver using a clean intensity-modulated signal at 10 Gb/s. As the optical filter bandwidth is varied, the decrease in sensitivity due to increased spontaneous-spontaneous beat noise can clearly be observed.

compared to the equivalent DWDM system. Note that there is still a sensitivity improvement of 10 dB with respect to the unamplified detection case.

6.4 DOPED FIBER-BASED CWDM AMPLIFIERS

Now we turn to a review of experimental work on broadband amplification with each of the three types of amplifiers. In order to amplify an appreciable number of CWDM channels in a rare earth doped fiber, a bandwidth significantly broader than the usual 1530 to 1565 nm band is needed. The first step in this direction was to use a split-band amplifier for the C- and L-band [16]. Because at lower inversion the gain in erbium-doped fiber shifts to longer wavelengths, the L-band (1565 to 1610 nm) was amplified by a longer EDF pumped to have lower gain per meter.

The next step is to use other elements to augment the gain band. A cascade of a TDFA and an EDFA is able to amplify four CWDM channels from 1460 to 1540 nm [17]. An appropriate pumping scheme can even extend the bandwidth to 100 nm [18–20].

Combining this technique with erbium in tellurite fiber that can amplify the four highest channels allows for the amplification of eight CWDM channels [21]. Figure 6.3 shows the configuration of such an amplifier. It consists of two band amplifiers that are combined with wavelength division multiplexing (WDM) couplers. The first half is a cascade of a TDFA and an EDFA, which take care of amplifying the 1460 to 1540 nm band, whereas the second half is an erbium-doped tellurite fiber amplifier (EDTFA) that amplifies the 1540 to 1620 nm band.

Amplifiers for CWDM

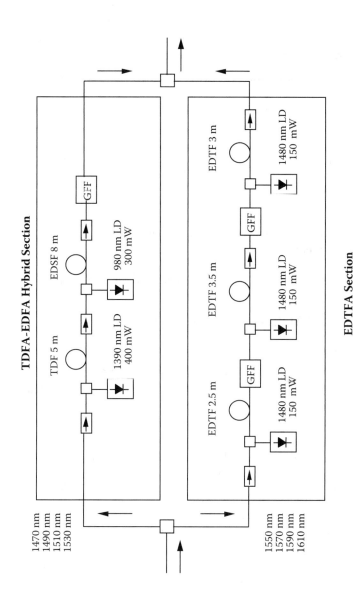

FIGURE 6.3 Hybrid EDFA–TDFA amplifier configuration: the four shorter wavelengths are amplified by the TDFA, whereas the remaining four CWDM wavelengths are amplified by the EDFA. (Taken from Sakamoto, T., Mori, A., and Shimizu, M., Proceedings OFC 2004, vol. 2, paper ThJ5. With permission.)

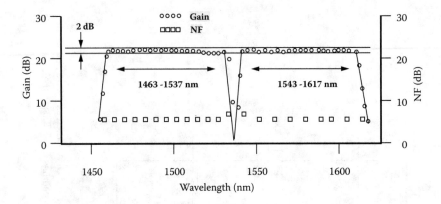

FIGURE 6.4 Gain after equalization and NF characteristics of hybrid EDFA–TDFA amplifier. (Taken from Sakamoto, T., Mori, A., and Shimizu, M., Proceedings OFC 2004, vol. 2, paper ThJ5. With permission.)

As Figure 6.4 shows, this combination provides a fairly flat gain of around 22 dB over 160 nm, good for eight CWDM channels, and an NF less than 8 dB. This is a high-quality, but also fairly expensive solution, as the amplifier consists of five doped fibers, five pump lasers, and three gain flattening filters, all spliced together.

6.5 SEMICONDUCTOR-BASED CWDM AMPLIFIERS

6.5.1 Device Specifics

The SOA [22] is more naturally suited to amplifying CWDM channels because of its broad bandwidth. Figure 6.5 displays the C-band coverage of a typical SOA, showing that the gain variation is less than 1 dB without any kind of gain flattening filter. The gain continues to drop smoothly on both sides, and 3-dB gain bandwidths of up to 80 nm are common.

The location of the gain peak is easily engineered by adjusting the composition of the active layer. Most SOAs have a quaternary $In_xGa_{1-x}As_yP_{1-y}$ active layer composed either of bulk material or of quantum wells. A definite relationship exists between x and y to lattice-match the material to the crystalline InP substrate. As a matter of fact, small variations from the lattice-matched condition help to improve the gain properties of quantum wells and to fine-tune the polarization independence of the amplifier. This leaves one degree of freedom, which serves to adjust the position of the gain band. A material composition close to InP provides gain at very short wavelengths (<1 μm), whereas in the limit of ternary material (InGaAs, with $y = 1$), the gain spectrum extends to 1.65 μm.

In practice, and quite logically, any wavelength that can be generated by an InGaAsP laser can be amplified by an InGaAsP SOA. With a gain bandwidth of 80 nm, four CWDM wavelengths, each occupying a slot of 20 nm, can be covered so that a full spectrum of 16-channel CWDM can be amplified by a set of four

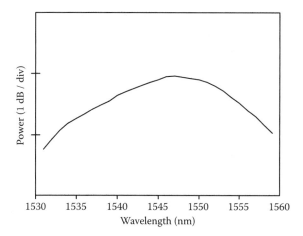

FIGURE 6.5 Amplified spontaneous emission spectrum of an SOA, which typically has a smooth parabola-like shape. (Taken from Spiekman, L.H., *Optical Fiber Telecommunications*, Academic Press, 2002. With permission.)

appropriately designed SOAs, for example, in an architecture as already shown in Figure 6.1.

The SOA bandwidth can be enhanced by employing more advanced active layer designs. It has been shown that using quantum wells a gain bandwidth of almost 100 nm can be reached [13,14]. This amplifier consisted of two quantum well structures separated by a thin barrier layer. Figure 6.6 shows the polarization-resolved gain of the device. The flat section in the center is considerably broadened with respect to the spectrum of Figure 6.5, and the 3-dB bandwidth extends from 1515 to 1615 nm.

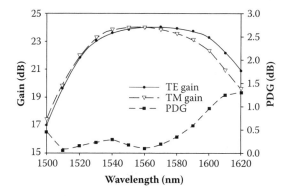

FIGURE 6.6 Measured TE and TM gains and PDG of an SOA chip using a coupled quantum well structure with broadband gain. (Taken from Park, S., Leavitt, R., Enck, R., Luciani, V., Hu, Y., Heim, P.J.S., Bowler, D., and Dagenais, M., *Photon. Technol. Lett.*, 17(5), 980–982, 2005. With permission.)

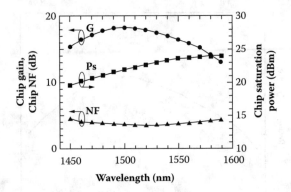

FIGURE 6.7 Dependence of gain, saturation output power, and NF on signal wavelength. Note that these parameters are referenced to the chip facets. (Taken from Morito, K., Tanaka, S., Tomabechi, S., and Kuramata, A., *Photon. Technol. Lett.*, 17(5), 974–976, 2005. With permission.)

6.5.2 Challenges and Solutions: Dynamics, Gain Clamping

In a broadband amplifier, it is important to maintain a good NF and a high output power over the entire gain band. A high output power is particularly important because of the gain dynamics of the SOA, as we will see subsequently. These parameters are plotted in Figure 6.7 for a device consisting of a thin multi-quantum well (MQW) active layer [23]. The 3-dB flatness of the gain is 120 nm, and a chip NF < 4.5 dB and a chip 3-dB saturation output power > +19.6 dBm are maintained over this entire band. Note that these figures are referenced to the chip. Packaging the chip (see Figure 6.8) adds a fiber coupling loss of typically 1 to 1.5 dB per side. Even then, the amplifier has an NF better than 6 dB and an output power better than +18 dBm. This is on par with present low-cost EDFA gain blocks, but over a significantly broader bandwidth.

High power SOA designs [24,25] raise a lot of interest. A fiber-coupled saturation output power as high as +22 dBm (160 mW) has recently been reached [26]. As said, this is important because of the dynamic behavior of the SOA. Statically, the gain versus power is described by a curve (Figure 6.9) that could

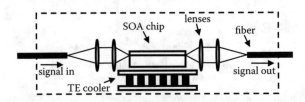

FIGURE 6.8 Typical configuration of a packaged SOA chip. The device has two fiber-chip couplings. Optional items like isolators may be placed in the light path. (Taken from Spiekman, L.H., *Optical Fiber Telecommunications*, Academic Press, 2002. With permission.)

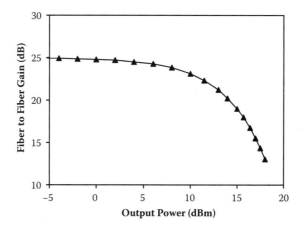

FIGURE 6.9 Gain versus output power curve of an SOA. The gain is quasi-linear in the small signal regime, but saturates at higher power levels. (Taken from Spiekman, L.H., *Optical Fiber Telecommunications*, Academic Press, 2002. With permission.)

also describe an EDFA. But dynamically, the picture is different. Where an EDFA reacts very slowly (~10 ms) to changes in power level, the reaction of an SOA is almost instantaneous (~100 ps).

The effect on gain of a short but intense optical pulse is shown in Figure 6.10. The pulse immediately compresses the gain, which recovers with a time constant

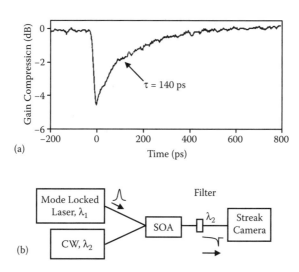

FIGURE 6.10 (a) Ultrafast gain compression and slower recovery of an SOA by an intense optical pulse. The gain recovery time is determined by the carrier lifetime. (b) Pump-probe setup with which the curve was measured. (Taken from Spiekman, L.H., *Optical Fiber Telecommunications*, Academic Press, 2002. With permission.)

comparable to the bitrate of a 10 Gb/s signal. The upshot of this is that the "1"-bits in a powerful intensity-modulated signal compress the gain of the SOA, and the gain recovers during the "0"-bits. This leads to ISI on the signal, and for WDM transmission additionally to cross-modulation between the channels.

The simplest way to avoid this, apart from not using intensity modulation, is to limit the total output power to the linear range of the amplifier. That this is not necessarily overly restrictive will be shown subsequently in a discussion of SOA WDM transmission experiments. However, a device solution is also available: gain clamping. By integrating a laser cavity on the chip, the SOA gain can be clamped by virtue of the lasing condition. This has been done by the following techniques:

1. Gain clamping can be done by integrating distributed Bragg reflectors (DBRs) in the waveguide path, on both ends of the chip [27]. These reflectors set up a lasing field at a wavelength outside of the bandwidth of interest for signal amplification. The lasing action fixes the gain at a value determined by the DBR loss so that the round-trip gain for the lasing wavelength is unity. Incoming signals will now always encounter a constant gain, irrespective of their power. Since the output power of the gain-clamped SOA is shared between the signals and the laser, output power is limited by the laser power: as soon as it goes below threshold, gain clamping is lost.
2. A disadvantage of the first type of gain-clamped amplifier is that a filter is required at the output to block the unwanted laser light. This disadvantage is not shared by a type of gain-clamped amplifier in which the lasing occurs vertically: By providing DBR mirrors above and below the waveguide, much like in a vertical cavity surface-emitting laser (VCSEL), the clamping light does not interfere with the signal amplification function of the device [28].
3. A third possibility for gain clamping is to use the interesting property of the SOA so that it can be integrated with other elements in a planar waveguide chip, and combine it with a Mach-Zehnder interferometer (MZI), using self-switching as a method to suppress ISI [29]. Note that this mechanism has only been demonstrated for single-channel amplification.

In the above list, patterning and cross-modulation between channels are the only nonlinear effects in the SOA that are discussed in this text. The important topic of four-wave mixing (FWM) is not reviewed since its mixing efficiency depends heavily on the wavelength spacing of the optical signals. Since in CWDM the channel spacing is large (7 nm min.), FWM can be neglected. A more complete overview of SOA nonlinearities can be found elsewhere [30].

The fact that the SOA is amenable to integration opens up a number of other interesting possibilities. CWDM (de)multiplexers implemented in planar waveguide circuits have been demonstrated in silica waveguides [31,32]. A similar waveguide design implemented in InP waveguides could be integrated with pre or postamplification SOAs, leading to lossless multiplexing or even multiplexing with gain.

Amplifiers for CWDM 187

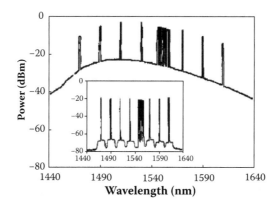

FIGURE 6.11 Input (inset) and output spectra of an SOA showing an 8-channel CWDM spectrum upgraded in one of the CWDM channels with eight DWDM channels. (Taken from Iannone, P., Reichmann, K., and Spickman, L., Proceedings OFC 2003, vol. 2, pp. 548–549, 2003. With permission.)

6.5.3 Fiber Transmission Using SOAs

Line extension using an SOA is straightforward. Amplification of four channels in the 80-nm 3-dB bandwidth of an ordinary SOA is the most obvious proposition [15], possibly integrating multiple SOAs with a planar band filter [33]. But even in a 160-nm spectrum, enough margin is available to show improvement for an 8-channel system using a single SOA [3]. In this same experiment, upgrading the capacity by replacing one of the CWDM channels with an 8-channel DWDM comb has been demonstrated (see Figure 6.11). Owing to the fact that the SOA is operated in the linear regime, such an upgrade can be done in-service, without disturbing the other channels by cross-gain modulation.

To see that line extension with an SOA does not have to be limited to a single additional span, refer to Figure 6.12, where a cascade of nine SOAs is shown. This experiment demonstrated 8×10 Gb/s DWDM transmission over six spans of fiber adding up to 240 km [10]. Note that in CWDM transmission with directly modulated lasers and no dispersion compensation, this reach would be practically limited by the accumulated dispersion, unless compensation techniques are applied after demultiplexing (since the dispersion accumulated for the shortest and the longest wavelength will be widely different).

Using the SOA not just as an amplifier, but as a functional element, architectures like the one shown in Figure 6.13 become possible: here, the SOA is used as a data modulator, with the wavelengths provided from the head end. This makes the end stations less wavelength-specific, as a single SOA can modulate any of a group of eight wavelengths [34].

A second example of CWDM transmission using SOAs is wireless-over-fiber transmission using a CWDM trunk feeding a ring as the transport architecture [35,36]. An SOA is used bidirectionally to amplify the trunk signal from the central

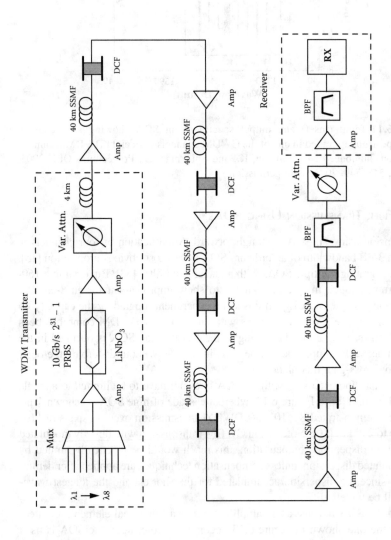

FIGURE 6.12 Transmission of eight WDM channels modulated at 10 Gb/s across 6 × 40 km spans of standard fiber using a total of nine SOAs. External modulation is used here to investigate the transmission over several cascaded SOA-amplified spans. (Taken from Spiekman, L.H., *Optical Fiber Telecommunications*, Academic Press, 2002. With permission.)

Amplifiers for CWDM

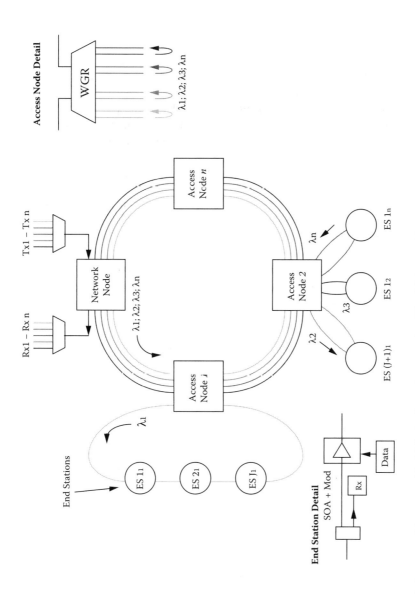

FIGURE 6.13 Access architecture with SOAs in the end user nodes. The SOAs provide the amplification needed in the metro ring, and double as modulators for the upstream data. (Taken from Spiekman. L.H., *Optical Fiber Telecommunications*, Academic Press, 2002. With permission.)

station feeding the ring, as well as to amplify signals originating from the wireless base-stations in the ring returning to the central station.

Finally, since SOAs have very fast dynamics, they are suited as nanosecond switching elements in a packet switch [37].

6.6 RAMAN AND HYBRID AMPLIFIERS

Erbium can be used to extend gain beyond the C-band toward the red (the L-band), but S-band EDFAs have small bandwidth (a single CWDM channel) and are not available commercially. Raman, on the other hand, is able to deliver gain in this band over a wide bandwidth, thanks to the property that gain moves with pump wavelength [38]. The same property allows fabrication of ultra-wide band Raman amplifiers simply by using multiple pump sources [39]. Figure 6.14 shows a flat gain profile thus achieved over 100 nm [40]. The flatness is better than ±0.5 dB, accomplished with asymmetric allocation of pump channels and without any gain equalization filters [41]. Figure 6.15 displays the pump unit used, consisting of 12 polarization-diversity wavelength channels and containing 24 pump lasers. Although clearly a very high quality amplifier, this complex device is not practical for use in low-cost CWDM networks.

A slightly more practical approach starts out with only two pump wavelengths for amplifying four CWDM channels, allowing for upgrading the amplifier with more pump lasers when more signal wavelengths have to be amplified [42]. Alternatively, a hybrid SOA-Raman approach can be chosen. Less Raman pump power is needed since the bulk of the gain is provided by the SOA, and the Raman amplifier serves to lower the NF of the combination and to further flatten the already flat SOA gain (same approach used for hybrid EDFAs in long haul).

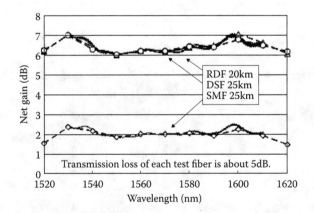

FIGURE 6.14 Flat gain profile (100 nm centered on C- and L-band) obtained by Raman-pumping fiber using a multi-wavelength source. (Taken from Namiki, S. and Emori, Y., *IEEE J. Sel. Topics Quantum Electron.*, 7(1), 3–16 (invited paper), 2001. With permission.)

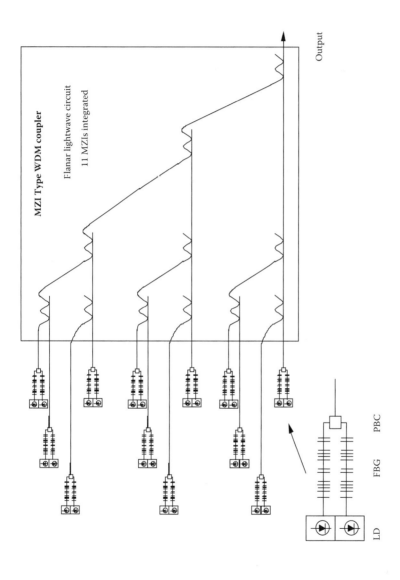

FIGURE 6.15 A 12-wavelength-channel WDM high-power pump LD unit. (Taken from Namiki, S. and Emori, Y., *IEEE J. Sel. Topics Quantum Electron.*, 7(1), 3–16 (invited paper), 2001. With permission.)

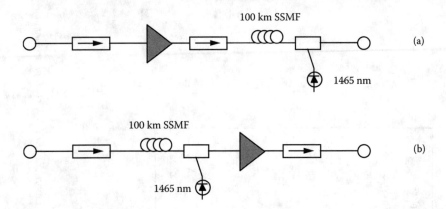

FIGURE 6.16 Two SOA-Raman hybrid schemes: (a) SOA + Raman and (b) Raman + SOA. (Taken from Chen, Y., Pavlik, R., Visone, C., Pan, F., Gonzales, E., Turukhin, A., Lunardi, L., Al-Salameh, D., and Lumish, S., Proceedings OFC 2002, pp. 390–391. With permission.)

As Figure 6.16 shows, the Raman gain can either follow the SOA or precede it [43]. Similarly, a hybrid consisting of Raman and rare earth doped fiber can be constructed, with the Raman section flattening and broadening the gain [44]. A hybrid SOA/Raman amplifier specifically constructed for CWDM amplification (see Figure 6.17) has been shown to extend four-channel transmission distance to 200 km [45]. Note that a Raman solution for all CWDM channels is not feasible in distributed Raman amplifiers since Raman pumps and signal channels are mixed.

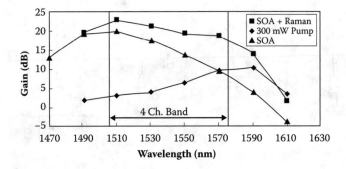

FIGURE 6.17 Gain versus wavelength for a hybrid Raman-SOA amplifier and its constituent SOA and Raman stages. (Taken from Iannone, P.P., Reichmann, K.C., Zhou, X., and Frigo, N.J., Proceedings OFC 2005, paper OthG3. With permission.)

6.7 AMPLIFIED CWDM TRANSMISSION LINE

Amplifiers can be introduced in a CWDM link in different configurations. Figure 6.18 shows booster amplification, preamplification, and both. In-line amplification is preferably not done because it requires a powered site along the transmission path. Booster amplification and preamplification can provide lossless (de)multiplexing, and especially make sense when a banded amplification concept is used, as the transmitters or receivers can be connected to the amplifiers using channel multiplexers, while the amplifiers can be connected to the line using band multiplexers. In a similar way, amplifiers can be introduced in add-drop demultiplexers.

When using preamplification, it has to be kept in mind that due to the relatively large bandwidth of noise reaching the receiver, the gain in link budget is less than the gain of the amplifier and less than in an optimally filtered system. This limitation does not occur when booster amplification is used, as in this case it is possible to maintain a high optical SNR.

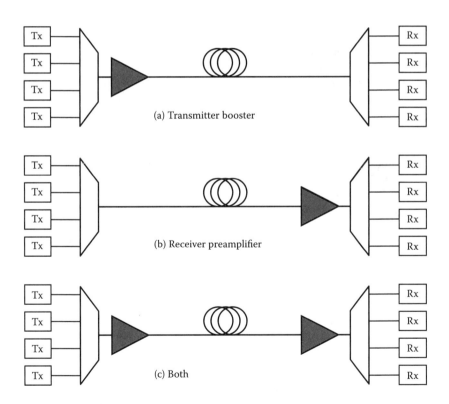

FIGURE 6.18 Various configurations of an amplified CWDM link.

On the other hand, when SOAs are used (which seems to make the most sense from a cost perspective) in a booster amplification setup, care has to be taken to limit the distortion due to gain compression. This limits the maximum usable output power out of the amplifier.

6.8 SUMMARY

Amplified CWDM systems have not yet taken root [46]. That may just be a matter of time, as system designers are pushing for more margin to increase bitrates or to go further, filling the gap between CWDM and DWDM. The technology is certainly available:

- in the form of fiber amplifiers that deliver high quality amplification, but at a relatively high cost and complexity; and
- in the form of SOAs that deliver more limited performance, mainly in terms of lower output power, but whose low-cost and simplicity are more in line with the goals of most CWDM deployments.

All amplifier technologies discussed here, rare earth doped fiber, semiconductor, and Raman, are capable of delivering gain in bands at least 80 nm (four channels) wide, and can be engineered for multiple such bands: rare earths can cover the eight highest channels, whereas SOAs and Raman can cover all channels. SOAs allow amplification of more than one channel (typically four) in a single device and might have a better-cost position with increasing channel count than simple repeaters based on small form factor pluggables (O-E-O). On the other hand, the latter are likely to be cheaper for single channel regeneration in CWDM systems.

The most important issues while designing an amplified system are the broad noise band into the receiver, and in the case of using a semiconductor booster, limiting the amount of gain compression. These issues can be overcome, and using amplification can significantly increase the link budget in CWDM systems. Whether this potential benefit is enough for designers to desire amplification will ultimately depend on economics.

REFERENCES

1. Shigematsu, M., Tanaka, M., Okuno, T., Hashimoto, J., Kawabata, Y., Takahashi, S., Nakanishi, H., Yamaguchi, A., Shibata, T., Inoue, A., Katsuyama, T., Nishimura, M., and Hayashi H., Amplified Coarse WDM System Employing Uncooled Fiber Bragg Grating Lasers with 600-GHz Channel Spacing, Proceedings OFC 2003, pp. 96–97, 2003.
2. Winzer, P.J., Fidler, F., Matthews, M.E., Nelson, L.E., Thiele, H.J., Sinsky, J.H., Chandrasekhar, S., Winter, M., Castagnozzi, D., Stulz, L.W., and Buhl, L.L., 10-Gb/s Upgrade of bidirectional CWDM systems using electronic equalization and FEC, *J. Lightwave Technol.*, 23(1), 203–210, 2005.

3. Iannone, P., Reichmann, K., and Spiekman, L., In-Service Upgrade of an Amplified 130-km Metro CWDM Transmission System Using a Single LOA with 140-nm Bandwidth, Proceedings OFC 2003, 2, 548–549, 2003.
4. Desurvire, E., *Erbium-Doped Fiber Amplifiers, Principles and Applications*, Wiley-Interscience 2002. ISBN 0-471-26434-2, 2002.
5. Arbore, M., Zhou, Y., Thiele, H.J., Bromage, J., and Nelson, L.E., S-band Erbium-doped Fiber Amplifiers for WDM Transmission between 1488 and 1508 nm, Proceedings OFC 2003, vol. 1, pp. 374–376, 2003.
6. Spiekman, L.H., Semiconductor Optical Amplifiers, in *Optical Fiber Telecommunications*, Kaminow, I. and Li, T., Eds., Vol. 4A, chap. 14. Academic Press, 2002. ISBN 0-123-95172-0.
7. Agrawal, G.P. and Dutta, N.K., *Semiconductor Lasers*, 2nd ed., Kluwer Academic Publishers, Dordrecht, The Netherlands, 1993. ISBN 0-442-01102-4.
8. Islam, M.N., *Raman Amplifiers for Telecommunications 1: Physical Principles*, Springer 2003. ISBN 0-387-00751-2.
9. Srivastava, A.K., Sun, Y., Zyskind, J.L., and Sulhoff, J.W., EDFA transient response to channel loss in WDM transmission system, *IEEE Photon. Technol. Lett.*, 9(3), 386–388, 1997.
10. Spiekman, L.H., Wiesenfeld, J.M., Gnauck, A.H., Garrett, L.D., van den Hoven, G.N., van Dongen, T., Sander-Jochem, M.J.H., and Binsma, J.J.M., 8 ×10 Gb/s DWDM transmission over 240 km of standard fiber using a cascade of semiconductor optical amplifiers, *Photon. Technol. Lett.*, 12(8), 1082–1084, 2000.
11. Olsson, N.A., Lightwave systems with optical amplifiers, *J. Lightwave Technol.*, 7(7), 1071–1082, 1989.
12. Masuda, H., Aozasa, S., and Shimizu, M., Ultra-wide-band hybrid amplifier consisting of two dispersion-compensating fibres for Raman amplification and thulium-doped fibre, *Electron. Lett.*, 38(11), 500–502, 2002.
13. Park, S., Leavitt, R., Enck, R., Luciani, V., Hu, Y., Heim, P.J.S., Bowler, D., and Dagenais, M., Semiconductor Optical Amplifier for CWDM Operating Over 1540–1620 nm, Proceedings CLEO 2004, vol. 1.
14. Park, S., Leavitt, R., Enck, R., Luciani, V., Hu, Y., Heim, P.J.S., Bowler, D., and Dagenais, M., Semiconductor optical amplifier for CWDM operating over 1540–1620 nm, *Photon. Technol. Lett.*, 17(5), 980–982, 2005.
15. Thiele, H.J., Nelson, L.E., Thomas, J., Eichenbaum, B., Spiekman, L.H., and van den Hoven, G.N., Linear Optical Amplifier for Extended Reach in CWDM Transmission Systems, Proceedings OFC 2003, vol. 1, pp. 23–24, 2003.
16. Sun, Y., Sulhoff, J.W., Srivastava, A.K., Abramov, A., Strasser, T.A., Wysocki, P.F., Pedrazzani, J.R., Judkins, J.B., Espindola, R.P., Wolf, C., Zyskind, J.L., Vengsarkar, A.M., and Zhou, J., A Gain-Flattened Ultra Wide Band EDFA for High Capacity WDM Optical Communications Systems, Proceedings ECOC 1998, vol. 1, pp. 53–54.
17. Sakamoto, T., Aozasa, S., Yamada, M., and Shimizu, M., High-gain hybrid amplifier consisting of cascaded fluoride-based TDFA and silica-based EDFA in 1458-1540 nm wavelength region, *Electron. Lett.*, 39(7), 597–599, 2003.
18. Yam, S.S.H., Akasaka, Y., Kubota, Y., and Inoue, H., 100-nm Cascaded Hybrid Doped Fiber Amplifier for Coarse Wavelength Division Multiplexing, Proceedings OFC 2004, paper JWA16.
19. Yam, S.S.H., Akasaka, Y., Kubota, Y., Inoue, H., and Parameswaran, K., Novel pumping schemes for fluoride-based thulium-doped fiber amplifier at 690 and 1050 nm (or 1400 nm), *Photon. Technol. Lett.*, 17(5), 1001–1003, 2005.

20. Yam, S.S.H., Akasaka, Y., Kubota, Y., and Inoue, H., Hybrid doped fiber amplifier with 100-nm bandwidth for coarse wavelength division multiplexing, *Optics Commun.*, 249, 539–542, 2005.
21. Sakamoto, T., Mori, A., and Shimizu, M., Rare-Earth-Doped Fiber Amplifier for Eight-Channel CWDM Transmission Systems, Proceedings OFC 2004, vol. 2, paper ThJ5.
22. Tombling, C., Michie, C., Andonovic, A., and Kelly, A.E., Recent Advances in Semiconductor Optical Amplifiers, Proceedings LEOS Annual Meeting 2003, vol. 2, pp. 892–893.
23. Morito, K., Tanaka, S., Tomabechi, S., and Kuramata, A., A broadband MQW semiconductor optical amplifier with high saturation output power and low noise figure, *Photon. Technol. Lett.*, 17(5), 974–976, 2005.
24. Borghesani, A., Fensom, N., Scott, A., Crow, G., Johnston, L., King, J., Rivers, L., Cole, S., Perrin, S., Scrase, D., Bonfrate, G., Ellis, A., Lealman, I., Crouzel, L., Chun, H.K., Lupu, A., Mahe, E., and Maigne, P., High Saturation Power (>16.5dBm) and Low Noise Figure (<6dB) Semiconductor Optical Amplifier for C-band Operation, Proceedings OFC 2003, vol. 2, pp. 534–536.
25. Morito, K., Ekawa, M., Watanabe, T., and Kotaki, Y., High-output-power polarization-insensitive semiconductor optical amplifier, *J. Lightwave Technol.*, 21(1), 176–181, 2003.
26. Morito, K. and Tanaka, S., Record high saturation power (+22 dBm) and low noise figure (5.7 dB) polarization-insensitive SOA module, *Photon. Technol. Lett.*, 17(6), 1298–1300, 2005.
27. Bachmann, M., Doussiere, P., Emery, J.Y., N'Go, R., Pommereau, F., Goldstein, L., Soulage, G., and Jourdan, A., Polarisation-insensitive clamped-gain SOA with integrated spot-size converter and DBR gratings for WDM applications at 1.55 µm wavelength, *Electron. Lett.*, 32(22), 2076–2078, 1996.
28. Francis, D.A., DiJaili, S.P., and Walker, J.D., A Single-Chip Linear Optical Amplifier, Proceedings OFC 2001, vol. 4 (post-deadline papers), paper PD13.
29. Patent, E.A., Van der Tol, J.J.G.M., Nielsen, M.L., Binsma, J.J.M., Oei, Y.S., Mørk, J. and Smit, M.K., Integrated SOA-MZI for pattern-effect-free amplification, *Electron. Lett.*, 41(9), 549–551, 2005.
30. Wiesenfeld, J.M., Gain dynamics and associated nonlinearities in semiconductor optical amplifiers, *Int. J. High-Speed Electron. and Syst.*, 7(1), 179–222, 1996.
31. Doerr, C.R., Cappuzzo, M., Gomez, L., Chen, E., Wong-Foy, A. and Laskowski, E., Planar Lightwave Circuit Eight-Channel CWDM Multiplexer, Proceedings OFC 2004, post-deadline papers, paper PDP11.
32. Doerr, C.R., Cappuzzo, M., Gomez, L., Chen, E., Wong-Foy, A., Ho, C., Lam, J., and McGreer, K., Planar lightwave circuit eight-channel CWDM multiplexer with <3.9-dB insertion loss, *J. Lightwave Technol.*, 23(1), 62–65, 2005.
33. Doerr, C.R., Pafchek, R., and Stulz, L.W., Integrated band demultiplexer using waveguide grating routers, *Photon. Technol. Lett.*, 15(8), 1088–1090, 2003.
34. Iannone, P.P., Reichmann, K.C., Smiljanic, A., Frigo, N.J., Gnauck, A.H., Spiekman, L.H., and Derosier, R.M., A transparent WDM network featuring shared virtual rings, *J. Lightwave Technol.*, 18(12), 1955–1963, 2000.
35. Ismail, T., Liu, C.P., Mitchell, J.E., Seeds, A.J., Qian, X., Wonfor, A., Penty, R.V., and White, I.H., Full-duplex wireless-over-fibre transmission incorporating a CWDM ring architecture with remote millimetre-wave LO delivery using a bi-directional SOA, Proceedings OFC 2005, vol. 4, paper OThG7.

36. Ismail, T., Liu, C.P., Mitchell, J.E., Seeds, A.J., Qian, X., Wonfor, A., Penty, R.V., and White, I.H., Transmission of 37.6-GHz QPSK wireless data over 12.8-km fiber with remote millimeter-wave local oscillator delivery using a bi-directional SOA in a full-duplex system with 2.2-km CWDM fiber ring architecture, *Photon. Technol. Lett.*, 17(9), 1989–1991, 2005.
37. Shacham, A., Small, B.A., Liboiron-Ladouceur, O., and Bergman, K., A fully implemented 12 x 12 data vortex optical packet switching interconnection network, *J. Lightwave Technol.*, 23(10), 3066–3075, 2005.
38. Bromage, J., Thiele, H.J., and Nelson, L.E., Raman Amplification in the S-band, Proceedings OFC 2002, pp. 383–385 (invited paper).
39. Namiki, S. and Emori, Y., Recent Advances in Ultra-Wideband Raman Amplifiers, Proceedings OFC 2000, vol. 4, pp. 98–99.
40. Namiki, S. and Emori, Y., Ultrabroad-band Raman amplifiers pumped and gain-equalized by wavelength-division-multiplexed high-power laser diodes, *IEEE J. Sel. Topics Quantum Electron.*, 7(1), 3–16 (invited paper), 2001.
41. Emori, Y., Tanaka, K., and Namiki, S., 100nm bandwidth flat-gain Raman amplifiers pumped and gain-equalised by 12-wavelength-channel WDM laser diode unit, *Electron. Lett.*, 35(16), 1355–1356, 1999.
42. Miyamoto, T., Tsuzaki, T., Okuno, T., Kakui, M., Hirano, M., Onishi, M., Shigematsu, M., and Nishimura, M., Highly-Nonlinear-Fiber-Based Discrete Raman Amplifier for CWDM Transmission Systems, Proceedings OFC 2003, vol. 1, pp. 20–21.
43. Chen, Y., Pavlik, R., Visone, C., Pan, F., Gonzales, E., Turukhin, A., Lunardi, L., Al-Salameh, D., and Lumish, S., 40-nm Broadband SOA-Raman Hybrid Amplifier, Proceedings OFC 2002, pp. 390–391.
44. Masuda, H., Review of Wideband Hybrid Amplifiers, Proceedings OFC 2000, vol. 1, pp. 2–4.
45. Iannone, P.P., Reichmann, K.C., Zhou, X., and Frigo, N.J., 200 km CWDM Transmission Using a Hybrid Amplifier, Proceedings OFC 2005, paper OthG3.
46. Iannone, P.P., Reichmann, K.C., and Spiekman, L.H., Amplified CWDM Systems, Proceedings LEOS Annual Meeting 2003, vol. 2, pp. 678–679.

7 CWDM — Upgrade Paths and Toward 10 Gb/s

Hans-Jörg Thiele and Peter J. Winzer

CONTENTS

7.1	Introduction	200
7.2	Overview of CWDM Capacity Upgrade Options	203
	7.2.1 Increasing Per-Channel Bit-Rates (Option 1)	204
	7.2.2 The Channel Overlay (Option 2c)	206
	7.2.3 Equalization and FEC for CWDM Transmission	209
	7.2.3.1 Forward Error Correction	209
	7.2.3.2 Electronic Equalization	210
7.3	Advances in CWDM Lasers	211
	7.3.1 Laser Output Power	211
	7.3.2 Laser Chirp, Extinction Ratio, and Chromatic Dispersion	213
	7.3.2.1 Directly Modulated Laser Chirp	213
	7.3.2.2 Chirp and Chromatic Dispersion	216
	7.3.2.3 Chirp and Extinction Ratio	218
	7.3.3 Increased Bit-Rate Operation	218
	7.3.3.1 Uncooled DMLs for 10 Gb/s	218
	7.3.3.2 Operating 2.5-Gb/s Rated DMLs at 10 Gb/s	219
	7.3.4 Other Laser Types for 10-Gb/s CWDM Systems	222
7.4	CWDM System Upgrade Demonstrations	223
	7.4.1 Example for A DWDM Overlay	223
	7.4.2 Example for Mixed Bit-Rate Transmission	227
	7.4.2.1 Upgrade Using 10-Gb/s DML	228
	7.4.2.2 Upgrade Using a 10-Gb/s EML	229
7.5	Mixed Fiber-Type Transmission	230
	7.5.1 10-Gb/s DML over AllWave and TrueWave-RS	230
	7.5.2 10-Gb/s DWDM Overlay with TrueWave-RS	232
	7.5.3 10-Gb/s EML with AllWave and Multiple Spans of TrueWave-RS	232
7.6	Full-Spectrum CWDM at 10 Gb/s	234

    7.6.1   Transmission over NZDSF...234
    7.6.2   FEC-Enabled Transmission over AllWave................................236
            7.6.2.1 FEC and Equalization in a Full-Spectrum
                     CWDM Experiment...236
            7.6.2.2 FEC and Equalization to Support Fully
                     Bidirectional CWDM Transmission 242
References 245

7.1 INTRODUCTION

Cost-effective short-haul but high-capacity optical transport systems are becoming increasingly important for metropolitan-area and access applications; examples are metro-feeders, inter- or intra-office links between routers and cross-connects, and storage-area networks. Access links of this kind are characterized by transmission distances of 10 to 100 km. Until recently, lighting a single wavelength channel (typically at 1310 nm), or in some applications two wavelengths (at 1310 and 1550 nm) was sufficient to support the required capacities over the deployed single-mode fiber infrastructure. Such two-wavelength systems were initially called coarse wavelength division multiplexing (CWDM) in order to distinguish them from the dense wavelength division multiplexing (DWDM) systems used for long-haul transport.

As more and more coarsely spaced wavelength channels were added in the access regime to support the growing need for capacity without having to deploy additional fiber at relatively high cost, the original two-wavelength systems were renamed "wide WDM," and the term CWDM was adopted and standardized in ITU-T G.694.2 for systems supporting up to 18 channels across the low-loss region of single-mode silica fiber [1,2]. Figure 7.1 shows the fiber loss as a function of wavelength. Toward the short-wavelength side, fiber loss is dominated by fundamentally unavoidable *Rayleigh scattering*, while fiber bend loss starts to set in at wavelengths beyond 1610 nm. Fiber bend loss depends significantly on fiber cabling and deployment. At around 1385 nm, we observe the "water-peak", a high-loss wavelength band due to the OH^- absorption of legacy standard single-mode fiber (SSMF). Modern manufacturing processes have managed to eliminate this absorption band (see Chapter 2), and low water peak fiber (LWPF) is widely offered today; the almost entirely suppressed water-peak of zero water peak fiber (ZWPF) is also shown in Figure 7.1 [3]. As a result of the suppressed water peak, the entire low-loss window from below 1270 to 1610 nm can be exploited for CWDM transmission. Note that significantly less fiber loss can be tolerated at the long-wavelength end of the standardized CWDM band than at the short-wavelength end since chromatic dispersion (CD) of legacy SSMF and standard ZWPF increases with wavelength, which gives rise to unacceptable dispersive pulse broadening at wavelengths beyond 1610 nm, as discussed in Section 7.3 of this chapter.

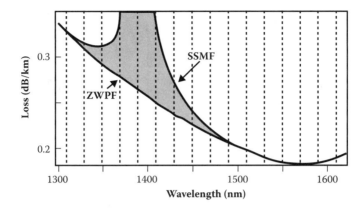

FIGURE 7.1 Attenuation coefficient of legacy standard single-mode fiber and advanced zero water peak single-mode fiber. (Taken from Winzer, P.J., Fidler, F., Thiele, H.J., Matthews, M., Nelson, L.E., Sinsky, J.H., Chandrasekhar, S., Winter, M., Castagnozzi, D., Stulz, L.W., and Buhl, L.L., *J. Lightwave Technol.*, 23(1), 203–210, 2005. With permission.)

In order to provide cost-effective solutions for CWDM systems, directly modulated, uncooled distributed feedback (DFB) lasers and low-cost thin-film multiplexing filters are used in CWDM systems. Since DFB sources have a temperature-induced wavelength drift of around 0.1 nm/°C, the standard CWDM channel spacing was set to 20 nm to allow for operation over a wide temperature range with existing low-cost laser and multiplexer technology and without the need for active cooling [ITU-T G.694.2]. Over the entire low-loss window of ZWPF, this allows for a maximum of 18 CWDM channels. Note, however, that the two shortest wavelength channels at 1270 and 1290 nm are typically omitted, owing to the high Rayleigh scattering loss at these short wavelengths. A CWDM system using all 16 channels from 1310 to 1610 nm is commonly referred to as a full-spectrum CWDM system, and the standardized CWDM center wavelengths of such a system are indicated by dotted lines in Figure 7.1.

A typical *N*-wavelength CWDM system, as standardized in ITU-T G.695, is shown in Figure 7.2. It either operates (a) unidirectionally or (b) bidirectionally over a single optical fiber. If operating bidirectionally, East and West traffic is carried on different wavelengths to avoid crosstalk problems. (We will see in Section 7.6 how such a system can be extended to carry bidirectional traffic on the same set of wavelengths.) CWDM systems, in contrast to their DWDM counterparts, are typically single-span, that is, they do not employ any kind of in-line optical amplification. The reason for the absence of optical amplifiers lies in the value proposition of optical amplification: optical amplification generally proves it if the optical amplifier can be shared among many WDM channels. However, in CWDM systems with wide channel spacings, this value proposition becomes questionable. As a result of the single-span operation, span lengths are

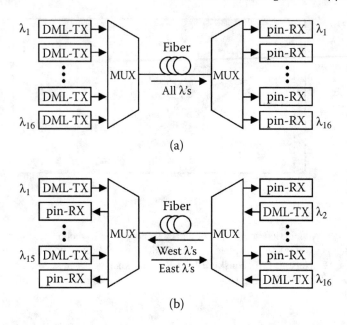

FIGURE 7.2 Typical CWDM system, as specified by ITU-T G.695, for $N = 16$ channels: (a) unidirectional and (b) bidirectional. (Taken from Winzer, P.J., Fidler, F., Thiele, H.J., Matthews, M., Nelson, L.E., Sinsky, J.H., Chandrasekhar, S., Winter, M., Castagnozzi, D., Stulz, L.W., and Buhl, L.L., *J. Lightwave Technol.*, 23(1), 203–210, 2005. With permission.)

limited to 40 to 80 km at 2.5 Gb/s, depending on the portion of the used CWDM band [ITU-T G.695].

Due to the absence of optical amplification with the exception of the applications discussed in Chapter 6, the laser output power has to be high enough to support the system's loss budget, consisting of transmission fiber, multiplexer loss, and additional splice and connector losses. For example, in an 80-km system, the fiber loss at 1310 nm (0.33 dB/km) amounts to 26.4 dB. Multiplexer losses may add another 5 dB. Assuming an avalanche photodiode (APD) receiver with a typical sensitivity of −31 dBm at 2.5 Gb/s, and a target bit error ratio (BER) of 10^{-9}, we arrive at a minimum laser launch power of −0.4 dBm. Directly modulated lasers (DMLs), which have modulated output powers of up to several dBm and are low-cost and widely available for direct modulation up to 2.5 Gb/s, thus make up for ideal CWDM sources. Adding commercial system margins for aging and component variations, or assuming lower-cost *pin*-type receivers, results in a reduced system reach.

Note that components for CWDM systems are generally specified with a broad range of parameter variations in order to keep the yield high and the cost low. As an example, Figure 7.3 shows the transmission spectra of two typical, nominally identical commercial 16-channel CWDM thin-film multiplexers. Figure 7.3 reveals a 3-dB spread in insertion loss across the CWDM band. Variations in multiplexer loss and laser output power are also reflected in the ITU

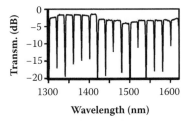 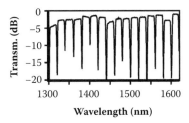

FIGURE 7.3 Insertion loss of typical thin-film CWDM multiplexers. (Taken from Thiele, H.J., Winzer, P.J., Sinsky, J.H., Stulz, L.W., Nelson, L.E., and Fidler, F., *Photon. Technol. Lett.*, 16(10), 2004. With permission.)

standard [ITU-T G.695], which for most systems specifies per-channel optical power levels between +3.5 and −4 dBm after multiplexing.

At the typical, SONET-oriented data rate of 2.5 Gb/s, a full-spectrum (16-channel) CWDM system has an aggregate capacity of 40 Gb/s [4]. Modulating the lasers at 1.25 Gb/s, and running Gigabit Ethernet (GbE) services on each wavelength, results in 16 GbE channels. While these capacities may be sufficient for some access and metro applications, future high-bandwidth services, such as remote storage, remote and distributed computing, or multimedia communications and gaming applications, may well lead to capacity bottlenecks in the not too distant future. There are several potential strategies to boost CWDM system capacities while still retaining the cost-effectiveness and the modular "pay as you grow" philosophy of CWDM. It is the aim of this chapter to discuss and compare these capacity upgrade options. We will first give an overview of potential upgrade solutions in Section 7.2. In Sections 7.3 through 7.6, we will then discuss aspects of these options in more technical detail.

7.2 OVERVIEW OF CWDM CAPACITY UPGRADE OPTIONS

There are basically two ways to upgrade the capacity of any WDM system, as visualized in Figure 7.4.

1. Increasing per-channel bit-rates
2. Increasing the number of WDM channels by
   a. extending the WDM band while keeping the channel spacing fixed
   b. reducing the channel spacing while keeping the WDM band fixed
   c. replacing individual channels of the original system by sub-bands of more densely packed upgrade channels (DWDM overlay)

In the case of CWDM systems, option 2(a) has to be ruled out since the standardized CWDM band already covers the entire low-loss region of optical transmission fiber, and any extension of this band would lead to unacceptable additional loss and/or dispersion. Furthermore, this approach would not be very effective since only 1/18th of additional capacity could be gained for every 20 nm of band expansion.

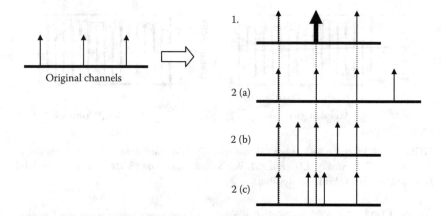

FIGURE 7.4 Upgrade paths for an existing WDM system, either by increasing the bit-rate (option 1) or additional wavelength channels (option 2a–c).

Upgrading a CWDM system using option 2(b) is also problematic since this would call for tighter specifications on the wavelength stability of all CWDM channels and would not lend itself to the smooth, cost-effective, and modular upgradability typical of existing CWDM systems. Nevertheless, the standardization of next-generation CWDM systems with closer channel spacings should not be completely ruled out. Candidate lasers could be fiber Bragg grating lasers (FBGLs), which have been studied for WDM applications due to their small temperature drift of less than 0.02 nm/°C [5]. Use of such uncooled lasers could therefore boost CWDM capacities by a factor of 5 without increasing per-channel bit-rates. These lasers are discussed in more detail in Section 7.3.4.

The most promising capacity upgrade paths for existing CWDM systems are options 1 and 2(c), which we will consider in more detail in this chapter. Both options lend themselves to smooth upgradeability since they allow capacity upgrades of individual channels without affecting the traffic on existing channels. This way, the upgrade process can start at a single channel, until eventually all channels across the CWDM band are upgraded.

7.2.1 Increasing Per-Channel Bit-Rates (Option 1)

Increasing per-channel bit-rates from, for example, 2.5 to 4 Gb/s or even 10 Gb/s can increase the aggregate CWDM capacity by a factor up to four. Increasing channel bit-rates is particularly attractive whenever the frequency-selective optical components of a deployed CWDM system (e.g., multiplexers and demultiplexers of an initially deployed 4-channel CWDM system) prevent an upgrade using additional CWDM wavelengths. The suitability of the increased bit-rate upgrade option hinges on three main considerations:

1. *Are uncooled, low-cost 10-Gb/s DMLs available at the upgrade wavelengths?*
   - Today, 10-Gb/s CWDM lasers are not commercially available across the full CWDM spectrum, although research experiments have demonstrated 10-Gb/s DMLs for use in CWDM systems [6] and have even shown the potential of 40-Gb/s DMLs for short-reach applications [7,8]. Most likely, upgrading CWDM channels to 10 Gb/s will start with channels around 1310 and 1550 nm, where appropriate uncooled DMLs already exist. Challenges in the development of 10-Gb/s DMLs for CWDM applications will be discussed in more detail in Section 7.3.
2. *Does the deployed system's link budget support 10-Gb/s operation?*
   - Fundamentally, receiver sensitivity scales linearly with bit-rate. Therefore, at least 6 dB of additional link budget is needed to accommodate an increase in per-channel bit-rate by a factor of 4. In particular at the short-wavelength end of the CWDM spectrum, where fiber attenuation is high, a deployed system may not directly support a bit-rate upgrade. To overcome this problem, one can resort to one of the following methods: (i) If available, higher-power DMLs can be used. Every additional dB in launched signal power reduces the gap in the link budget by 1 dB. (ii) If the existing system uses pin-type receivers at 2.5 Gb/s, an upgrade with 10-Gb/s APD receivers can be sufficient to close the gap in the link budget, while retaining all system margin. (iii) If more margin is needed, or if the 2.5 Gb/s system uses APDs already, the use of forward error correction (FEC) at 10 Gb/s can help to close the link. The use of FEC for capacity upgrades is discussed in detail in Section 7.2.3 and experimentally validated in Section 7.6.2.
3. *Does the deployed system's CD support 10-Gb/s operation?*
   - Even if the deployed system's link budget allows operation at higher per-channel bit-rates, CD may not. Since the amount of tolerable CD scales quadratically with bit-rate, an increase in bit-rate by a factor of 4 results in a decrease in dispersion tolerance by a factor of 16. If operating over standard-dispersion fiber (such as SSMF or standard LWPF), this can pose severe problems at the longer-wavelength end of the CWDM spectrum, where dispersion is highest; Figure 7.5 shows the CD of standard-dispersion fiber across the CWDM band. Since the tolerable (2 dB power penalty) CD for 10-Gb/s DMLs is on the order of 100 ps/nm, system reach at 1610 nm [20 ps/(nm.km) dispersion] is limited to some 5 km. In contrast, at 2.5 Gb/s direct modulation, dispersion does not pose problems up to about 75 km of standard fiber [4]. To counteract the effect of CD, one of the following methods may be used for upgrading long-wavelength channels to higher bit-rates:

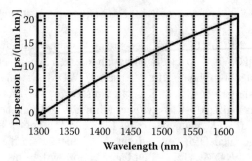

FIGURE 7.5 Chromatic dispersion of standard-dispersion fiber (SSMF and standard LWPF).

a. If traffic can be interrupted,
– the system can be moved to a nonzero dispersion-shifted fiber (NZDSF) with much lower dispersion values, if such a fiber can be cost-effectively deployed [9], or
– a broadband dispersion-compensating module (DCM) can be used to compensate CD, prior to demultiplexing; DCMs are available either as dispersion-compensating fiber (DCF) or as grating-based devices, and can be inserted if their loss can be accommodated in the system's link budget.
b. If traffic is not to be interrupted,
– single-channel DCMs may be inserted after the demultiplexer for the long-wavelength channels. Since this option is typically not cost-effective, one can also use
– advanced electronic equalization and FEC in the receiver to combat dispersion. This method, which lends itself to high-density integration and potentially low-cost implementation, is discussed in more detail in Section 7.2.3 and applied to 10-Gb/s CWDM systems in Section 7.6.2.

7.2.2 The Channel Overlay (Option 2c)

In this approach, the capacity upgrade of the CWDM system is carried out by adding additional wavelengths within the low-loss bandwidth of the transmission fiber. This technique is particularly useful when the criteria for introducing new, higher bit-rate optical and electrical CWDM components are not met, for example, due to the limited availability of 10-Gb/s DMLs, an insufficient link budget, or high accumulated dispersion. Using this upgrade option, the basic CWDM grid is maintained, and sub-bands of new, densely packed DWDM upgrade channels are added to form an overlay on the existing CWDM channel grid. This upgrade option is called "DWDM over CWDM" and has certain cost advantages over other approaches, where the channel grid structure of the CWDM system is completely changed. In a 16-channel full-spectrum CWDM system operating at 2.5 Gb/s per channel, one of the CWDM wavelengths, for example, the channel at 1550 nm, can be substituted with a subset of DWDM channels at 2.5 or 10 Gb/s, which all lie within the passband of the existing CWDM multiplexer of typically 12 nm (see Chapter 4). Therefore, the upgrade capacity is mainly determined by how many additional DWDM channels can be fed through one of the ports of a CWDM multiplexer [10–12]. Assuming a 100-GHz channel spacing within the DWDM sub-band, the CWDM port centered at 1550 nm can be upgraded with 15 new DWDM channels.* Therefore, even by replacing only one single CWDM channel with a full DWDM sub-band, a total capacity increase by a factor of almost 2 is possible: by substituting with a DWDM sub-band the 2.5-Gb/s CWDM channel at 1550 nm, where most of the available DWDM equipment exists, a total capacity of 77.5 Gb/s can be obtained. In the hypothetical case of replacing all 16 CWDM channels with 16 fully occupied DWDM sub-bands, up to 224 DWDM channels would be possible, although this value may never be reached since DWDM lasers at wavelengths other than the S-, C-, and L-band are hardly available. A high-capacity DWDM system with a continuous band of channels within the C-band may be the more appropriate solution if such high capacities are actually required.

In general, the suitability of this DWDM overlay option hinges on the following considerations:

Are suitable lasers for the DWDM channel overlay available?
- Narrow DWDM channel spacing of typically 100 GHz requires cooled DMLs to avoid filtering penalties due to the chirp-broadened spectrum of directly modulated sources, wavelength drift, and DWDM channel crosstalk [13]. Although temperature-stabilized lasers are generally more expensive and require extra footprint and power consumption, they are the key elements in pursuing overlay upgrades of this kind. The attractiveness of smoothly growing CWDM system capacity without traffic interruption may justify the extra cost associated with these sources. DWDM sources are mostly available in the C-band today, which would allow the replacement of up to three CWDM channels by DWDM sub-bands.

Can the additionally required passive optical components be accommodated in the link budget?
- As shown in Figure 7.6, a DWDM overlay needs a DWDM multiplexer for each sub-band to feed the DWDM upgrade channels into one of the ports of the existing CWDM multiplexers. At the receiver side, DWDM demultiplexers are needed to separate the DWDM channels at the receiver. Although the span loss between the CWDM multiplexers is

---

* Note that CWDM standards define a *wavelength* grid, while DWDM systems use a *frequency* grid. Therefore, the number of 100-GHz spaced channels fitting within a 12-nm window varies between 21 and 14 when going from 1310 to 1610 nm.

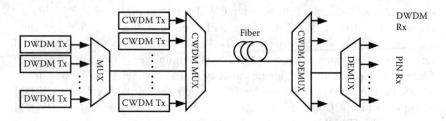

FIGURE 7.6 A DWDM overlay substituting a single CWDM channel.

unaffected by the DWDM overlay, the combined insertion loss of the two DWDM multiplexers of approximately 5 to 6 dB requires an increased launch power for the DWDM channels. Luckily, cooled DMLs can typically provide the extra output power compared to uncooled CWDM lasers. Note that DWDM multiplexers for directly modulated sources require more flat-top characteristics than components used in conjunction with externally modulated channels, which feature lower-chirp and thus a narrower-modulated bandwidth. In order to allow for more relaxed component requirements of the DWDM multiplexers, resulting in a lower cost DWDM upgrade, 200 GHz DWDM channel spacing may be more cost-effective for the DWDM channels, albeit at the expense of reduced upgrade capacity.

Are any other modifications of the system necessary?
- If the bit-rate of the DWDM upgrade channels is the same as that of the underlying CWDM system, CD will not be of concern for the DWDM overlay. Only if option 2(c) is combined with option 1, that is, the DWDM overlay is implemented with increased per-channel bit-rates, the feasibility of the upgrade with respect to CD and link budget have to be assessed, as discussed along with option 1 mentioned earlier. Some tradeoffs between more channels at lower bit-rates and fewer channels at high bit-rates will be discussed in Section 7.4.

Note that the DWDM upgrade channels can potentially operate at higher bit-rates than the original CWDM channels. Exchanging a single CWDM wavelength at 2.5 Gb/s by a DWDM sub-band of 15 channels operating at a per-channel rate of 10 Gb/s results in a highly efficient capacity upgrade of more than a factor of 60. Obviously, whether going to higher bit-rates in the DWDM overlay is supported in a particular system depends on the issues discussed in Section 7.2.1, most notably on the system's link budget (including the DWDM multiplexers) and on the fiber's CD. Fortunately, the problem is somewhat relaxed for the DWDM overlay as compared to increasing CWDM bit-rates since one is more flexible in choosing the most appropriate CWDM wavelength for the DWDM overlay, that is, a wavelength band that simultaneously minimizes loss and dispersion. Also, DWDM sources are cooled to maintain wavelength stability, thus offering higher output power.

7.2.3 EQUALIZATION AND FEC FOR CWDM TRANSMISSION

As will be discussed later in Section 7.4, the low dispersion tolerance of DMLs at 10 Gb/s prevents uncompensated transmission over CWDM link of legacy standard-dispersion fiber (SSMF or standard LWPF/ZWPF, see Figure 7.5). As an alternative to using scarcely installed low-dispersion NZDSF for CWDM transmission (discussed in Section 7.6.1), electronic equalization and FEC can be used to combat dispersion penalties and to achieve transmission distances typical of existing 2.5-Gb/s CWDM systems.

7.2.3.1 Forward Error Correction

Over the past few years, FEC has become an invaluable tool to increase system margins [14]. FEC schemes operate by introducing redundancy into the transmitted signal, which allows for the correction of a certain number of errors after detection. The redundancy required by FEC manifests itself in a bit-rate overhead. Typical FECs for terrestrial multi-Gb/s optical communication systems have an overhead of 7%, increasing to 25% and beyond for strong FEC used predominantly in submarine systems.

To characterize the correction capabilities of an FEC, one typically specifies the correction curve, which translates the raw channel BER at the FEC input (pre-FEC BER) to a corrected BER at the FEC output, assuming white noise statistics. An example for the correction curve of two common FEC schemes with 7% bit-rate overhead is shown in Figure 7.7. The BER at the FEC input that

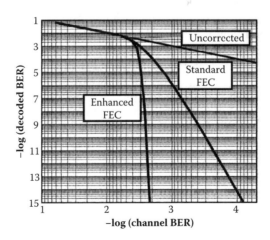

FIGURE 7.7 Correction curves of the enhanced FEC scheme used in our experiment. For comparison, standard RS(255,239) FEC and the FEC-free (uncorrected) case are also shown. (Taken from Winzer, P.J., Fidler, F., Thiele, H.J., Matthews, M., Nelson, L.E., Sinsky, J.H., Chandrasekhar, S., Winter, M., Castagnozzi, D., Stulz, L.W., and Buhl, L.L., *J. Lightwave Technol.*, 23(1), 203–210, 2005. With permission.)

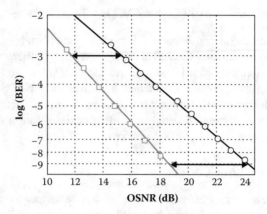

FIGURE 7.8 BER vs. OSNR for NRZ-OOK at 10.7 Gb/s. *Squares*: back-to-back; *circles*: after 60 km of SSMF; *arrows*: BER-dependent dispersion penalties.

is required to achieve a decoded BER of better than 10^{-16} is often referred to as the FEC's correction threshold, and is typically between 10^{-3} and 10^{-5}.

Alternatively, an FEC scheme may be specified by its coding gain, which is the gain in required optical signal-to-noise ratio (OSNR) for achieving the same target BER with FEC as for an uncoded system [14]. However, when assessing the benefits of FEC in a particular system scenario, coding gains alone are insufficient: first, the FEC overhead implies a higher data rate, which changes the impact of certain propagation impairments. For example, the tolerance to CD shrinks with the square of the bit-rate, which implies a 14% reduction in dispersion tolerance for a typical FEC overhead of 7% and reduces the coding gain. Secondly, many impairments (including CD) lead to higher penalties at good BER than at poor BER (see Figure 7.8) [15], or even to error floors, which together with the steep correction curve of FECs results in enhanced coding gains in the presence of impairments [16]. Thirdly, the occurrence of burst errors can degrade FEC characteristics for certain kinds of impairments, such as coherent crosstalk (see Section 7.6.2) [17,18].

7.2.3.2 Electronic Equalization

Studied for over 10 years for lightwave systems [19–24], electronic equalization at 10 Gb/s has recently experienced a boost by progress in high-speed integrated circuits, and has proven to be a powerful tool to increase dispersion tolerance [21,25–29]. Equalization techniques range from multi-tap feed-forward and decision feedback structures [25,29], to sophisticated schemes employing multiple thresholds in conjunction with FEC [21,26]. Even advanced maximum-likelihood sequence detectors (MLSEs) [14,27,28] have been demonstrated at 10 Gb/s [30].

7.3 ADVANCES IN CWDM LASERS

In the previous section, we discussed general strategies for upgrading the capacity of CWDM systems. We pointed out that low-cost components are key in successfully deploying CWDM systems, and that DMLs present a serious bottleneck as far as high per-channel bit-rates are concerned. In this section, we review recent developments in CWDM laser technology and discuss how the key laser parameters impact CWDM system design. More details and also general information about transmitters, especially transceivers designed to operate in CWDM systems, are found in Chapter 3. Here we merely focus on the aspects that are critical for CWDM capacity upgrades.

Although vertical cavity surface emitting laser with reasonable performance are available, we focus on DFB lasers since these devices offer a better performance, for example, higher output power. Driven by the need for low overall system cost, the main requirements for CWDM lasers are:

- Operation across a wide temperature range since thermoelectric cooling (TEC) is to be avoided in CWDM systems in order to keep the transponder power consumption, the complexity of the control electronics, and the transponder footprint as low as possible.
- High output power across the operating temperature range in order to meet the system's link budget, especially at the loss-dominated short-wavelength end of the CWDM spectrum.
- Low laser chirp, so that CD over legacy SSMF does not set unreasonably low limits to transmission distances, especially at the highly dispersive long-wavelength end of the CWDM band.
- Good modulation performance at high bit-rates to avoid excessive back-to-back penalties.

7.3.1 LASER OUTPUT POWER

Uncooled DMLs are commonly used in CWDM transmission links due to their high modulated output power and their relatively low cost. Although the average modulated output power can reach +6 dBm or more at room temperature, CWDM lasers are typically rated for operation between 0 and +3 dBm. This is because high-temperature operation and aging reduce the laser's output power, and have to be accounted for in CWDM specifications.

As an example, Figure 7.9 shows the dependence of continuous wave (CW) laser output power on the drive current for an uncooled 1550-nm CWDM laser at different case temperatures between 0 and 85°C. It can be seen that the laser drive current has to be substantially increased in order to maintain constant laser output power across a wide temperature range. Note that for the design of laser control electronics, the higher laser current increases the laser chip temperature even further. Figure 7.9 also shows that the laser's slope efficiency (i.e., the slope of the P-I curves) decreases considerably with increasing temperature,

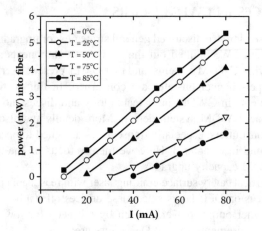

FIGURE 7.9 Measured variation of output power for uncooled CWDM laser.

which implies higher output voltage swing requirements for the high-speed data driver amplifier than would be needed for cooled laser operation.

Note that it is important to distinguish between chip temperature and case temperature when assessing the performance of the uncooled lasers. The chip temperature can be significantly higher than the case temperature due to imperfect heat transfer to the case. For example, when the ambient temperature is increased to 85°C, the optical subassembly, that is, the laser chip with the attached fiber, must work at temperatures in excess of 95°C [31]. Typically, commercial lasers are specified using the case temperature.

Due to the typical quantum structure and conduction band offset, the conventional InGaP/InP active layers are often considered to be responsible for poor temperature characteristics for the DFB lasers at higher temperatures. Hence, work has been done to find multi quantum well (MQW) structures with better temperature characteristics. AlGaInAs was considered the best candidate, and 10-Gb/s operation up to 85°C has been demonstrated [32,33]. However, Al-based MQW layers are not compatible with the conventional InGaAsP manufacturing process. Recently, progress has been made with multi-junction buried heterostructure MQW with phosphorus (P)-based MQW active structures. DMLs operating error-free up to 100°C with 5 dB extinction ratio (ER) have been demonstrated; even at 90°C, a laser threshold of 30 mA and an output power of up to 13 mW was obtained [34]. Other work has focused on high-temperature operation of MQW ridge-waveguide DFBs based on aluminum-doped InGaAlAs. Laser devices with 10 mW output power at 115°C were demonstrated, and transmission of the DML signal at 10 Gb/s was shown at these elevated temperatures [35]. More recently, this AlGaInAs structure achieved 5 mW of modulated output power and 7 dB extinction at 10 Gb/s modulation and at 120°C [36]. Key parameters were a high relaxation oscillation frequency and high differential gain of the laser device; no indications of aging were observed. Although most of these

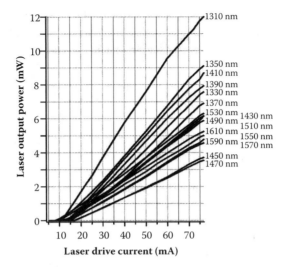

FIGURE 7.10 Widely varying static characteristics of the nonselected DMLs covering the full 16-channel CWDM band from 1310 to 1610 nm. (Taken from Thiele, H.J., Winzer, P.J., Sinsky, J.H., Stulz, L.W., Nelson, L.E., and Fidler, F., *Photon. Technol. Lett.*, 16(10), 2004. With permission.)

examples are confined to the 1310-nm region, these results are also relevant for high-speed and high-temperature operation of upgraded CWDM transmitters.

Figure 7.10 shows the room temperature characteristics of 16 unselected, commercially available CWDM lasers rated for 2.5 Gb/s operation across the full CWDM band (1310 to 1610 nm). It can be seen that the laser output powers vary between 3 and 11 mW for 70 mA drive current. The threshold currents vary from 8 to 17 mA, and the slope efficiencies vary from 0.06 to 0.2 mW/mA. Device variations of this magnitude are typical of low-cost CWDM sources.

As mentioned in Section 7.2.1, receiver sensitivity scales linearly with bitrate, which implies that future DMLs have to operate not only at higher bit-rates but also at higher output powers in order to accommodate comparable link losses without having to resort to more advanced receivers, such as APD-based receivers, or to FEC. At present, a complete set of uncooled DMLs covering the full CWDM band from 1270 (1310) to 1610 nm is only commercially available for 2.5 Gb/s operation.

7.3.2 Laser Chirp, Extinction Ratio, and Chromatic Dispersion

7.3.2.1 Directly Modulated Laser Chirp

Direct modulation of the laser current not only imprints the desired transmit data onto the emitted optical intensity, but also leads to a modulation of the carrier density within the laser cavity and in turn to a modulation of the refractive index of the active region. This in turn leads to the modulation of the frequency of the emitted light (laser chirp). The modulation behavior of DMLs is described by the coupled rate equations of the carriers and the optical field in the laser [37,38]. The analytic solution yields an expression for the frequency chirp Δv of the single mode laser:

$$\Delta v(t) = \frac{\alpha}{4\pi}\left(\frac{1}{P}\frac{dP}{dt} + \kappa \cdot \Delta P(t)\right) \quad (7.1)$$

The first term of Equation 7.1 represents dynamic chirp, which is a frequency shift at the pulse transitions, that is, between the "1" and "0" states. Dynamic or transient chirp dominates for a fast rise time or if the laser is operated close to threshold. It has been shown that the dynamic chirp can be reduced by adjusting laser design parameters, such as the optical confinement factor [39]. The second term in Equation 7.1 represents adiabatic chirp, and captures the wavelength difference between the steady-state emission frequency for the "1" and "0" bits. Adiabatic chirp is proportional to the power difference ΔP between the two states; reducing the ER therefore mitigates adiabatic chirp. The linewidth enhancement factor α, which drives the overall magnitude of laser frequency chirp, is a measure of the coupling strength between intensity and frequency modulation (FM) of the laser diode. It is defined as [40,41]

$$\alpha = -(dn/dN)/(dg/dN). \quad (7.2)$$

Here, n is the refractive index of the laser cavity, N is the carrier density, and g is the gain of the laser material, with dg/dN being also known as the differential gain. The case $\alpha > 0$ ($\alpha < 0$) is referred to as positive (negative) chirp; DMLs always exhibit positive chirp. The linewidth enhancement factor quantifies the ratio of refractive index change to the change in gain g for some modulation-induced variation in carrier density. When the injected current is suddenly increased, the carrier density rises before the light output can increase. This temporary jump in carrier density leads for that transition to a temporary reduction in the refractive index in the active region, which shortens the optical path and results in a blue-shift of the wavelength at the leading edge of an optical pulse. At turn-off, the carrier density decreases and leads to a red-shift of the wavelength at the trailing edge of a pulse. The magnitude of the frequency shift depends on the magnitude and speed of the current change, the laser pre-bias condition, and the oscillation damping [42]. A laser with large damping necessarily has large adiabatic chirp, and since oscillations are small, transient chirp is small. In contrast, a laser with low damping has low adiabatic chirp, but pronounced relaxation oscillations due to low damping produce a large output power derivative, and hence significant transient chirp. Increasing the damping factor can be accomplished by decreasing either the cavity volume or the photon lifetime, or by increasing the nonlinear gain suppression factor.

We proceed to discuss the combination of the two chirp components (adiabatic and transient chirp) for two nonselected, commercially available CWDM lasers operating at 2.5 Gb/s with 8 dB extinction. Figure 7.11 shows both the

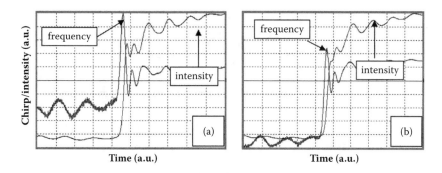

FIGURE 7.11 Frequency-chirp and intensity waveforms measured for two nonselected CWDM lasers at (a) 1510 nm and (b) 1530 nm.

modulated optical intensity and the time-resolved frequency variation for the same part of a pseudo-random bit sequence (PRBS) measured with a delay interferometer [43]. It can be seen that the laser of Figure 7.11a is dominated by transient chirp, shown by a pronounced frequency peak at the pulse transition. In contrast, the laser of Figure 7.11b has a high damping factor and thus a larger adiabatic chirp indicated by only a small frequency overshoot at the pulse transition but a larger frequency difference between the "0" and "1" state than in Figure 7.11a. The different performance of the lasers is due to variations in the device parameters, for example, detuning, doping, optical confinement, and facet phase.

According to Equation 7.2, the α-factor (and hence the DML chirp) can be minimized by increasing the differential gain dg/dN, which in practice is achieved by using multiple quantum well active regions, p-doping the active region, and increasing the wavelength detuning between the material gain peak λ_{gain} and the DFB emission wavelength λ_{DFB} [44]. In order to understand the impact of wavelength detuning on the properties of the laser, the gain curve at three different carrier densities N, the resulting differential gain dg/dN temperatures, and the α-factor and differential gain for a typical DFB laser are depicted in Figure 7.12. Without the DFB structure, the laser would operate at the gain peak λ_{gain} of the laser material. However, in a DFB structure, the emitted wavelength is essentially controlled by the grating period and can therefore differ from the gain peak. Since lower energy states always fill before higher energy states with increasing laser pumping, the differential gain dg/dN is higher for higher photon energies (equivalent to shorter emission wavelengths). Therefore, as the detuning $\lambda_{DFB} - \lambda_{gain}$ becomes more negative (i.e., the laser emits at wavelengths shorter than the gain peak), the differential gain dg/dN is enhanced, with only little change in dn/dN, resulting in reduced α-factors of ~3 for InGaAsP-type lasers. Therefore, the DFB laser wavelength should be slightly shifted toward wavelengths shorter than the gain peak for low-chirp operation. However, lasing off the gain peak also implies a reduced laser output power. Particularly for uncooled lasers, a suitable trade-off between output

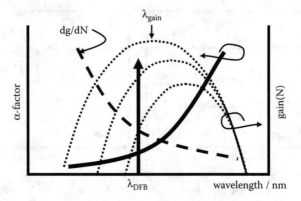

FIGURE 7.12 Principle of wavelength detuning in DFB lasers; variation of α-factor, differential gain and gain curve of the device.

power and chirp has to be found since the gain peak is reduced and shifts to longer wavelengths with increasing temperature (see Figure 7.12).

7.3.2.2 Chirp and Chromatic Dispersion

As a consequence of the positive laser chirp, pulses of the modulated output signal broaden when transmitted over positive-dispersion fiber, such as SSMF above 1310 nm (see Figure 7.5). When the chirped signal propagates through the positive dispersion fiber, the blue-shifted parts of the pulses travel faster than red-shifted parts, that is, the leading pulse edge travels faster than the falling pulse edge. The pulse is thus broadened. When the chirped signal propagates through a negative dispersion fiber, such as NZDSF (see Section 7.4), the blue-shifted portions of the pulse travel slower than red-shifted portions. The pulse is therefore compressed up to a point where CD takes over and lets the pulse broaden again [45]. Dispersive pulse broadening gives rise to transmission impairments through the spreading of energy from the "1"-bits into the "0"-bits, which reduces the eye aperture. Chirp-free nonreturn to zero (NRZ) at 10 Gb/s, received by simple pin-detection without[*] optical preamplification, allows for a CD of about 1400 ps/nm at a 2-dB power penalty, whereas chirped DML sources at 10 Gb/s are limited to about 100 ps/nm, equivalent to some 5 km of standard-dispersion fiber at 1610 nm. At 10 Gb/s, transmission distances of up to 38.5 km at 1550 nm have only been reported for selected low-chirp lasers and using spectral filtering [46]. In contrast, due to the quadratic scaling of dispersion-limited distances with bit-rate, dispersion does not pose problems up to about 80 km of standard fiber at 2.5 Gb/s using direct modulation; transmission distances of 75 km have been demonstrated across the full CWDM band [4]. Like for DWDM sources, the dispersion tolerance of CWDM

---

[*] Note that optical pre-amplification, due to "0"-bit beat noise, reduces the dispersion tolerance to about 900 ps/nm at 10 Gb/s.

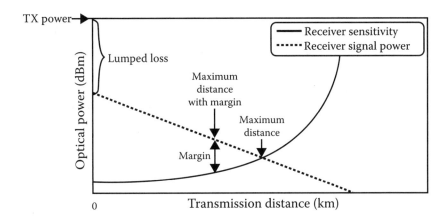

FIGURE 7.13 Signal power and receiver sensitivity as a function of transmission distance.

lasers is typically specified for a 1- or 2-dB receiver power penalty. Commercial DMLs at 2.5 Gb/s with a dispersion tolerance of 1600 to 1800 ps/nm for a 2-dB power penalty are available today. Figure 7.13 depicts a schematic of the power evolution in a CWDM system where the received signal power decreases with distance due to lumped component loss and accumulated fiber loss. In contrast, the receiver sensitivity increases with distance while the difference between both traces represents the available power margin and the intersection point defines the maximum transmission distance of the CWDM laser. Note that the channel performance is dependent on the design parameters of the individual lasers: while the received signal power is determined by the laser output power, the receiver sensitivity at a given distance, expressed by the dispersion tolerance parameter, is a result of the interaction between laser chirp and fiber dispersion.

One has to keep in mind that the penalty curve becomes quite steep at high penalties and therefore CWDM systems might not be strictly bound to a 2-dB power penalty limit, particularly for the long-wavelength channels. However, robust system design should still require penalties to stay within moderate bounds.

An example for the measured performance of a full set of 16 CWDM lasers is shown in Figure 7.14. The experiment, described in more detail in Section 7.6.1, was carried out over 40 km of NZDSF with reduced water peak. The fiber had a zero-dispersion wavelength of 1465 nm, with negative dispersion below that wavelength and positive dispersion above [9]. The measured sensitivity penalty (circles) directly reflect the interplay of laser chirp and dispersion: Negative dispersion compresses the pulses and leads to a sensitivity improvement compared to back-to-back performance (a "negative penalty"), whereas positive dispersion yields penalties of up to 5 dB at the high-wavelength end of the CWDM band. The margin shown in Figure 7.14 reflects the difference between the available power at the receiver and the required receiver input power necessary to satisfy the prescribed BER.

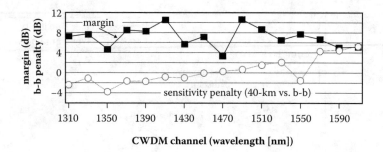

FIGURE 7.14 Power margin and sensitivity penalty for 16 CWDM channels at 10 Gb/s for 40-km NZDSF transmission link. (Taken from Thiele, H.J., Winzer, P.J., Sinsky, J.H., Stulz, L.W., Nelson, L.E., and Fidler, F., *Photon. Technol. Lett.*, 16(10), 2004. With permission.)

7.3.2.3 Chirp and Extinction Ratio

As mentioned along with Equation 7.1, DML chirp can be reduced by decreasing the modulated signal's ER. Doing so, however, also reduces receiver sensitivity, as can easily be seen from looking at the eye opening penalty due to finite signal extinctions: The eye opening of an NRZ signal with perfect signal extinction is given by $E = 2P_{av}$, where P_{av} is the average optical signal power. Reducing the signal's ER r leads to an eye opening of $E(r) = 2P_{av}(1 - 1/r)/(1 + 1/r)$. We thus find for the power penalty as a function of signal extinction,

$$\text{Power penalty} = (1 + 1/r)/(1 - 1/r). \tag{7.3}$$

Modulation of a DML in the presence of dispersion therefore has to balance performance degradations due to ER, as given by Equation 7.3, and chirp, as visualized in Figure 7.15, comparing the contribution of ER and chirp penalty for a DML transmitted over 100 km SSMF [37]. Here, the optimum ER is identified to be around 10 dB. Practical values are $r < 10$ dB where receiver sensitivity improves for higher extinction, and dispersion tolerance is better for lower ER.

7.3.3 INCREASED BIT-RATE OPERATION

7.3.3.1 Uncooled DMLs for 10 Gb/s

In principle, DMLs can be operated at higher bit-rates to achieve capacity-enhanced CWDM transmission. However, this is challenging, since DMLs require a laser relaxation oscillation frequency of about twice the bit-rate, that is, approximately two oscillations per bit period. In order to move to 10-Gb/s CWDM operation, oscillation frequencies of around 20 GHz would be required for the entire operating temperature range: Due to the electrical filtering commonly used at the receiver (0.7 times the bit-rate), these oscillations imprinted on the modulated

CWDM — Upgrade Paths and Toward 10 Gb/s

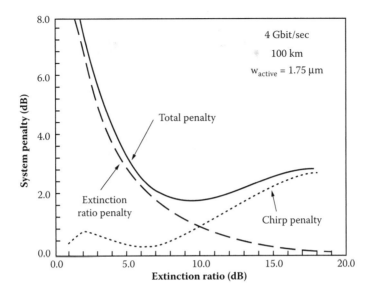

FIGURE 7.15 Power penalty as a function of ER (100-km link, 4 Gb/s, laser active region w = 1.75 μm wide). (Taken from reference Corvini, P.J. and Koch, T.L., *J. Lightwave Technol.*, 5(11), 1591–1595, 1987. With permission.)

signal power are filtered out by the receiver, and the resulting eye appears wide open as shown in Figure 7.16 for an uncooled 2.5-Gb/s CWDM laser at 1550 nm.

From the laser rate equations, the laser oscillation frequency is found to be proportional to

$$f_r \propto \sqrt{\frac{dg/dN \cdot P}{\tau_p}}, \tag{7.4}$$

where important design parameters defining f_r are dg/dN as the differential gain, the optical power P corresponding to the saturation current, and the photon lifetime τ_p [38]. Although there have been demonstrations of cooled lasers exceeding $f_r = 20$ GHz at room temperature, this typically requires high modulation currents of 80 mA and more [46], which is the current limit of many standard laser drivers. Uncooled operation of lasers further complicates the problem of sufficiently high f_r since differential gain and saturation current decrease at elevated temperatures. Therefore, benefit of high differential gain for high-speed DML design is twofold, first, due to a reduced chirp factor α and, secondly, an increased relaxation oscillation frequency resulting in improved dispersion tolerance and eye-opening, respectively.

7.3.3.2 Operating 2.5-Gb/s Rated DMLs at 10 Gb/s

To date, uncooled CWDM lasers at 10 Gb/s have only been demonstrated for selected wavelengths in the 1.3 and 1.55 μm wavelength region, and transmission

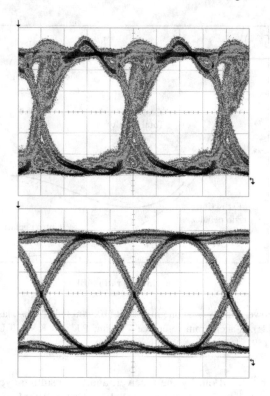

FIGURE 7.16 CWDM laser output signal at 2.5 Gb/s. *Top*: detected with 20-GHz PIN diode; *below*: with additional electrical filtering (cut-off at 2.488 GHz).

over a low water-peak NZDSF fiber has been shown [6,9]. In order to study 10-Gb/s CWDM systems before properly rated lasers across the full CWDM band become commercially available, one can try to operate a full bank of 2.5-Gb/s rated DMLs at 10 Gb/s. Taking this approach a step further, one could even think of initially using widely available and low-cost 2.5-Gb/s rated DMLs for short-reach 10-Gb/s CWDM capacity upgrades. This approach is enabled by optimizing the mounting and driving conditions of the 2.5-Gb/s rated uncooled DWDM lasers, as shown in Reference [9]. A similar approach, running 10-Gb/s cooled DMLs at 40 Gb/s, has been demonstrated for a single-wavelength, and more recently also for up to four WDM channels [7,8].

As reported in reference [9], 16 uncooled DMLs operating across the full CWDM band, and rated for 2.5-Gb/s operation, were mounted at the end of a 50-Ω microstrip line, in series with a 47-Ω surface mount chip resistor placed as close to the laser's coaxial package as possible, as shown in Figure 7.17. This way, the impedance matching of the laser was improved, which significantly flattened the DMLs' modulation response at frequencies up to 4 GHz, and enabled 10-Gb/s modulation. This increase in bandwidth comes at a reduced modulation efficiency due to the series resistor.

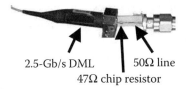

2.5-Gb/s DML   50Ω line
47Ω chip resistor

FIGURE 7.17 To improve modulation performance, each laser in a coaxial package was mounted at the end of a 50-Ω microstrip line, in series with a 47-Ω surface mount chip resistor.

Figure 7.18 shows the small-signal modulation characteristics of all 16 DMLs after improving the laser mounting as described earlier. Measurements were taken at 25°C. The figure also shows the characteristics of a commercial 10-Gb/s rated, cooled DML at 1550 nm as a reference. At frequencies below 4 GHz, both the 2.5-Gb/s DMLs and the 10-Gb/s DML exhibit similar performance once the mounting is improved for the lower-speed devices. A fast roll-off at about 4 GHz is consistently found for all 2.5-Gb/s DMLs. This is an intrinsic property of the laser chips, and can therefore not be mitigated by careful laser mounting.

Note that the measured data shown in Figure 7.18 were obtained for small-signal modulation using a network analyzer. In transmission systems, the lasers are typically driven harder, that is, with a large bias current and/or large RF amplitude. Therefore, small signal measurements can only give a first insight into the lasers' performance; under large-signal modulation, saturated rise and fall times as well as relaxation oscillations are equally important to assess modulation performance. Since the

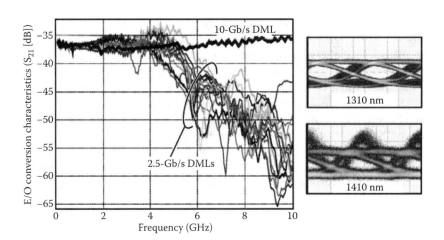

FIGURE 7.18 *Left*: Modulation characteristics of the DMLs using the improved mounting of Figure 7.17. The characteristics of a 10-Gb/s rated DML is shown for reference. *Right*: Two eye diagrams of typical lasers at 10 Gb/s. (Taken from Thiele, H.J., Winzer, P.J., Sinsky, J.H., Stulz, L.W., Nelson, L.E., and Fidler, F., *Photon. Technol. Lett.*, 16(10), 2004. With permission.)

relaxation oscillation frequency increases with higher laser bias currents (Equation 7.4), some DMLs yielded better results when driven with low data signal amplitudes but at relatively high bias currents, resulting in a poor (3 to 4 dB) ER. In contrast, other devices preferred to be driven with the "0"-level closely approaching the lasing threshold and using high drive amplitudes. This resulted in better ERs at the expense of significant amplitude overshoot. Figure 7.18 also shows two typical back-to-back eye diagrams obtained for both described driving conditions when optimizing the driving for maximum transmission distance, as discussed in Section 7.6.

Although the 10-Gb/s measurements on a set of 16 nonselected DMLs rated for 2.5-Gb/s were taken over a wide temperature range, the approach of pushing data rates beyond their specifications has not yet been investigated in detail for an appreciably large ensemble of lasers, which would be necessary for actually deploying this approach in the field.

7.3.4 Other Laser Types for 10-Gb/s CWDM Systems

The use of 10-Gb/s directly modulated DFB lasers in CWDM systems is restricted by the low dispersion tolerance caused by laser chirp. Other types of DMLs can overcome this limitation, such as external fiber Bragg grating lasers (FBGLs). These lasers have been discussed in the literature for a variety of different applications, ranging from pump lasers and pulse generators with wide frequency ranges to low-cost sources for optical access networks. These lasers are built by antireflection coating the fiber-pigtailed facet of a Fabry-Perot (FP) laser and forming an external cavity by writing a Bragg grating into the laser's fiber pigtail. Several systems experiments using these lasers have been carried with these lasers: Four-channel WDM transmission at 2.5 Gb/s over 300 km of fiber has been reported [47]. More recently, up to 600 km uncompensated reach at 2.5 Gb/s was reported, a distance sufficient for most applications of metro and regional networks [48]. The FBGLs have also been operated at higher bit-rates, for example, at 10 Gb/s where devices with a small signal-bandwidth in excess of 15 GHz were used, having a line width of less than 40 kHz at a maximum output power of 2 dBm and 48 dB side-mode suppression ratio. The advantages of these devices were a low FM response of 10 MHz/mA, resulting in a low laser chirp and less than 0.02 nm/°C wavelength drift [49]. Note that conventional DFB-DMLs feature an FM response of typically 250 MHz/mA, implying severe chirping. The FBGLs have been used for some metro applications, and could also have the potential to become a solution for uncooled DWDM or CWDM with narrower channel spacing and increased capacity. Some considerations for the future improvement of these devices and for their application to CWDM systems are the following:

- The maximum modulation rate B is driven by the round-trip time $\Delta\tau$ of the signal within the laser cavity.

$$B = \frac{1}{\Delta\tau} = \frac{c}{2} \frac{1}{(n_c L_c + n_f L_f)}, \qquad (7.5)$$

where n_c and n_f denotes the refractive index of the FP laser chip and the external fiber cavity. While the chip length L_c is typically 500 μm, L_f is much larger and defined by the distance between the fiber Bragg grating and the fiber-chip interface. A 10-GHz modulation bandwidth therefore requires to minimize the cavity length and spacing of the fiber as close as 30 μm from the chip, which becomes challenging for manufacturing.
- Controlling the laser emission wavelength by means of an external grating can simplify manufacturing since a pluggable external fiber grating could be used to define the wavelength of identical FP chips.
- Mode-hopping of uncooled FBGLs during temperature cycling leads to error bursts in the receiver and has to be avoided. One obvious solution is the extension of the mode-hopping free temperature range, although this requires at least a partial temperature control with a small thermoelectric heater or cooler.
- The first step toward a CWDM upgrade using FBGLs is described in reference [50] where the small temperature drift of a FBGL is utilized to create a CWDM-type multi-wavelength system with lasers spaced by 600 GHz (4.8 nm spacing around 1550 nm) and therefore lie within the gain bandwidth of standard EDFAs, which can be used for reach extension or to improve the link budget. However, a full-spectrum CWDM-type system based completely on FBGLs at 2.5 Gb/s has not been demonstrated yet but would have the potential to increase the transmission capacity due to a significantly increased channel count.
- Due to their lower temperature drift, uncooled FBGLs can enable wavelength spacing closer than the 20-nm spacings specified for today's CWDM systems, and can thus re-establish the value proposition of optical amplifiers. Many CWDM channels could fall within the amplification bandwidth of, for example, a compact semiconductor optical amplifier or a fiber-based amplifier, making possible multi-span transmission systems.

7.4 CWDM SYSTEM UPGRADE DEMONSTRATIONS

In this section, we discuss several concrete experimental demonstrations of CWDM system upgrades, using techniques outlined in our general discussions mentioned earlier. Some aspects of the experiments reported here are taken from reference [10].

7.4.1 EXAMPLE FOR A DWDM OVERLAY

First, we present an example for a capacity upgrade of an existing 16-channel CWDM system, where a sub-band of new DWDM channels was added to the transmission system. Figure 7.19 shows the experimental setup of the upgraded CWDM system. The original 16-channel CWDM system operated at 2.5 Gb/s per channel over 60 km of ZWPF (AllWave fiber). This fiber has positive dispersion over the full CWDM band, with a zero-dispersion wavelength around 1310 nm.

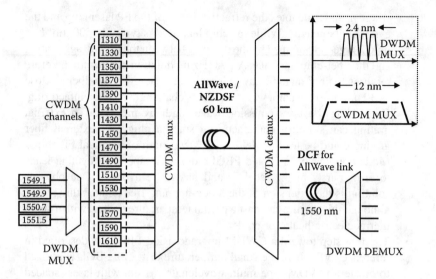

FIGURE 7.19 DWDM upgrade with four DWDM channels replacing CWDM channel number 13 at 1550 nm in a combined DWDM/CWDM transmission link. *Inset*: example of overlap between CWDM and DWDM multiplexer spectrum for the 1550-nm channel. (Taken from Thiele, H.J., Nelson, L.E., and Das, S.K., *Electron. Lett.*, 39(17), 2003. With permission.)

Figure 7.20 shows the power penalties of all 16 original CWDM lasers at 2.5 Gb/s. As expected, the penalties obtained for the short-wavelength channels (1310 to 1350 nm) are negligible due to the low CD in this band. System performance at the short-wavelength end of the CWDM band is only limited by fiber

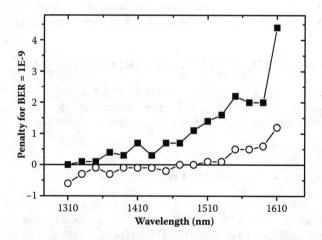

FIGURE 7.20 Measured power penalties (at 10^{-9} BER) for the 2.5-Gb/s reference CWDM system, channels modulated with 8-dB ER and $2^{23}-1$ PRBS word length, transmission over 60 km of: AllWave fiber (*squares*) and low water-peak NZDSF (*circles*).

loss due to Rayleigh scattering. The transmission penalties increase toward longer wavelengths, from 1.5 dB at 1550 nm to values as high as 4 dB for the 1610-nm channel due to increasing fiber dispersion. These penalties are a result of the interaction of the accumulated positive link dispersion and DML chirp. It is evident that the actual penalties do not increase in a uniform way since the α-factor is different for each laser. Variations in α are most pronounced at longer wavelengths, where CD is highest. Comparing the penalties of different CWDM lasers to their respective chirp (see Figure 7.11), DMLs with a large amount of adiabatic chirp achieve lower transmission penalties compared to lasers with substantial transient chirp. The penalties are considerably reduced when replacing the transmission fiber with an NZDSF with 1465-nm zero-dispersion wavelength and 0.047 ps/(km·nm^2) dispersion-slope (see Figure 7.26). Note that unlike the conventional NZDSF used for transmission in Section 7.5, the spectral attenuation of the prototype NZDSF is similar to the attenuation of ZWPF fiber (see Figure 7.1). Due to the reduced accumulated dispersion for all 2.5-Gb/s channels, the penalty falls below 0.5 dB for all but the 1590- and 1610-nm channel. Note that the interaction of negative dispersion and laser chirp leads to negative penalties up to −0.6 dB for the shortest wavelength at 1310 nm.

In the capacity upgrade process (see Figure 7.19), channel 13 of the CWDM system (at 1550 nm) was replaced by a sub-band of four DWDM channels operating at 10 Gb/s each and having a channel spacing of 100 GHz. The DWDM demultiplexer had a transmission bandwidth of 50 GHz, which was sufficiently wide to introduce no penalty due to filtering of the 10-Gb/s DML signals. After transmission, the DWDM channels were separated from the CWDM channels using the CWDM demultiplexer, and simultaneously post-compensated with a single spool of DCF. Due to the scaling of dispersion-limited distance with bit-rate (see Section 7.3.2), postcompensation was necessary at 10 Gb/s in order to support link distances of 60 km over standard-dispersion AllWave fiber. In the case of the NZDSF, however, no postcompensation was necessary. Since the DCF is a component that is shared by all channels of the DWM overlay, the per-channel cost benefit increases with the number of DWDM upgrade channels. In order to find the optimum length of post-compensating DCF for the AllWave span, the receiver sensitivity at BER = 10^{-9} was measured for different amounts of CD. The results are shown in Figure 7.21 for two different DML ERs (circles: 6 dB; squares: 8 dB). The back-to-back performance (0 ps/nm CD) is indicated by the crossed boxes. A performance improvement (i.e., a negative sensitivity penalty due to interaction of laser chirp and fiber dispersion) was measured over a range of 370 ps/nm for 8 dB extinction, increasing to 420 ps/nm for 6 dB extinction, corresponding to 24 km of transmission fiber.

Due to the positive DML chirp, the optimum total link dispersion was found to be around −280 ps/nm, and a DCF with −1290 ps/nm dispersion was chosen for the DWDM upgrade, adding to the desired total dispersion when including the dispersion of the transmission fiber. The measured accumulated dispersion of the combined AllWave fiber — DCF link in the DWDM upgrade path is

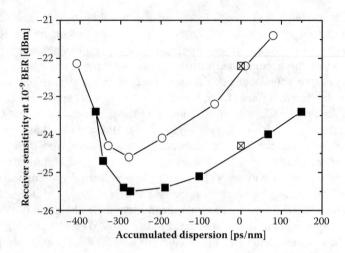

FIGURE 7.21 Impact of accumulated dispersion on performance of the 1551-nm 10-Gb/s DWDM channel over a post-compensated 60-km AllWave fiber span: *circles*: 6-dB ER; *squares*: 8-dB ER; *crosses*: back-to-back. (Taken from Thiele, H.J., Nelson, L.E., and Das, S.K., *Electron. Lett.*, 39(17), 2003. With permission.)

summarized in Figure 7.22. The total accumulated link dispersion at 1550 nm is −276 ps/nm and varies only by a small amount across the four DWDM channels.

The DCF used in this set-up had a high figure of merit (FOM = ratio of fiber dispersion to attenuation) of 317 ps/nm/dB [51]. The resulting low DCF loss of only 4.1 dB at 1550 nm, together with 11.4 dB for the 60-km ZWPF span, 5 dB for the CWDM muxes/demuxes and approximately 7 dB for the DWDM mux and demux filter allowed unrepeatered operation of the DWDM overlay; in this experiment, the 10-Gb/s DWDM channels had a back-to-back sensitivity between −22 and −25 dBm, as compared to 31 dBm, on average, for the 2.5-Gb/s CWDM

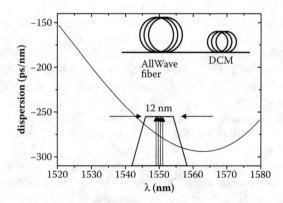

FIGURE 7.22 Variation of accumulated dispersion for the 1550-nm DWDM channels in the CWDM-DWDM transmission link.

TABLE 7.1
Transmission Penalties for 1550-nm DWDM Sub-Band

DML	[nm]	P[dBm]	Penalty [dB] AllWave + DCF	NZDSF
1	1549.1	7.4	−1.5	1.3
2	1549.9	8.4	−1.4	2.0
3	1550.7	8.1	−1.5	1.5
4	1551.5	7.9	−1.8	1.1

channels. Finally, the power penalties at BER = 10^{-9} were measured for the new DWDM channels operating at 10 Gb/s for both types of transmission fiber. All cooled DMLs were operated with 8 dB extinction over the AllWave link, and power levels launched into the system were exceeding 7 dBm. For the NZDSF link, the extinction was lowered to 6 dB to reduce the transmission penalties. For the DCF-compensated DWDM band, negative penalties were measured due to the negative net dispersion of the over-compensated link. On the contrary, the NZDSF link supported a DWDM band without additional compensation to the lowered accumulated dispersion (Table 7.1).

In this example it was shown that the growth in transmission capacity for metro-access networks can be met by an upgraded CWDM system. This can be achieved by the introduction of a DWDM overlay at 10 Gb/s increasing the capacity limit of conventional 16 × 2.5-Gb/s CWDM. This overlay technique promises a high capacity increase wihout sacrificing the original low-cost architecture of CWDM.

7.4.2 EXAMPLE FOR MIXED BIT-RATE TRANSMISSION

Although 10-Gb/s DWDM lasers have been discussed in the previous section for upgrading a CWDM system to higher capacity, the main focus of that work was the CWDM–DWDM architecture. In this section, individual CWDM channels are upgraded to 10 Gb/s without the introduction of new wavelengths, creating a mixed transmission system with both 2.5 Gb/s and 10 Gb/s CWDM channels. Unlike before, the goal in this upgrade is to avoid any additional optical components such as multiplexers, DCF, or even a change of transmission fiber. The question as to which channels of the 16-wavelength CWDM system can be easily upgraded in bit-rate hinges on the following considerations:

- *What is the wavelength of the upgraded channel?*
  – At short wavelengths, the small accumulated dispersion of the All-Wave fiber and high fiber loss allow the use of lasers with relaxed requirements on chirp but requires a high output power. In contrast, lasers at longer wavelengths experience less fiber loss but experience higher accumulated dispersion.

- *What low-cost lasers are available for the upgrade?*
  – The two most likely candidates for low-cost 10-Gb/s CWDM channels are DMLs with a low dispersion tolerance and high output power, or continuously operating DFBs integrated with an electro-absorption modulator [electro-absorption modulated lasers (EMLs)]. The EMLs have a higher dispersion tolerance than the DMLs (equivalent to 80-km SSMF at 1550 nm), but a major drawback is their relatively low modulated output power of typically 0 dBm, limited by the electro-absorption modulator's high insertion loss.

7.4.2.1 Upgrade Using 10-Gb/s DML

The 2.5-Gb/s lasers at 1310 and 1330 nm can directly be replaced with 10-Gb/s DML sources since the accumulated dispersion of G.652 fiber, such as AllWave fiber, is low in that wavelength range. In an experimental demonstration, the 1310-nm channel is upgraded with a cooled 10-Gb/s DML operating at a nominal wavelength of 1307.1 nm and an average modulated output power of 5 dBm at 8-dB extinction [10]. The second upgraded channel at 1337.2 nm and 5.5-dBm output power could still pass the 1330-nm port of the CWDM MUX. These two laser devices were not selected, and both output power and wavelength could be further improved. The 10-Gb/s DMLs were transmitted over 60 km of AllWave fiber together with the existing 2.5-Gb/s CWDM channels (see Figure 7.19), and detected with an APD-based 10-Gb/s receiver.

As shown in Figure 7.23, the upgraded 10-Gb/s channel at 1307.1 nm has a small negative penalty after 60 km since the fiber's zero-dispersion wavelength

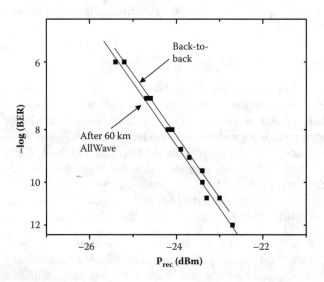

FIGURE 7.23 Penalty measurement of 1310-nm CWDM channel after 60 km, upgraded with a 10-Gb/s cooled DML. (Taken from Thiele, H.J., Nelson, L.E., and Das, S.K., *Electron. Lett.*, 39(17), 2003. With permission.)

lies around 1310 nm, higher than the laser wavelength (see Figure 7.5). Hence, the channel is limited by fiber loss, and the upgrade from 2.5 to 10 Gb/s can be achieved without dispersion compensation. If uncooled 10-Gb/s DMLs of sufficiently high output power are available, they could also be used in this upgrade.

The transmission penalty of the second 10-Gb/s DML at 1337 nm was 2 dB due to the higher positive dispersion at this wavelength (72 ps/nm for the 60-km link). A 10-Gb/s upgrade with DMLs for even longer wavelengths is challenging for this 60-km link due to the increasing dispersion penalties. A larger number of CWDM channels can be upgraded in shorter transmission links, for example, for 20-km distances typically found in access networks, where the accumulated dispersion across the band is reduced by two-thirds compared to the 60-km link considered here. In this shorter link, all CWDM channels at wavelengths below the water-peak region could be operated at 10 Gb/s without dispersion compensation.

7.4.2.2 Upgrade Using a 10-Gb/s EML

At longer wavelengths in the CWDM spectrum, the use of 10-Gb/s DMLs is prevented by the high accumulated dispersion, and lower-chirp sources such as EMLs have to be used. In the experimental demonstration discussed here, an EML is emulated by using a separate 10-Gb/s electro-absorption modulator, following a cooled 10-dBm DFB laser operating at 1569.1 nm. In this case, the transmission penalty for the 10-Gb/s channel at 1569.1 nm was 3.5 dB, with no evidence of an error floor.

Finally, the different upgrade techniques reported in Section 7.4.1 and Section 7.4.2 were combined, as shown in Figure 7.24 for the 60-km AllWave

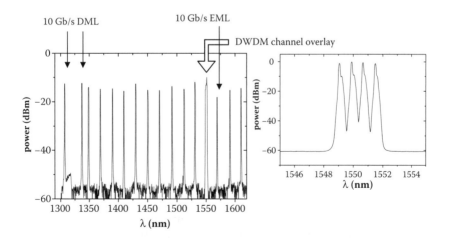

FIGURE 7.24 *Left*: Optical spectrum of the 100-Gb/s capacity mixed bit-rate CWDM system over 60 km AllWave fiber. *Right*: 4 × 10-Gb/s channels of the DWDM overlay.

fiber link. The remaining channels continue to operate at the original bit-rate of 2.5 Gb/s. The three combined upgrade techniques increase the total capacity of the CWDM system from initially 16 channels at 2.5 Gb/s to 7 channels at 10 Gb/s with 12 other channels remaining at 2.5 Gb/s, resulting in 100-Gb/s total system capacity.

7.5 MIXED FIBER-TYPE TRANSMISSION

In the examples discussed earlier, the CWDM capacity upgrade was based on an existing 2.5-Gb/s CWDM system running over a single type of fiber, for example, AllWave fiber. In this section, we investigate how the performance of the 10-Gb/s sources is influenced by the use of a combination of different fiber types and, hence, accumulated dispersion. The set-up shown in Figure 7.25 combines all three previously discussed approaches for enhanced CWDM capacity: a single 10-Gb/s DML, a DWDM overlay, and a 10-Gb/s EML. The 2.5-Gb/s CWDM channels were omitted for simplicity, as reported in reference [52].

All 10-Gb/s sources were transmitted over a common 50-km G.655-type fiber link (conventional NZDSF fiber, TrueWave-RS); the 1310-nm DML and the EML are additionally transmitted over mixed SSMF and NZDSF fiber sections representative of an access and a metro regional network. Figure 7.26 shows the dispersion characteristics of both fiber types across the CWDM spectrum. Since the zero-dispersion wavelength of NZDSF (around 1450 nm) falls in the center of the CWDM transmission band, the accumulated dispersion remains low for all channels and no dispersion compensation is necessary, even for the DWDM sub-band at 1550 nm. The DWDM overlay, the single DML at 1310 nm and the EML were driven with a 10-Gb/s PRBS bit-sequence of $2^{23}-1$ word length. After transmission over the NZDSF link and, for some channels, over the additional fiber spans shown in Figure 7.25, all channels were detected with an APD-based receiver (10-Gb/s sensitivity of -25 dBm for BER = 10^{-9}).

7.5.1 10-Gb/s DML over AllWave and TrueWave-RS

A single, 1307.2-nm DML with 4.0-dBm output power was transmitted over a section of the edge network represented by the 50-km span of TrueWave-RS fiber (-8 ps/nm/km at 1310 nm, approximately 4 ps/nm/km at 1550 nm), a distance comparable to the previous CWDM links. The add-drop sections for this signal consisted of two 10-km laterals of AllWave fiber representing the access section of the network. This cooled DML could be substituted with a low-cost uncooled 10-Gb/s DML at 1310 nm (0 dBm output) when the combined AllWave and TrueWave fiber length was reduced to approximately 50 km. Uncooled devices with increased output power should overcome this limitation in transmission length and can directly replace the cooled DML.

CWDM — Upgrade Paths and Toward 10 Gb/s

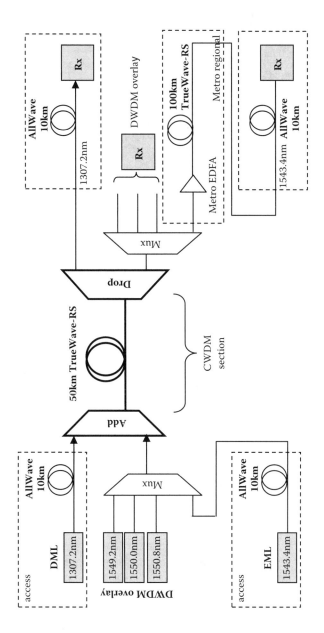

FIGURE 7.25 Experimental set-up for 10-Gb/s uncompensated transmission over a mixed fiber plant.

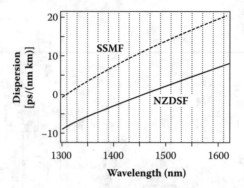

FIGURE 7.26 Dispersion of SSMF (*dashed*) and NZDSF (*solid*) across the CWDM spectrum. Dotted vertical lines indicate nominal CWDM wavelengths.

7.5.2 10-GB/S DWDM OVERLAY WITH TRUEWAVE-RS

Three 100-GHz-spaced DMLs between 1549.2 and 1550.8 nm were operated at 10 Gb/s with modulated output powers between 5.3 and 6 dBm. The narrow channel spacing requires the use of cooled DMLs for wavelength stability. Due to the linewidth broadening of DMLs, a multiplexer with flat-top characteristic was used (3-dB bandwidth ≈ 0.6 nm). The combined 10-Gb/s signals were directly launched into the 50-km TrueWave-RS fiber section. Note that the driving conditions of the lasers, for example, 8-dB ER, do not allow a direct comparison with the 10-Gb/s penalty results of Section 7.4.1.

7.5.3 10-GB/S EML WITH ALLWAVE AND MULTIPLE SPANS OF TRUEWAVE-RS

A single 1543.4-nm EML with −1 dBm output power and ER = 10 dB was transmitted over the access (AllWave fiber), edge (50 km TrueWave-RS) and metro regional sections (amplified TrueWave-RS span) without compensation. This scenario emulates the case when express channels are transmitted over a longer distance within the optical network and routed through the 50-km section of a CWDM link. After separation from the other co-propagating short-reach channels such as the DWDM band or the 1310-nm channel, the EML is transmitted over an amplified TrueWave-RS span. Due to the lower output power of the EML and the additional uncompensated 100-km fiber span, a single-stage, nongain flattened metro EDFA with 6-dBm output power was used for amplification.

Figure 7.27 shows the transmitted spectrum of the four lasers (EML and DMLs) in the 1550-nm region before demultiplexing. As expected, although a NZDSF fiber with a small effective area of 55 μm^2 and 100-GHz-spaced DMLs were used, nonlinearities such as four-wave-mixing (FWM) and cross-phase modulation (XPM) did not play a role for the comparatively short distances considered here: less than 4% peak-to-valley variation for XPM-induced distortion

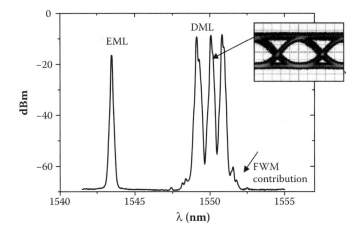

FIGURE 7.27 Optical spectrum of 10-Gb/s EA-EML and DML overlay channels after 50-km TrueWave-RS fiber. *Inset*: back-to-back filtered eye diagram of center DML. (Taken from Thiele, H.J., Das, S.K., Boncek, R., and Nelson, L.E., Proceedings OECC 2003, paper 16E1-4, pp. 717–718. With permission.)

on a CW channel, and FWM crosstalk below −50 dB [52]. The penalties are dominated by fiber dispersion and laser chirp. Generally, no impact of nonlinearities was observed for all upgraded CWDM links discussed in this chapter.

Table 7.2 summarizes the transmission results for all segments of this sample network. All channels were error-free (BER < 10^{-9}) after transmission. The 1307-nm DML exhibited a negative penalty due to the negative NZDSF dispersion. Positive transmission penalties were measured in the 1550-nm region as a result of positive DML chirp combined with positive fiber dispersion. The 1549.2-nm DML shows that a lower α-factor corresponds to a reduced transmission penalty, as discussed in Section 7.3.2. Lowering the ER of the other two DMLs to 6 dB reduced their penalties by 1 dB. Although the use of 1550-nm DMLs is challenging due to their intrinsic positive chirp, the reduced dispersion of TrueWave-RS fiber enables their use in metro edge or CWDM applications with 50-km uncompensated reach at 10 Gb/s. The additional benefit of low laser chirp is demonstrated with the EML over sections of AllWave and TrueWave-RS fiber, extending out to metro regional networks.

TABLE 7.2
Overview of 10-Gb/s Transmission Results

λ [nm]	1307.2	1549.2	1550.0	1550.8	1543.2
α	2.8	2.3	4.0	3.8	0.5
Distance [km]	10 +50 +10	50	50	50	10 + 50 +100 + 10
ER [dB]	8	8	8	8	10
penalty at 10^{-9} BER	−0.7	1.3	3.8	2.7	4.5

7.6 FULL-SPECTRUM CWDM AT 10 Gb/s

In the previous two sections, we discussed CWDM capacity upgrades by adding new channels or by substituting a small number of existing CWDM channels with new devices running at higher bit-rates. As a result of this partial upgrade, a transmission system with mixed bit-rates at both 2.5 and 10 Gb/s was obtained. In this section, we move to the next step, considering systems with all channels operating at 10 Gb/s. In order to demonstrate full-spectrum 10-Gb/s CWDM transmission despite the unavailability of 10-Gb/s rated laser sources across the entire CWDM band, we employed 16 commercially available, uncooled DMLs rated for 2.5-Gb/s operation, but modulated them at 10-Gb/s data rates, as already described in Section 7.3.3. Note that these DMLs were not pre-selected, and thus had widely varying static as well as dynamic characteristics (see Figure 7.10).

7.6.1 Transmission over NZDSF

In the following transmission experiment, 16 commercially available uncooled DMLs were used, rated for 2.5-Gb/s modulation, at CWDM wavelengths from 1310 to 1610 nm (see Figure 7.2a). The lasers were simultaneously driven by 9.95-Gb/s PRBSs of length $2^{31} - 1$ via a 2:16 electrical splitting network. The 20-nm-spaced optical data signals were combined using a thin-film filter-based CWDM multiplexer (see Figure 7.3) before propagating over 40 km of a low water peak NZDSF with 1465-nm zero-dispersion wavelength, as discussed in Section 7.4.1. A CWDM demultiplexer was used to separate the channels before entering an APD receiver (RX) with a 10-Gb/s sensitivity of –25 dBm at 1550 nm and at a BER of 10^{-9}.

Figure 7.28 shows the BER for all 16 channels as a function of the optical power at the receiver after CWDM propagation over 40 km of NZDSF and at 25°C. Laser bias currents and drive voltages as well as the receiver's decision threshold were adjusted for optimum BER after transmission. No indications of an error floor were found, and no sensitivity difference was observed when turning off the neighboring channels, indicating negligible WDM channel crosstalk. The large scatter in the DMLs' characteristics (Section 7.3, Figure 7.10 and Figure 7.18) and chirp behavior (see Figure 7.11) is reflected in the ~8 dB of sensitivity variation among the 16 channels.

Figure 7.29a summarizes the receiver sensitivities (BER = 10^{-9}) after 40 km of CWDM propagation for all channels at 25°C (solid squares). The solid triangles show the available power at the receiver for each received channel. The difference between the markers indicates a substantial amount of available power margin, for example, 10 dB at 1490 nm; the values for system margin are represented in Figure 7.29b by solid squares. The open circles in both Figure 7.29a and 7.29b denote the back-to-back sensitivities and the propagation-induced sensitivity penalties, respectively, at BER = 10^{-9}, obtained using the same laser driving conditions as in the 40-km case. Although these conditions are not optimized for back-to-back operation, we used them in order to demonstrate full

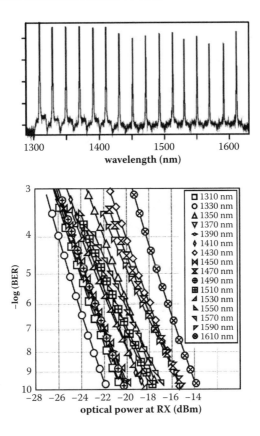

FIGURE 7.28 *Top*: 16×10-Gb/s CWDM spectrum measured after multiplexing; *below*: BER measurements for all 10-Gb/s channels at $T = 25°C$ after 40-km of simultaneous transmission over low water peak NZDSF. (Taken from Thiele, H.J., Winzer, P.J., Sinsky, J.H., Stulz, L.W., Nelson, L.E., and Fidler, F., *Photon. Technol. Lett.*, 16(10), 2004. With permission.)

functionality without the need for dynamic laser reconfiguration. The negative penalties below 1450 nm are due to the interaction of positive laser chirp and negative dispersion of the fiber at those wavelengths.

When speaking about uncooled lasers, it is imperative to demonstrate operation not only at 25°C, but also at elevated temperatures, where DML characteristics tend to degrade rapidly (see Figure 7.9). Figure 7.30 shows, for all CWDM channels, the maximum laser case temperature that still allowed detection with BER = 10^{-9} after 40 km of 10-Gb/s CWDM transmission. With the exception of a single device (laser at 1470 nm), a temperature of 75°C is well supported, and selected devices reach operation temperatures beyond 90°C. The eye diagrams for all lasers were recorded at the highest possible temperature supporting BER = 10^{-9} and show the different settings for the DMLs when operating at 10 Gb/s. Some have a low ER while others yield optimum performance with a large ER and slight pulse distortion.

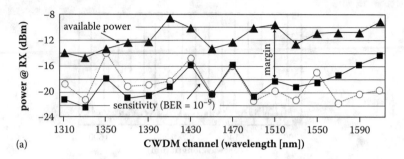

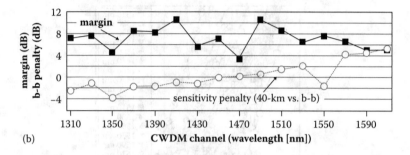

FIGURE 7.29 (a) Receiver sensitivity for all CWDM channels (■), maximum available receiver power (5), and power for BER = 10^{-9} after 40 km (O). (b) Calculated power margin after 40-km fiber propagation (■) and sensitivity penalty with respect to the back-to-back case (O). (Taken from Thiele, H.J., Winzer, P.J., Sinsky, J.H., Stulz, L.W., Nelson, L.E., and Fidler, F., *Photon. Technol. Lett.*, 16(10), 2004. With permission.)

In conclusion, the key factors for enabling uncompensated full-spectrum CWDM transmission at 10 Gb/s in this experiment were both the laser sources and the transmission fiber. Although the lasers were designed to work at 2.5 Gb/s, small modifications of the connectors and optimized driving conditions improved the small-signal bandwidth. In addition, the low accumulated dispersion of the prototype low water peak NZDSF helped to maintain a transmission link length comparable to the distances used for conventional 2.5-Gb/s CWDM systems.

7.6.2 FEC-Enabled Transmission over AllWave

7.6.2.1 FEC and Equalization in a Full-Spectrum CWDM Experiment

This subsection describes an experimental demonstration of a full-spectrum 10-Gb/s CWDM system operating over standard-dispersion fiber and using equalization and FEC, which is discussed in Section 7.2.3 and in more detail in reference [53]. The CWDM sources were nonselected 2.5-Gb/s DMLs, mounted carefully for improved RF performance, and operated at 10 Gb/s, as described in Section 7.3.3 and Section 7.6.1. The chipset used at the receiver consisted of a CMOS-based FEC device and a SiGe-based, highly integrated equalization circuit

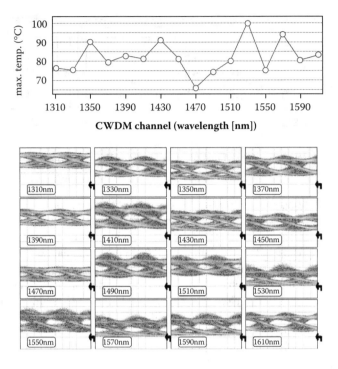

FIGURE 7.30 *Top*: Maximum achievable laser case temperature allowing BER = 10^{-9} after 40-km fiber propagation. *Below*: back-to-back eye diagrams for all CWDM channels at temperature corresponding to BER = 10^{-9}. (Taken from Thiele, H.J., Winzer, P.J., Sinsky, J.H., Stulz, L.W., Nelson, L.E., and Fidler, F., *Photon. Technol. Lett.*, 16(10), 2004. With permission.)

[21], as shown in Figure 7.31. The equalization chipset incorporates a combination of several previously studied equalization techniques [54–57], such as adaptive thresholding, feed forward equalization (FFE), decision feedback equalization (DFE), and a reduced-complexity version of MLSE. The signal is first passed through a variable-gain stage, a variable peaking circuit for bandwidth equalization, and a multi-tap FFE. A DFE in combination with multiple adaptive thresholding provides quantized information for subsequent soft-decision decoding by simplified, MLSE-like processing and FEC feedback. Full clock and data recovery as well as SFI-4 compliant demultiplexing are part of the chipset. A parallel sub-rate feedback channel between the SiGe device and the FEC chip provides rapid automatic equalizer adaptation over a broad range of signal conditions. The FEC scheme has a 7.14% bit-rate overhead, including framing, and complies with ITU standard G.709. The FEC is built on interleaved Reed-Solomon (RS) and Bose-Chandhuri-Hocquenghem (BCH) codes, and is able to correct channel BERs of $< 2 \cdot 10^{-3}$ to values $< 10^{-15}$, as visualized in the BER conversion curves shown in Figure 7.7 [21].

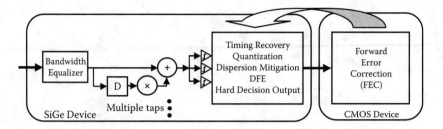

FIGURE 7.31 Block diagram of the integrated electronic equalization and FEC chip set. This is the device for the receiver.

Figure 7.32 shows the back-to-back sensitivities (circles) of the 16 10-Gb/s CWDM channels using a standard 10-Gb/s PIN-type photoreceiver without electronic equalization. All 16 DMLs were driven with an FEC-precoded PRBS of length $2^{31} - 1$ using NRZ modulation at 10.664 Gb/s. This data rate corresponds to a 9.953-Gb/s OC-192 information bit-rate and a 7.14% FEC overhead. The sensitivities are taken at a channel BER of 10^{-3}, yielding a corrected (post-FEC) BER of better than 10^{-15}. Note that the CMOS-based FEC board was included in all our BER measurements. This way, all deteriorating effects of burst errors on FEC performance were fully included in our measurement results. As in the experiment described in Section 7.6.1, there is significant device-dependent scatter in receiver sensitivities (see Figure 7.28 and Figure 7.29).

The squares in Figure 7.32 represent the back-to-back sensitivities of the 16 CWDM channels with electronic equalization in combination with the same photoreceiver front-end that was used for the measurements without equalization (circles). Note that the equalizer is able to improve the back-to-back sensitivities of our limited bandwidth 2.5-Gb/s rated DMLs by up to 4 dB.

Figure 7.33 shows, as a typical example, the receiver sensitivity as a function of the transmission distance over standard-dispersion fiber for the 1550-nm CWDM

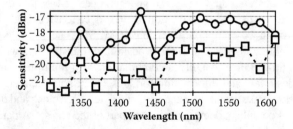

FIGURE 7.32 Back-to-back sensitivities of all 16 CWDM channels (channel BER = 10^{-3}; corrected BER < 10^{-16}). Drive conditions optimized for maximum transmission distance. *Circles*: without equalization; *squares*: with equalization. (Taken from Winzer, P.J., Fidler, F., Thiele, H.J., Matthews, M., Nelson, L.E., Sinsky, J.H., Chandrasekhar, S., Winter, M., Castagnozzi, D., Stulz, L.W., and Buhl, L.L., *J. Lightwave Technol.*, 23(1), 203–210, 2005. With permission.)

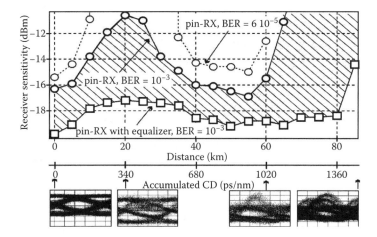

FIGURE 7.33 Receiver sensitivity (*solid*: BER = 10^{-3}; *dashed*: BER = $6 \cdot 10^{-5}$) vs. transmission distance and accumulated CD for the 1550-nm CWDM laser. *Circles*: unequalized receiver; *squares*: equalized receiver; *hatched area*: equalization gain. Optical eye diagrams are shown at various distances. (Taken from Winzer, P.J., Fidler, F., Thiele, H.J., Matthews, M., Nelson, L.E., Sinsky, J.H., Chandrasekhar, S., Winter, M., Castagnozzi, D., Stulz, L.W., and Buhl, L.L., *J. Lightwave Technol.*, 23(1), 203–210, 2005. With permission.)

channel without (circles) and with (squares) electronic equalization. Both solid curves were taken at a channel BER of 10^{-3} and with enabled FEC to ensure block error correctability. The dotted curve shows sensitivities at BER = 6×10^{-5}. All curves used the same laser driving conditions, optimized for maximum reach. The improvement in receiver performance beyond 20 km reflects the effect of self-steepening (Section 7.3.2) caused by the interplay of CD and adiabatic laser chirp, and leading to pulse recompression [58,59]. The effect of equalization, visualized by the hatched area in Figure 7.33, is twofold: first, equalization provides a back-to-back sensitivity improvement of almost 4 dB, which compensates for the limited 10-Gb/s performance of our 2.5-Gb/s rated DMLs. For the compressed pulses at around 50 km, we find an equalization gain of 2 to 3 dB. Secondly, at distances where the pulses are broadened by dispersion (at distances around 20 km and in excess of 60 km), the equalizer gain exceeds 7 dB. Using equalization and FEC, error-free transmission over 85 km was achieved, corresponding to an accumulated CD of 1445 ps/nm at 1550 nm. Note from the dotted curve in Figure 7.33 that transmission between 15 and 35 km of standard-dispersion fiber was not possible at a channel BER of 6×10^{-5} due to severe error flooring. Since this channel BER corresponds to the correction threshold of standard RS(255,239) FEC (see Figure 7.7), we conclude that only the combination of electronic equalization and enhanced FEC enables 10-Gb/s CWDM to continuously attain appreciable distances as needed in flexible metro/access networks.

It is important to note from Figure 7.33 that electronic equalization significantly reduces the variations in CD penalty over distance. In paricular, the poor

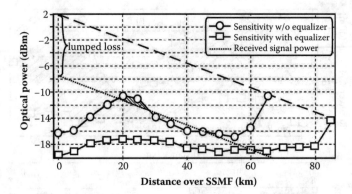

FIGURE 7.34 Receiver sensitivity at BER = 10^{-3} vs. transmission distance for the 1550-nm CWDM laser (*circles*: unequalized receiver; *squares*: equalized receiver). The dashed line represents the optical power evolution as a function of transmission distance, while the dotted line includes any lumped optical loss and represents the available optical power at the receiver. The hatched area indicates unattainable transmission distances caused by high penalty around 20 km in the unequalized case. (Taken from Winzer, P.J., Fidler, F., Thiele, H.J., Matthews, M., Nelson, L.E., Sinsky, J.H., Chandrasekhar, S., Winter, M., Castagnozzi, D., Stulz, L.W., and Buhl, L.L., *J. Lightwave Technol.*, 23(1), 203–210, 2005. With permission.)

receiver sensitivity around 20 km is substantially improved. This feature of electronic equalization significantly facilitates system design, as visualized in Figure 7.34, showing the measured 1550-nm receiver sensitivities from Figure 7.33. Also shown is the signal power evolution as a function of distance (dashed), for a signal launch power of +2 dBm and a fiber loss coefficient of 0.19 dB/km. Any additional lumped loss, for example, due to multiplexers, connectors, couplers, or splices, shifts down the dashed curve, resulting in the dotted curve in Figure 7.34, representing the available optical power at the receiver as a function of transmission distance. System margins may also be added to further lower the dotted curve. As explained along with Figure 7.13, the intersection of the available power with the receiver sensitivity then determines the maximum transmission distance. Typically, and as depicted in the schematic drawing in Figure 7.13, the solution to finding the maximum transmission distance is unique, that is, the available power and the receiver sensitivity intersect only once. However, if the receiver sensitivity exhibits non-monotonic behavior as a function of distance, as found for the unequalized receiver (circles), more than one intersection may occur, which in the equalizer-free example of Figure 7.34 excludes transmission distances between 10 and 30 km (hatched area in Figure 7.34), but otherwise enables distances of up 45 km. Such a situation significantly complicates system design, unless one is willing to accept the lowest distance intersection as the maximum allowed transmission distance. In general, the criterion for the sensitivity penalty to yield unattainable transmission distances (and thereby complicate system

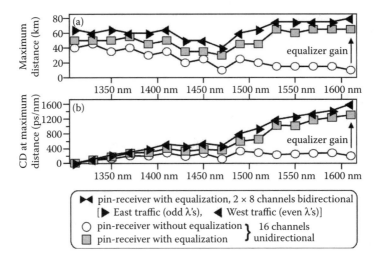

FIGURE 7.35 (a) Maximum transmission distance for all CWDM channels and (b) CD at maximum distance shown in (a). (Taken from Winzer, P.J., Fidler, F., Thiele, H.J., Matthews, M., Nelson, L.E., Sinsky, J.H., Chandrasekhar, S., Winter, M., Castagnozzi, D., Stulz, L.W., and Buhl, L.L., *J. Lightwave Technol.*, 23(1), 203–210, 2005. With permission.)

design) reads [53]

$$-\frac{\partial P_{sens}}{\partial z} > \alpha, \qquad (7.6)$$

where P_{sens} denotes receiver sensitivity in decibels, z is the coordinate in propagation direction in kilometers, and α stands for the fiber attenuation coefficient in dB/km. As evident from Figure 7.33 and Figure 7.34, electronic equalization substantially reduces the variations in receiver sensitivity, and thus facilitates system design.

Figure 7.35 summarizes the results reported in reference [53] for a 16-channel full spectrum bidirectional CWDM system according to Figure 7.2b, where adjacent channels propagated in opposite directions on a standard dispersion ZWPF (see Figure 7.1 and Figure 7.5), as well as for a 16-channel unidirectional system with reduced laser output power The black triangles show the results of the bidirectional transmission experiment, using electronic equalization and FEC; the triangles point in the direction of traffic flow (East and West traffic). The significant performance variations among channels are typical for nonselected DMLs (see Figure 7.10). In the long-wavelength region (high dispersion but low loss), the equalizer proves particularly valuable, enabling transmission distances in excess of 40 km for all channels and up to 80 km for the best-performing channel; more than 50% of all channels go beyond 65 km. The results for the unidirectional 16-channel system are represented by the open circles (no equalization) and the gray squares (with equalization). Electronic equalization is seen to boost transmission distances from 10 km up to 65 km. As is evident from Figure 7.35b, the unequalized receiver only supports ~ 300 ps/nm of CD, whereas the equalizer allows up to 1600 ps/nm.

7.6.2.2 FEC and Equalization to Support Fully Bidirectional CWDM Transmission

The previous section discussed a bidirectional transmission experiment, where eight CWDM channels propagated in each direction along a fiber. In contrast, in a fully bidirectional system, all 16 CWDM wavelengths are simultaneously used for both transmission directions. Such fully bidirectional systems, which make the most efficient use of installed fiber infrastructure, have been frequently discussed, both for amplified multi-span transmission and for single-span access scenarios [60,61]. So far, however, their deployment has been prevented by beat noise due to in-band crosstalk, arising from connector reflections and Rayleigh backscatter. This places severe restrictions on the tolerable link loss, and thus on the achievable transmission distances.

Figure 7.36 illustrates this limitation: transmitter TX_B sends information to be received by receiver RX_A. The signal power after coupler B is denoted $P_{TX,B}$. Due to (lumped and distributed) span loss L, the signal power just before coupler A is $L\,P_{TX,B}$. At the same time, a reflected power $R\,P_{TX,A}$ will be present at that point, caused by any (lumped and distributed) reflections R of the counter-propagating signal within the span, as shown in the figure. The ratio of signal power to interferer power at the receiver [signal-to-interference ratio (SIR)] can then be expressed in dB as [61]

$$SIR_{[dB]} = L_{[dB]} - R_{[dB]} - \Delta P_{[dB]}, \qquad (7.7)$$

where $\Delta P = P_{TX,A} / P_{TX,B}$ is the power divergence between same-wavelength transmitters. Typically, R is around −32 dB [61], and SIR > 25 dB is required for less than 1-dB power penalty at a BER of 10^{-9} with a pin-type photoreceiver and NRZ modulation [61–65]. If the interferers are due to a single discrete reflection (e.g., a connector) rather than due to multiple reflections (e.g., Rayleigh backscatter), the tolerable SIR can be somewhat lower [63]. Allowing for a transmitter power divergence of 5 dB, Equation 7.7 predicts a maximum tolerable span loss of 2 dB,

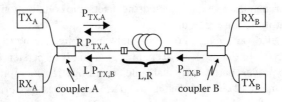

FIGURE 7.36 Fully bidirectional fiber communication system.

which translates into a transmission distance of some 10 km at 1550 nm. (Note that the span loss L excludes CWDM multiplexers and couplers.)

Using FEC, the requirements on SIR can be significantly relaxed, and the bidirectional transmission distance can be extended. This beneficial aspect of FEC [17,18,66] owes to the fact that in-band crosstalk, like many other optical signal degradations [15], leads to reduced sensitivity penalties at poor channel BER (where FEC operates) than at low channel BER (where uncoded systems operate): At low BER, for example, at BER 10^{-9}, it only takes a few highly distorted bits within a bit pattern to significantly affect the overall BER of the pattern or to even produce a noticeable error floor, whereas at poor BER, for example, at 10^{-3}, the incremental contribution of a few distorted bits to the overall BER of the pattern is negligible. For the case of in-band crosstalk, the BER-dependence of power penalties is visualized in Figure 7.37a, showing the BER as a function of received signal power for a range of different SIRs, measured with a 10-Gb/s rated DML at 1550 nm (eye diagram of Figure 7.37b, ~7.5-dB extinction) according to the setup in Figure 7.37c. In the measurements, we used two optical attenuators to set SIR and received signal power, respectively. A polarization controller was used to achieve worst-case interference conditions by co-polarizing signal and interferer, which were decorrelated by a 10-km long spool of NZDSF. Since in-band crosstalk constitutes a source of noise rather than distortion, we used our pin-receiver front-end, without equalization but with enabled FEC.

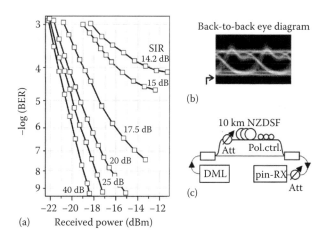

FIGURE 7.37 (a) Measured BER as a function of received signal power for different SIRs using a 10-Gb/s rated DML at 1550 nm. (b) Back-to-back eye diagram of the 10-Gb/s laser without interferer. (c) Measurement setup, incorporating two attenuators to set SIR and received power, a polarization controller to align signal and interferer, and 10 km of NZDSF to decorrelate signal and interferer. (Taken from Winzer, P.J., Fidler, F., Thiele, H.J., Matthews, M., Nelson, L.E., Sinsky, J.H., Chandrasekhar, S., Winter, M., Castagnozzi, D., Stulz, L.W., and Buhl, L.L., *J. Lightwave Technol.*, 23(1), 203–210, 2005. With permission.)

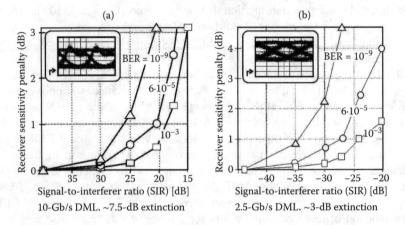

FIGURE 7.38 10-Gb/s receiver sensitivity penalties vs. SIR at different target BERs, measured at 1550 nm using the setup of Figure 7.37c. (a) 10-Gb/s rated DML with ~7.5-dB extinction and (b) 2.5-Gb/s rated DML with ~3-dB extinction. (Taken from Winzer, P.J., Fidler, F., Thiele, H.J., Matthews, M., Nelson, L.E., Sinsky, J.H., Chandrasekhar, S., Winter, M., Castagnozzi, D., Stulz, L.W., and Buhl, L.L., *J. Lightwave Technol.*, 23(1), 203–210, 2005. With permission.)

Figure 7.38 shows the BER-dependent sensitivity penalties as a function of SIR, both for the (a) 10-Gb/s rated DML and for the (b) 2.5-Gb/s rated DML at 1550 nm. Both DMLs exhibit a significantly increased tolerance to in-band crosstalk at FEC channel error ratios [17,18,66]. In Figure 7.38a, the 1-dB penalty is pushed from 26 to 19 dB when going from BER = 10^{-9} to 10^{-3}, which in the above example increases the tolerable span attenuation to 8 dB (or 40 km at 1550 nm). Since the data ER has a strong impact on the beat-noise-induced penalties caused by in-band crosstalk [63], the poor extinction (~3 dB) of the 2.5-Gb/s rated DML operated at 10 Gb/s (see Figure 7.38b) artificially reduces the allowable SIR, and is not too representative of realistically deployable 10-Gb/s systems.

To prove that fully bidirectional CWDM transmission over meaningful distances is possible with the help of FEC, provided that 10-Gb/s rated DMLs are available, we carried out the single-channel version of the experiment shown in Figure 7.39, using a 10-Gb/s rated DML at 1550 nm and an ER of ~7.5 dB. After going through a CWDM multiplexer, we split off a fraction of the signal by means of a broadband optical coupler prior to entering the span. At the receiver side, we used a second optical coupler to feed in the signal, emulating the counter-propagating channel. Appropriate selection of the couplers in combination with an attenuator at the input to the span set the ratio of signal powers entering the span from each side. A polarization controller was used to align the reflections from backward traffic with the West-East propagating detected signal for maximum impairment. Although we used various types of angled and straight fiber connectors, the dominant source of in-band crosstalk was Rayleigh backscatter from backward traffic. At a transmission distance of 60 km, we were able to close the link,

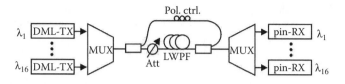

FIGURE 7.39 Experimental setup to generate backward traffic and emulate the worst-case scenario for a fully bidirectional CWDM system.

using the pin receiver with equalization and FEC, and measuring SIR = 17.6 dB for equal launch powers in both traffic directions. However, due to the lack of availability of full-band CWDM sources at 10 Gb/s with reasonably good ERs, a full-band demonstration of bidirectional CWDM transport and equal signal powers in each direction could not be done.

REFERENCES

1. Draft New Recommendation ITU-T G694.2: *Spectral Grids for WDM Applications*: SG 15, 2002.
2. Sugie, T., Nakamura, H., Tsujikawa, K., and Nonoyama, Y., Wide Passband WDM (WWDM) Technology and its Application to Access Networks for Broadband Services, Proceedings NOC 2001, pp. 37–43.
3. Chang, K.H., Kalish, D., and Pearsall, M., New Hydrogen Aging Loss Mechanism in the 1400 nm Window, Proceedings OFC'99, paper PD-22, 1999.
4. Das, S.K., Mysore, S.M., Villa, R.A., Thomas, J.J., Thiele, H.J., Jiao, L., and Nelson, L., 40 Gb/s (16 x 2.5 Gb/s) Full Spectrum Coarse WDM Transmission over 75km Low Water Peak Fiber for Low-Cost Metro and Cable TV Applications, Proceedings NFOEC'02, pp. 881–887.
5. Ziari, M., Mathur, A., Jeon, H., Booth, I., and Lang, R.J., Fiber-grating Based Dense WDM Transmitters, Proceedings NFOEC' 97, San Diego, CA, pp. 503–512.
6. Sogawa, I., Kaida, N., Iwai, K., Takagi, T., Nakayabashi, T., and Sasaki, G., Study on Full Spectrum Directly Modulated CWDM Transmission of 10 Gb/s per Channel over Water-peak suppressed Non-zero Dispersion Shifted Fiber, Proceedings ECOC'02, Copenhagen, paper 8.2.1
7. Sato, K., Kuwahara, S., Miyamoto, Y., and Shimizu, N., 40 Gb/s direct modulation of distributed feedback laser for very-short-reach optical links, *Electron. Lett.*, 38(15), 816–817, 2002.
8. Sato, K., Kuwahara, S., Hirano, A., Yoneyama, M., and Miyamoto, Y., 4 x 40 Gbit/s Dense WDM Transmission over 40-km SMF Using Directly Modulated DFB Lasers, Proceedings ECOC'04, Stockholm, paper We 1.5.7.
9. Thiele, H.J., Winzer, P.J., Sinsky, J.H., Stulz, L.W., Nelson, L.E., and Fidler, F., 160-Gb/s CWDM capacity upgrade using 2.5-Gb/s rated uncooled directly modulated laser, *Photon. Technol. Lett.*, 16(10), 2004.
10. Thiele, H.J., Nelson, L.E., and Das, S.K., Capacity-enhanced coarse WDM transmission using 10 Gbit/s sources and DWDM overlay, *Electron. Lett.*, 39(17), 2003.

11. Roka, R., The Utilization of the DWDM/CWDM Combination in the Metro/Access Networks, Proceedings on Mobile Future and Symposium on Trends in Communications, 2003. SympoTIC '03, pp. 160–162.
12. Antosik, R., Super Channel Architectures for In-service Capacity Expansion of CWDM/DWDM Systems, Proceedings ICTON 2003, paper We.C.8, pp. 84–86.
13. Tomkos, I., Hesse, R., Antoniades, N., and Boskovic, A., Impact of filter concatenation effects on the performance of metropolitan area optical networks utilizing directly modulated lasers, *Photon. Technol. Lett.*, 13(9), 2001.
14. Mizuochi, T., Kubo, K., Yoshida, H., Fujita, H., Tagami, H., Akita, M., and Motoshima, K., Next Generation FEC for Optical Transmission Systems, Proceedings OFC'03, paper ThN1.
15. Winzer, P.J., Essiambre, R.-J., and Chandrasekhar, S., Dispersion-tolerant Optical Communication Systems, Proceedings ECOC'04, paper We2.4.
16. Liu, X., Zheng, Z., Kaneda, N., Wei, X., Tayahi, M., Movassaghi, M., Radic, S., Chandrasekhar, S., and Levy, D., Enhanced FEC OSNR gains in dispersion-uncompensated 10.7-Gb/s duobinary transmission over 200-km SSMF, *Photon. Technol. Lett.*, 15(8), 1162–1164, 2003.
17. Chandrasekhar, S., Buhl, L.L., and Zhu, B., Performance of Forward Error correction Coding in the Presence of In-band Cross-talk, Proceedings OFC'02, paper WP1.
18. Radic, S., Vukovic, N., Chandrasekhar, S., Velingker, A., and Srivastava, A., Forward error correction performance in the presence of Rayleigh-dominated transmission noise, *Photon. Technol. Lett.*, 15(2), 326–328, 2003.
19. Winters, J.H. and Gitlin, R.D., Electrical signal processing techniques in long-haul fiber optic systems, *IEEE Trans. Commun.*, 38(9), 1439–1453, 1990.
20. Buchali, F., Bulow, H., and Kuebart, W., Adaptive Decision Feedback Equalizer for 10 Gb/s Dispersion Mitigation, Proceedings ECOC'00, vol. 2, pp. 101–102.
21. Castagnozzi, D., Digital Signal Processing and Electronic Equalization (EE) of ISI, Proceedings OFC'04, paper WM6.
22. Haunstein, H.F., Sticht, K., Dittrich, A., Lorang, M., Sauer-Greff, W., and Urbansky, R., Implementation of Near Optimum Electrical Equalization at 10 Gbit/s, Proceedings ECOC'00, vol. 3, pp. 223–224.
23. Buchali, F. and Bulow, H., Adaptive PMD compensation by electrical and optical techniques, *J. Lightwave Techn.*, 22(4), 1116–1126, 2004.
24. Kanter, G.S., Capofreddi, P., Behtash, S., and Gandhi, A., Electronic Equalization for Extending the Reach of Electro-absorption Based Transponders, Proceedings OFC '03, vol. 2, pp. 476–477.
25. Woodward, S.L., Huang, S.Y., Feuer, M.D., and Borodisky, M., Demonstration of an electronic dispersion compensator in a 100-km 10-Gb/s ring network, *Photon. Technol. Lett.*, 15(6), 867–869, 2003.
26. Winzer, P.J., Fidler, F., Matthews, M.J., Nelson, L.E., Chandrasehkar, S., Buhl, L.L., Winter, M., and Castagnozzi, D., Electronic Equalization and FEC Enable Bidirectional CWDM Capabilities of 9.6 Tb/s-km, Proceedings OFC'04, paper PDP7.
27. Fludger, C.R.S., Whiteaway, J.E.A., and Anslow, P.J., Electronic Equalization for Low Cost 10 Gb/s Directly Modulated Systems, Proceedings OFC'04, paper WM7.
28. Cavallari, M., Fludger, C.R.S., and Anslow, P.J., Electronic Signal Processing for Differential Phase Modulation Formats, Proceedings OFC'04, paper TuG2.
29. Feuer, M.D., Huang, S.Y., Woodward, S.L., Coskun, O., and Borodisky, M., Electronic dispersion compensation for a 10-Gb/s link using a directly modulated laser, *Photon. Technol. Lett.*, 15(12), 1788–1790, 2003.

30. Faerbert, A., Langenbach, S., Stojanovic, N., Dorschky, C., Kupfer, T., Schulien, C., Elbers, J.-P., Wernz, H., Griesser, H., and Glingener, C., Performance of a 10.7 Gb/s Receiver with Digital Equaliser Using Maximum Likelihood Sequence Estimation, Proceedings. ECOC'04, paper Th4.1.5.
31. Silver, M., Booij, W.E., Malik, S., Galbraith, A., Uppal, S., McBrien, P.F., Berry, G.M., Ryder, P.D., Chandler, S.J., Atkin, D.M., Chan, S., Harding, R., and Ash, R.M., Wide temperature -20°C–95°C operation of an uncooled 2.5Gb/s 1300-nm DFB laser, *IEEE Photon. Technol. Lett.*, 14(6), 741–743, 2002.
32. Morton, P.A., Shtengel, G.E., Tzeng, L.D., Yadvish, R.D., Tanbun-Ek, T., and Logan, R.A., 38.5 km error free transmission at 10 Gbit/s in standard fibre using a low chirp, spectrally filtered, directly modulated 1.55 μm DFB laser, *Electron. Lett.*, 33(4), 310–311, 1997.
33. Takiguchi, T., Hanamaki, Y., Kadowaki, T., Tanaka, T., Takemi, M., Mihashi, Y., Omura E., and Tomita, N., 1.3μm AlGaInAs Lasers with 12.0 GHz Relaxation Oscillation Frequency at 85°C for Gigabit Ethernet System, Proceedings ECOC 2000, vol. 1, pp. 127–128.
34. Berry, G., Burns, G., Charles, P., Crump, P., Davies, A., Ghin, R., Holm, M., Kompocholis, C., Massa, J., Ryder, P., Taylor, A., Agresti, M., Bertone, D., Fang, R.Y., Gotta, P., Magnetti, G., Meneghini, G., Paoletti, R., Rossi, G., Valenti, P., and Meliga, M., 100 deg C, 10 Gb/s Directly Modulated InGaAsP DFB Lasers for Uncooled Ethernet Applications, Proceedings OFC 2002, talk ThF1, pp. 415–416.
35. Nakahara, K., Tsuchiya, T., Kitatani, T., Shinoda, K., Taniguchi, T., Fujisaki, S., Kikawa, T., and Nomoto, E., 1.3μm InGaAlAs Directly Modulated MQW RWG DFB Lasers Operating over 10Gb/s and 100°C, OFC 2004, paper ThD1.
36. Takagi, K., Shirai, S., Tatsuoka, Y., Watatani, C., Ota, T., Takiguchi, T., Aoyagi, T., Nishimura, T., and Tomita, N., 120°C 10-Gb/s uncooled directly modulated 1.3μm AlGaInAs MQW DFB laser diodes, *Photon. Technol. Lett.*, 16(11), 2415–2417, 2004.
37. Corvini, P.J. and Koch, T.L., Computer simulation of high-bit-rate optical fiber transmission using single-frequency lasers, *J. Lightwave Technol.*, 5(11), 1591–1595, 1987.
38. Kaminov, I.P. and Koch, T.L., *Optical Fiber Telecommunications IV-A*, Academic Press, 2002, chap. 12.
39. Shimizu, J., Yamada, H., Murata, S., Tomita, A., Kitamura, M., and Suzuki, A., Optical confinement factor dependence of K-factor, differential gain, and nonlinear gain in 1.55-μm MQW and strained MQW lasers, *Electron. Devices*, 38(12), 2698–2705, 1991.
40. Henry, C. H., Theory of the linewidth of semiconductor lasers, *J. Quantum. Electron.*, QE-18, 259–264, 1982.
41. Cartledge, J.C. and Burley, G.S., The effects of laser chirping on lightwave system performance, *J. Lightwave Technol.*, 7(3), 568–573, 1989.
42. Petermann, K., *Laser Diode Modulation and Noise*, Kluwer Academic Publishers, Dordrecht, 1991.
43. Saunders, R.A., King, J.P., and Hardcastle, I., Wideband chirp measurement technique for high bit-rate sources, *Electron. Lett.*, 30(16), 1336–1337, 1994.
44. Yamanaka, T., Yoshikuni, Y., Yokoyama, K., Lui, W., and Seki, S., Theoretical study on enhanced differential gain and extremely reduced linewidth enhancement factor in quantum-well lasers, *J. Quantum Electronics*, 29(6), 1609–1616, 1993.
45. Agrawal, G.P., *Nonlinear Fiber Optics*, 2nd edition, Academic Press, 1995.
46. Morton, P.A., Tanbunek, T., Logan, R.A., and Chand, N., Packaged 1.55-Mu-M DFB laser with 25 GHz modulation bandwidth, *Electron. Lett.*, 30(24), 2044–2046, 1994.

47. Hamakawa, A., Kato, T., Sasaki, G., and Shigehara, M., A Four Channel Multi-wavelength Fiber Grating External Cavity Laser Array, Proceedings OFC'96, paper ThM3, pp. 297–299.
48. Timofeev, F.N., Bayvel, P., Mikhailov, V., Gambini, P., Wyatt, R., Kashyap, R., Robertson, M., Campbell, R.J., and Midwinter, J.E., Low Chirp, 2.5Gb/s Directly Modulated Fiber Grating Laser for WDM Networks, Proceedings OFC'96, paper ThM1, pp. 296–298.
49. Paoletti, R., Meliga, M., Rossi, G., Scofet, M., and Tallone, L., 15-GHz modulation bandwidth, ultralow chirp, 1.55μm directly modulated hybrid distributed bragg reflector (HDBR) laser source, *Photon. Technol. Lett.*, 10(12), 1691–1693, 1998.
50. Shigematsu, M., Tanaka, M., Okuno, T., Hashimoto, J., Kawabata, Y., Takahashi, S., Nakanishi, H., Yamaguchi, A., Shibata, T., Inoue, A., Katsuyama, T., Nishimura, M., and Hyashi, H., Amplified Coarse WDM System Employing Uncooled Fiber Bragg Grating Lasers with 600 GHz Channel Spacing, Proceedings OFC'03, paper MF77, pp. 96–97.
51. Gruener-Nielsen, L. and Edvold, B., Status and Future Promises For Dispersion Compensating Fibers, Proceedings ECOC'02, paper 6.1.1.
52. Thiele, H.J., Das, S.K., Boncek, R., and Nelson, L.E., Economics and Performance of 10-Gb/s Metro Transport Over Mixed Fiber Plant of G.655 NZDF and G.652.C Zero Water-peak Fibers, Proceedings OECC 2003, paper 16E1-4, pp. 717–718.
53. Winzer, P.J., Fidler, F., Thiele, H.J., Matthews, M., Nelson, L.E., Sinsky, J.H., Chandrasekhar, S., Winter, M., Castagnozzi, D., Stulz, L.W., and Buhl, L.L., 10-Gb/s upgrade of bidirectional CWDM systems using electronic equalization and FEC, *J. Lightwave Technol.*, 23(1), 203–210, 2005.
54. Proakis, J.G., *Digital Communications*, McGraw-Hill, 1995.
55. Winters, J.H. and Gitlin, R.D., Electrical signal processing techniques in long-haul fiber-optic systems, *IEEE Trans. Comm.*, 38(9), 1439–1453, 1987.
56. Bulow, H. and Thielecke, G., Electronic PMD Mitigation: From Linear Equalization to Maximum Likelihood Detection, Proceedings OFC'01, paper WAA3-3.
57. Haunstein, H.F., Sticht, K, Sauer-Greff, W., and Urbansky, R., Design of Near Optimum Electrical Equalizers for Optical Transmission in the Presence of PMD, Proceedings OFC'01, paper WAA4-1.
58. Corvini, P.J. and Koch, T., Computer simulation of high-bit-rate optical fiber transmission using single-frequency lasers, *J. Lightwave Tech.*, 5(11), 1591, 1987.
59. Mohrdiek, S., Burkhard, H., Steinhagen, F., Hillmer, H., Losch, R., Schlapp, W., and Gobel, R., 10-Gb/s standard fiber transmission using directly modulated 1.55-μm quantum-well DFB lasers, *Photon. Technol. Lett.*, 7(11), 1357–1359, 1995.
60. Radic, S., Chandrasekhar, S., Srivastava, A., Kim, H., Nelson, L., Liang, S., Tai, K., and Copner, N., Dense interleaved bidirectional transmission over 5x80 km of nonzero dispersion-shifted fiber, *Photon. Technol. Lett.*, 14(2), 218–220, 2002.
61. Das, S.K. and Harstead, E.E., Beat interference penalty in optical duplex transmission, *J. Lightwave Technol.*, 20(2), 213, 2002.
62. Goldstein, E.L., Eskildsen, L., and Elrefaie, A.F., Performance implications of component crosstalk in transparent lightwave networks, *Photon. Technol. Lett.*, 6(5), 657–660, 1994.
63. Rasmussen, C.J., Liu, F., Pedersen, R.J.S., and Jorgensen, B.F., Theoretical and Experimental Studies of the Influence of the Number of Cross-talk Signals on the

Penalty Caused by Incoherent Optical Cross-talk, Proceedings OFC'99, talk TuR5, pp. 258–260, 1999.
64. Liu, F., Rasmussen, C.J., and Pedersen, R.J.S., Experimental verification of a new model describing the influence of incomplete signal extinction ratio on the sensitivity degradation due to multiple interferometric cross-talk, *Photon. Technol. Lett.*, 11(1), 137–139, 1999.
65. Fludger, C.R.S., Zhu, Y., Handerek, V., and Mears, R.J., Impact of MPI and modulation format on transmission systems employing distributed Raman amplification, *Electron. Lett.*, 37(15), 970–972, 2001.
66. Kaman, V., Zheng, X., Pusarla, C., Keating, A.J., Helkey, R.J., and Bowers, J.E., Mitigation of optical crosstalk penalty in photonic cross-connects using forward error correction, *Electron. Lett.*, 39(8), 678–679, 2003.

8 CWDM in Metropolitan Networks

Jim Aldridge

CONTENTS

8.1 Metro Network 252
    8.1.1 Definitions 252
    8.1.2 CWDM in the Metro 253
    8.1.3 CWDM Network Building Blocks 254
        8.1.3.1 Multiplexer and Demultiplexer 254
        8.1.3.2 Add/Drop Multiplexer 256
        8.1.3.3 Cross-Connect 257
        8.1.3.4 Transceiver 257
    8.1.4 Rings Using CWDM 258
8.2 Key Issues for Optical Network Engineers 259
8.3 Metro CWDM Applications 263
    8.3.1 Disaster Recovery 263
    8.3.2 Adding CWDM to Single Wavelength Networks 264
    8.3.3 Scalability with DWDM Over CWDM 265
    8.3.4 Future Applications 266
References 267

Coarse wavelength division multiplexing (CWDM) already started to show up in the metro access network in late 1999 [1] before finally the new ITU-T G.694-2 standard went into effect in 2002 [2]. The metro segment was an interesting new telecommunication market that was growing rapidly and needed a low cost "plug and play" solution. What made the metro especially interesting was that it was rich in a wide variety of communication protocols that all evolved in parallel as customer demand for bandwidth grew. Further, the metro network is a lucrative target for service providers who could provide the bandwidth quickly to medium to large businesses. The long haul network had been built and many users in cities around metropolitan areas wanted an internet access on-ramp. Dense wavelength division multiplexing (DWDM) was the first consideration for providing the needed capacity. DWDM had been a proven technology to build out the long haul section of the network and was therefore the obvious first choice. However, DWDM proved too costly, even at short distances with no amplifiers. Then CWDM was introduced as a new possibility for low-cost multi-wavelength transmission. Small service providers hailed CWDM as the answer over DWDM. CWDM was regarded as a "poor man's" way of building a metro ring. Validation of the CWDM market came soon when multitudes of vendors appeared with CWDM pluggable transceivers and passive optical equipment.

8.1 METRO NETWORK

8.1.1 DEFINITIONS

To understand the metro network, a greater picture of the whole network is helpful. The entire fiber network is made up of three main segments. Figure 8.1 shows these three main segments of the fiber network in North America today: Long-haul, core or metro regional and metro access network. Each main segment is defined by several distinctions: distance, number of wavelengths, data rate, topology, protocols, number of access points (nodes) or immediate customers and equipment. Furthermore, each area of the fiber optical network has its own set of special requirements.

Long-haul networks extend traffic point-to-point greater than 400 km, carrying 64 to 160 wavelengths of DWDM. The long-haul has the fat pipes in the network tasked to shuttle data across the country to central hubs. The equipment used consists of SONET platforms that carry data at 10 or 40 Gb/s. The data path requires amplifiers to regenerate the optical signal from hub to hub. Today, the long haul has been mostly built out and expects low to moderate growth in the future.

Core networks, also known as regional networks, are the backbone of the fiber network. Core networks are made of mesh rings that carry 16 to 32 DWDM

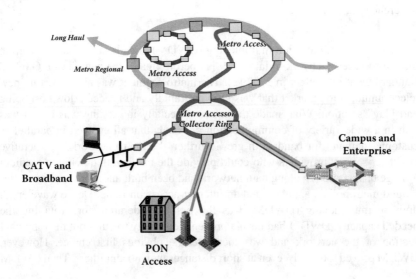

FIGURE 8.1 A fiber optics network with long-haul, metro regional and metro access.

wavelengths and connect central offices together. Traffic in the core is likely to be shipped to any of the routes out of each core location, therefore requiring the network to be configured as a mesh. Distances in the core are 50 to 300 km in distance. The equipment used consists of SONET or Ethernet platforms that carry data at 2.5 to 10 Gb/s. Recently, the core network is experiencing growth as the result of a need for more bandwidth in access networks from Video on Demand (VoD) and internet protocol television (IPTV).

The *metro access network* is the on-ramp point for end users. It is the largest area of potential growth and innovation. The metro has the characteristics of short distance coupled with the requirement of multi-service equipment capable of a wide variety of protocols. Topologies consist of short rings and point-to-point networks carrying 4 to 16 wavelengths. Data rates vary from 100 Mb/s to 2.5 Gb/s. Protocols used are T1, SONET, Fiber Channel, Ethernet, ATM, and ESCON.

CWDM blends well into the metro network. Its usefulness is threefold. First, because installing fiber is too expensive in the metro, CWDM allows expansion of bandwidth by simply aggregating additional wavelengths into an existing pair of fibers between two buildings. Secondly, multiple protocols can be transported transparently by using a CWDM solution. Finally and most importantly, CWDM is a low-cost "pay as you grow" technology. In Chapter 9 and 10, the role of CWDM in the access network is discussed for cable television (CATV) and fiber to anywhere (FTTx)/passive optical network (PON) applications, respectively.

8.1.2 CWDM in the Metro

CWDM was, from the start, designed for the metropolitan area network. Enterprises and service providers are keen on CWDM because of its capability to scale as the network grows. Further, to implement a CWDM network is a trivial exercise. CWDM allows for a low-cost WDM architecture with simple interconnections to make it easy to construct point-to-point and ring networks [3–5]. The topologies used to build optical are in general:

- Mesh with multiple links interconnected at nodes
- Ring with links and add/drop functions
- Point-to-multipoint for broadcasting
- Point-to-point

The key topologies used particularly in the metro networks are ring and point-to-point. Mesh and multipoint, which are used mostly in core networks, are costly to implement and, therefore, are seldom used in metro networks. A CWDM point-to-point network consists of a simple connection from one building to another building via a pair of fiber optic cables. A CWDM ring involves connecting three or more buildings together via separate links, as schematically shown in Figure 8.2.

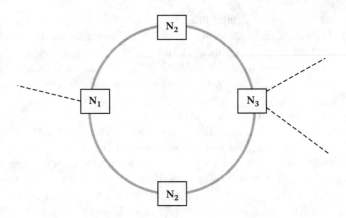

FIGURE 8.2 CWDM ring topology using four nodes $N_1...N_4$ interconnected via links.

The basic building blocks required in the nodes of Figure 8.2 to construct CWDM network rings and point-to-point links are the MUX (multiplexer), the DEMUX (demultiplexer), and the OADM (add/drop multiplexer), which are all discussed in Chapter 4. Also, pluggable transceivers that are investigated in Chapter 3 are needed at the end points of each optical path. While those two chapters cover the function of the network elements in much detail, we remain here on a network level and solely focus on the function of those components necessary for CWDM networking.

8.1.3 CWDM Network Building Blocks

8.1.3.1 Multiplexer and Demultiplexer

The passive CWDM MUX building block shown in Figure 8.3 aggregates wavelengths of light from several transmitter sources (TX) and transmits the combined light into one fiber. Each wavelength of light remains unchanged and transparent in the presence of neighboring wavelengths. Typically, CWDM MUX and DEMUX modules are designed with a minimum of 4 channels to a maximum of 16 channels. Eight channel modules are commonly used in a configuration for all channels from 1470 to 1610 nm. Typical signal loss through a TFF-based

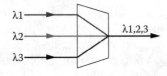

FIGURE 8.3 CWDM MUX at Tx side of the link shown with three different channels having a center wavelength $\lambda_1...\lambda_3$.

CWDM in Metropolitan Networks

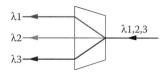

FIGURE 8.4 CWDM DEMUX at Rx side of the link shown with three different channels having a center wavelength $\lambda_1 \ldots \lambda_3$.

MUX or DEMUX is about 2 to 4 dB for each wavelength. Isolation between adjacent channels is a minimum of 30 dB (see chapter 4).

At the receive end, the demultiplexer does the reverse of the multiplexer by separating out the different combined light sources (Rx), as shown in Figure 8.4. The properties of the DEMUX are similar to the MUX. In the case of bidirectional transmission, each site contains both MUX and DEMUX components.

The typical metro point-to-point link would employ a MUX and DEMUX pair to terminate the wavelengths at each end. This also applies to the CWDM ring since it can be divided into several interconnected point-to-point links. Figure 8.5 shows a typical point-to-point metro link with 50 km of fiber. A pair of fibers, one fiber for the wavelengths going West-East and one fiber for the wavelengths going East-West, carries the data traffic between the two end-points. The ports of the multiplexer are connected to wavelength-selective CWDM transceivers located in the communications equipment. Note that commonly two separate fibers are used for this bidirectional transmission meaning that the total equipment needed is twice as much as in the case of unidirectional transmission, for example, West to East. Therefore, the choice of CWDM wavelengths for each direction is

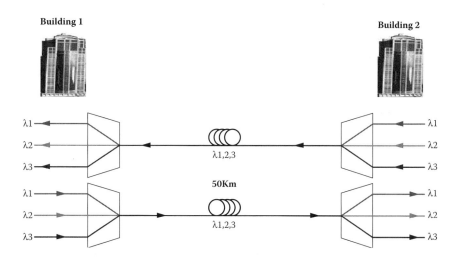

FIGURE 8.5 Bidirectional CWDM point-to-point connection between two buildings.

8.1.3.2 Add/Drop Multiplexer

The second basic building block is the passive CWDM OADM. The OADM is located in a ring network and drops (Rx) wavelengths from the ring and adds (Tx) wavelengths to the ring or straight line transmission link. All other wavelengths pass through the OADM unaffected as express channels. The OADM example shown in Figure 8.6a adds and drops a single wavelength. When all channels $\lambda_1 \ldots \lambda_N$ carry traffic, only the dropped channel λ_1 can be replaced with a new channel having the same wavelengths. Due to the colored ports of the multiplexer components used, the correct CWDM wavelengths must be known for the add/drop operation. Typical signal loss through the add/drop ports is

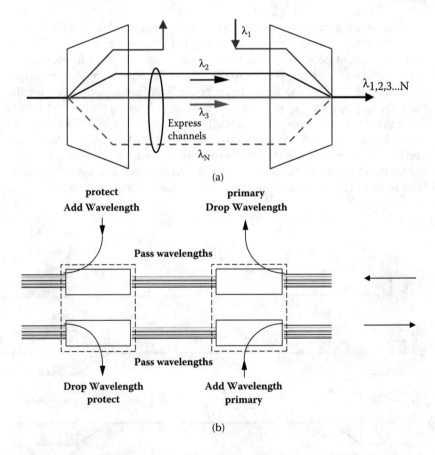

FIGURE 8.6 (a) OADM schematic with a single add/drop channel plus express channels, the OADM is functionally composed of a single DEMUX-MUX pair; (b) usage of a dual OADM in a protected ring configuration.

CWDM in Metropolitan Networks

1.5 to 2.5 dB. The pass through loss from East to West is typically 1 to 2 dB per port. Isolation between adjacent channels is a minimum of 30 dB while a superior product would have 50 dB of isolation. CWDM OADMs are typically made with either single- or four-channel configurations.

The use of the OADM is illustrated in Figure 8.6b for a protected ring network. In a protected ring, the protected fiber connection is accompanied by a second fiber where the redundant information propagates in the opposite direction. The main idea behind this scheme is that in the case of a fiber cut the information is still accessible at the sites via the second fiber. Therefore, the add/drop will exist twofold, as a primary OADM and a protecting OADM. We have seen that OADMs provide access to a single or even more wavelengths of a wavelength-multiplexed system increasing the possibility of networking. Although this greatly improves the flexibility for CWDM, the insertion loss of these devices poses a challenge on the design of rings as CWDM uses no optical amplification to overcome losses.

8.1.3.3 Cross-Connect

While in the previous example the multiplexer have been used in pairs to achieve add/drop function, also higher order nodes with more than two fibers are possible by using a multiplexer at the end of each fiber and then an appropriate switch fabric to provide the connections on a wavelengths basis among the different incoming fibers. The complexity and cost is high so that we are not considering those network elements in this context.

8.1.3.4 Transceiver

The previously discussed building blocks for creating CWDM networks were passive and based on wavelength-selective filters. Instead of using discrete uncooled laser components and PIN receivers for transmission at the different CWDM wavelengths, these optical components are increasingly integrated into transceivers. These compact transceivers are particularly useful when operating on bidirectional links since each site comprises a transmitter as well as a receiver. Laser, receiver diode, and relevant electronics for driving the laser and shaping the received signal are integrated in a small form factor module with a standardized interface. The transceiver modules are colored, that is, the transmitter unit has a characteristic wavelength between 1310 and 1610 nm while the receiver will work for any of those wavelengths. The transceivers can directly be driven with the data to be transmitted that is modulated onto the particular wavelengths.

CWDM transceivers typically use directly modulated DFB lasers operating at 2.5 Gb/s and PIN receivers with a receiver module and decision circuit. The modulated output power ranges from 0 to 3 dBm, although it could be reduced at elevated temperatures since no active cooling of the devices is available due to lower-cost design. The (PIN) receiver sensitivity of the transceivers is around $-24\ldots 26$ dB so that a link budget of at least 24 dB should be available, which can be used to accommodate both the insertion loss of components (multiplexer, fiber) as well as penalties due to the interaction of fiber dispersion and laser chirp. At lower bit-rates, the link budget is increasing up to 32 dB at 1.25 Gb/s. Therefore, even with nodes present, a ring circumference of 50 km is possible.

8.1.4 Rings Using CWDM

A full ring with a head end from where all channels are originating can be constructed with the MUX/DEMUXs and OADM building blocks discussed before. In the network diagram shown in Figure 8.7, a four-node ring is constructed with OADMs in the building node locations and a pair of MUX/DEMUXs at the head end. The wavelengths (clockwise: West-East, counterclockwise: East-West) originate at the head end or central office (node 1) and for simplicity only a single wavelength path is shown here. Each OADM adds/drops one wavelength in each direction (East and West) from the ring and passes on the remaining wavelengths

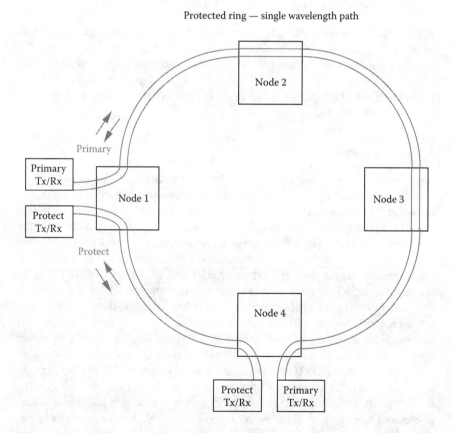

FIGURE 8.7 Add/drop multiplexer, multiplexer and transceiver used to create a CWDM ring with four nodes and protection.

CWDM in Metropolitan Networks

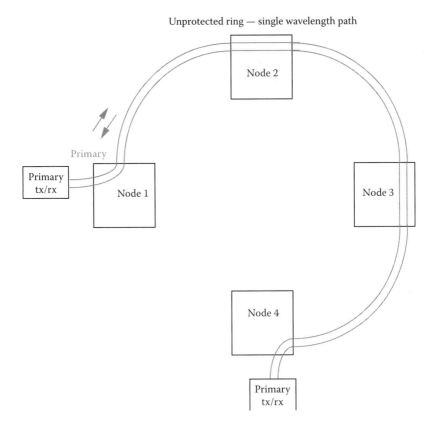

FIGURE 8.8 Add/drop multiplexer, multiplexer and primary transceiver used to create a CWDM ring with four nodes and no protection.

to the next OADM. For example, a wavelength λ starts in the West-East direction from the head end through the East MUX/ DEMUX of the node, then passes through two OADMs in building 1 and 2 (node 2 and 3), and is finally dropped at building 3 (node 4). In the opposite East-West direction from the head end, the same λ starts westbound from the West MUX/DEMUX and is added/dropped immediately at building 3 with no intermediate OADMs. The head end dual MUX/DEMUX setup thus allows for two paths into each building (plus reverse directions) creating redundancy in case of a path failure. If this redundancy feature or protection is not needed, a less costly single MUX/DEMUX at the head end in one direction would look like Figure 8.8.

8.2 KEY ISSUES FOR OPTICAL NETWORK ENGINEERS

Service provider and enterprise network engineers have several key issues in building metro networks. Among those issues is scalability, protection, transparency, and transport connectivity. The plan for most metro network engineers is that their network will need more bandwidth in the future and, because of cost, adding more new fiber in the metro network is not a consideration. CWDM is a scalable technology that allows for addition of more bandwidth with a "pay as you grow" philosophy. The challenges for upgrading CWDM transmission links are addressed in Chapter 7. Here, we want to focus on some issues important on the network level:

- Scalability by increased channel count
- Network protection by redundancy
- Transport connectivity design requirements
- Transparency for multiple transmission protocols

Figure 8.9 illustrates the simple concept of *scaling CWDM* to multiple wavelengths. A point-to-point metro link can start with just two CWDM wavelengths λ_1 and λ_2 plugged into a 16-channel CWDM passive MUX/DEMUX. As more bandwidth is needed, additional CWDM wavelengths λ_3 through λ_{16} can be added for expansion where channels are often upgraded in groups of 4 to 4, 8 or 16 wavelengths; such as in a Storage Area Networks (SANs) application requiring rapid bandwidth capability for disaster recovery applications. Other applications include wireless backbone networks and VoD networks. This is also discussed in Section 3.

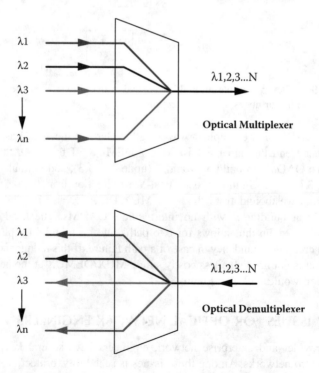

FIGURE 8.9 MUX/DEMUX scaling from 1 to 16 wavelengths for full-spectrum CWDM.

CWDM in Metropolitan Networks

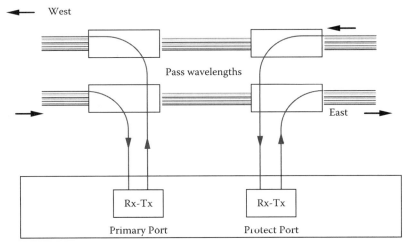

FIGURE 8.10 Node with OADM connected Ethernet switch.

As discussed earlier, *network protection* (see Figure 8.7) involves creating redundant paths through a network so that when a network failure occurs traffic does not stop. A network failure such as a fiber cut somewhere in a ring network can be a disaster if a redundant traffic path is not created in the network. The CWDM OADM basic is shown in Figure 8.10 connected to an Ethernet switch in a network ring with a path sending data in West-East direction (clockwise) and a separate path sending data East-West (counterclockwise). This clever OADM implementation is a simple way to provide protection in a ring network. Ethernet switches provide alternate paths for data to travel via spanning-tree-type algorithms.

Figure 8.11 now illustrates an example of how protection would work via a CWDM OADM Ethernet switch combination with a fiber cut in the network. In the figure, the fiber cut occurs on the West path around the ring. The Ethernet switch detects the data loss from the CWDM transceiver from the West side and then routes data from the switch in the other direction around the ring. Traffic bandwidth is half until the fiber cut is repaired. The cost of this protection solution is the cost of one transceiver and an extra port in the OADM node.

Another aspect of a metro network (Table 8.1) is the *transparency* for a variety of protocols that may be offered throughout the network. Instead of converting each signal separately to have a common underlying transmission protocol, it is often more cost effective to assign each signal to a unique wavelength and aggregate these wavelengths in a CWDM system using a low-cost multiplexer. This flexibility comes from the fact that each wavelength can be treated optically as unique and transparent from all of the others. Figure 8.12 illustrates an example of aggregating multiple protocols from various sources and equipment each having a different protocol.

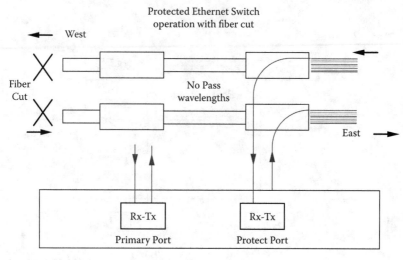

FIGURE 8.11 OADM connected to Ethernet with network fiber cut, compared to Figure 8.10.

Transport connectivity design requirements are common network issues that involve serious consideration. Among those issues critical to metro service providers are link budget and passive optic isolation. In the metro network, link budget and attenuation losses are the key specifications for establishing the maximum point-to-point distance. The link budget is the launch power of the fiber optic transceiver plus the receiver sensitivity. For example, the launch power and receiver sensitivity for most CWDM transceivers is +1.0 dBm and −29.0 dBm (APD receiver at 2.5 Gb/s), respectively, producing a link budget of 30 dB between transmitter and receiver.

TABLE 8.1
Overview of Common Network Properties and Usage of CWDM

	Long-haul	Metro Regional	Metro Access
Length scale (km)	>400	50–300	<50
Bit-rate (Gb/s)	10–40	2.5–10	0.1–2.5
Channel count	>40	<32	4–16
Transport	SONET	SONET, Ethernet	T1, SONET, Fiber Channel, Ethernet, ATM, ESCON
Topology	Point-to-point, nodes, OXC	Rings, point-to-point, (R)OADMs	Feeder (short link)
CWDM	No	Yes (partially)	Yes
Cost	High	Medium	Low

CWDM in Metropolitan Networks

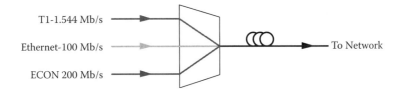

FIGURE 8.12 Transparency between signals with different protocols. At the far end, not shown in this figure, is a DEMUX that separates the wavelengths after transmission with the individual protocol traffic unaffected by the other signals in the same path.

The attenuation losses in the path between transmitter and receiver are defined by the losses of the fiber optic cable, passive fiber optic devices, and distribution connection points. In the following example, insertion losses of a CWDM optical channel in a MUX, DEMUX, or OADM with add/drop are 2.5 dB per device. The loss through an OADM pass through is 1.5 dB. Therefore, the typical passive fiber optic loss for the optical path in the four-node network of Figure 8.7 from the head end (node 1) to building 3 (node 4) is 8 dB (2.5 dB + 1.5 dB + 1.5 dB + 2.5 dB). By adding another 8 dB for distribution connection patch panels (2 dB per building for four buildings), the total losses, including both passive fiber optic and distribution connection patch panels, is 16 dB (8 dB + 8 dB). Networks with good fiber optic cable typically have an attenuation of 0.25 dB per kilometer not considering the wavelengths around 1310 nm where the loss is higher.

Hence, the maximum distance (for CWDM wavelengths around 1550 nm) from the head end node to node 4 is accomplished by a two-step process. First, subtract the attenuation losses (16 dB) from the total link budget of a CWDM transceiver (30 dB) to find the budget for fiber cable loss: 30 dB − 16 dB = 14 dB. Second, to get the maximum distance by dividing the budget for the fiber cable loss by the fiber cable loss coefficient (0.25 dB/km): 14 dB/0.25 dB/nm = 56 km.

8.3 METRO CWDM APPLICATIONS

As we have seen so far, CWDM optics is a simple and cost-effective approach to scale bandwidth and extend the size of networks in the metro segment. In the special applications shown in the following examples utilizing CWDM optics solves a few basic problems in metro networks.

8.3.1 DISASTER RECOVERY

Recently, SANs have had a high bandwidth requirement in remote data replication for disaster recovery. While this type of redundancy can be implemented within a facility, remote site redundancy adds an additional layer of protection from disasters and outages affecting an entire site. If a data storage server becomes unavailable, the secondary replications at an offsite facility make it easy to do point-in-time update replication. As shown in Figure 8.13, we can

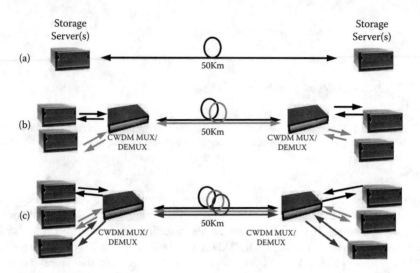

FIGURE 8.13 Path to extended storage networks with CWDM: (a) single fiber connection, (b) two channels, and (c) three channels multiplexed.

use CWDM to extend a SAN over distance. Figure 8.13a shows one storage array for remote backup to another storage array separated by 50 km. Figure 8.13b and 8.13c illustrate that adding a CWDM MUX/DEMUX at both ends scales the system to more bandwidth. Multiple storage servers can be interconnected where each pair is receiving a dedicated CWDM channel. This is a significant savings over the approach shown in Figure 8.13a where separate fiber links are used. Note that with CWDM pluggable optics and a CWM MUX/DEMUX the time for backup can be tailored to a "pay as you grow" approach. Hence, more wavelengths have to be added if we choose to increase storage space or want to decrease the required time of backup to the remote storage facility.

8.3.2 Adding CWDM to Single Wavelength Networks

Adding additional bandwidth to a SONET OC-192 link operating with a single 1310-nm wavelength can be an expensive and difficult task. Using CWDM pluggable transceivers, plus a CWDM MUX/DEMUX fitted with a 1310-nm port, the bandwidth addition transforms into a simple low-cost upgrade. Figure 8.14 illustrates the original SONET reference link with a 1310-nm wavelength.

The graceful upgrade is accomplished by inserting a CWDM MUX/DEMUX with a 1310-nm port into the link at both ends, as shown in Figure 8.15. The 1310-nm wavelength then passes through the CWDM infrastructure transparently with the CWDM wavelengths. Adding the multiplexers cannot be done without traffic interruption and also the link budget must be sufficient to accommodate the insertion loss of the multiplexer/demultiplexer pair. However, future upgrades

CWDM in Metropolitan Networks

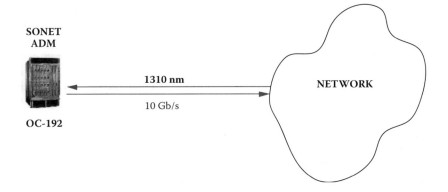

FIGURE 8.14 SONET link with a single wavelength at 1310 nm.

can be carried out without traffic interruption of existing channels, such as the 1310-nm SONET channel, provided there are enough free ports at the CWDM multiplexer.

8.3.3 Scalability with DWDM Over CWDM

It would seem that CWDM's scalability is limited to 16 wavelengths. Therefore, services for only 16 nodes in a ring are possible. However, scalability is possible to more than 64 wavelengths and a correspondingly larger number of users by expanding the basic architecture with DWDM. This approach promises a larger scalability than increasing the bit-rates of the CWDM channels, another option discussed in Chapter 7. The idea of using DWDM in the CWDM environment

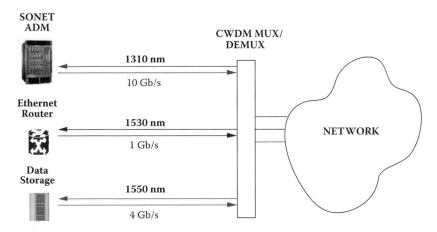

FIGURE 8.15 Adding CWDM wavelengths to a single wavelength link. Here, a common solution is shown for adding IP (Ethernet router) or fiber channel data services to a SONET link without a major expensive upgrade.

is rather simple: in the DWDM world, channels are spaced at 0.4 or 0.8 nm, whereas in CWDM, the wavelengths are 20 nm apart. Using the CWDM 20-nm grid, the channels in a CWDMUX/DEMUX are designed with a 12-nm passband, wide enough to fit 8 to 10 DWDM wavelengths that can substitute the original CWDM channel. Due to the narrow spacing and spectral usage of the entire passband of the filter-components, the DWDM channels require temperature-stabilized laser sources to avoid any wavelength drift, leading to channel crosstalk and signal drop. The principle of DWDM bands can be seen for Figure 8.15 where we added new services to the SONET link. That scaled link used three CWDM wavelengths for SONET, Ethernet, and fiber channel for data storage on the 1550-nm CWDM wavelength. If, in particular, the storage application requires more storage capacity and faster bandwidth, additional wavelengths could be added down the 1550-nm channel by inserting a DWDM MUX/DEMUX. Figure 8.16 illustrates this concept for the known link by adding DWDM on to CWDM to expand bandwidth. This is one of the clever approaches underlining what CWDM has given to the metro network world. It can be characterized by the statement "Build your network for your present requirements with a flexible architecture and expand as you grow."

8.3.4 Future Applications

Since the late 1990s, CWDM has delivered for everyone in the metro space the capability to build low-cost, transparent and scalable networks. CWDM optics

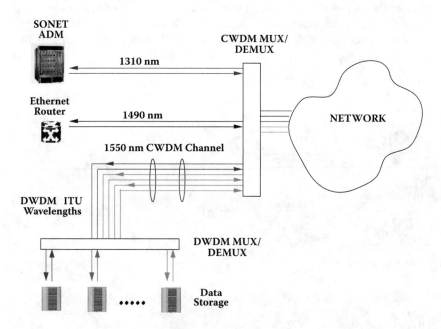

FIGURE 8.16 Using DWDM (lower part of figure) multiplexed on CWDM to provide a capacity upgrade for the network of Figure 8.15.

supplied in the form of pluggable transceivers and passive optic multiplexers have transformed the way metro networks are being built. The advances in device fabrication and the integration of CWDM with other parts of the network have helped to make CWDM a well-established player with a complete set of standards and many vendors from which to choose devices.

As the metro continues to grow and mature, two shifts are occurring in metro networks:

- First, capacity-demanding applications such as VoD are now driving the requirement for more bandwidth to the end-user. For now, this new trend is only going to be satisfied by 10G single wavelength transceivers. As the bandwidth demand increases further by numerous applications and a growing user base, CWDM 10G transceivers will start to show up to aggregate multiple 10G CWDM wavelengths onto the existing fiber infrastructure. As bandwidth demand increases even further, 10G DWDM on 10G CWDM solutions may start to emerge.
- Secondly, the price of CWDM equipment (transceivers and passive optical devices) is dropping significantly. This price erosion of CWDM solutions will eventually enable even the smallest enterprise to build CWDM networks. As a result, in the case of fiber exhaust, the CWDM solution will have a great advantage over adding more fiber to the network. With advances in design and manufacturing of DWDM fiber optical equipment, the price erosion seen in CWDM equipment will also start to happen in DWDM segment. A low-cost form of DWDM may soon compete with CWDM at the 1.25 to 2.5-G speeds. If the price gap between DWDM and CWDM mainly defined by the transceivers shrinks to the tipping point, then CWDM may be at risk to become an obsolete technology.

REFERENCES

1. Campbell, J., *Coarse WDM Makes Waves in Metro/Access Markets*, Laser Focus World, PennWell Publishing, pp. 24–27, 2000.
2. ITU-T recommendation G.694.2: Spectral grids for WDM applications: CWDM wavelength grid.
3. Levinson, F., *Coarse WDM Comes of Age for Local Carriers*, WDM Solutions, January 2003.
4. Aldridge, J., *The Best of Both Worlds*, Lightwave Europe, September 2002.
5. Kincade, K., CWDM Breathes Life into Metro, Access, and Enterprise Applications, *Laser Focus World*, March 2003.

9 CWDM in CATV/HFC Networks

Jim D. Farina

CONTENTS

- 9.1 Introduction...269
- 9.2 CATV Architecture...270
  - 9.2.1 HFC Forward Path Optics...273
  - 9.2.2 HFC Reverse Path Optics..273
  - 9.2.3 Analog Direct Modulation 274
- 9.3 Applications for CWDM..276
  - 9.3.1 Forward Path..276
  - 9.3.2 Forward Path Narrowcast Overlay................................276
  - 9.3.3 Return Path..278
  - 9.3.4 Limitations Due to SRS 279
- 9.4 CWDM in FTTx and Future Network Designs...........................283
- References 283

9.1 INTRODUCTION

Since the introduction of low-loss single mode fiber, there has been a constant effort to exploit the bandwidth of the medium for both digital and analog transmission systems. This has led to transmission systems operating around 1310 nm in the zero dispersion region of standard single mode fiber (SSMF) and later also to transmission in the C-band where the fiber loss reaches a minimum. Multi-channel dense wavelength division multiplexing (DWDM) transmission has increased the bandwidth requirement around 1550 nm extending to the neighboring S- and L-bands for additional channels. While DWDM uses narrow channel spacing of typically 0.4 to 0.8 nm, coarse wavelength division multiplexing (CWDM) follows a different path; here, the channel spacing is increased to 20 nm and the bandwidth covers the entire low-loss region of single mode fiber from 1270 or 1310 nm to 1610 nm [1]. The idea behind wide channel spacing is the use of low-cost uncooled lasers exhibiting a wavelength drift with temperature and relaxed requirements for manufacturing passive components such as multiplexers. One of the simplest and most cost-effective techniques for modulating a light source is to directly modulate the laser drive current with the desired waveform. This continues to be the most common approach for low-cost optical communication systems while external modulation is used for achieving higher transmission performance in trunk applications where transmission distance noise and distortion requirements are beyond the capabilities of direct modulation. In this approach, a modulator, external to the laser device, is used to impart the intensity modulation on the continuous wave (CW) output of the laser.

Before 1980, most cable television (CATV) systems were coax-based, but by the early 1980s the CATV industry began using directly modulated 1310-nm vestigial sideband (VSB)/ amplitude modulation (AM) links for fiber-based distribution super trunks. Hybrid fiber coax (HFC) networks combine both optical fiber and coaxial cable lines. Optical fiber runs from the cable head end to neighborhoods of 500 to 2000 subscribers. Coaxial cable runs from the optical fiber feeders to each subscriber. Hybrid networks provide many of fiber's reliability and bandwidth benefits at a lower cost than a pure fiber network. As of late 1996, about 7% of cable systems in the U.S. had been upgraded to HFC. By transporting a high quality replica of the head end signals, this system architecture reduced the number of cascaded radio frequency (RF) amplifiers required in the local distribution area. Later, also multi-channel fiber transmission was used for increased bandwidth, but the CATV networks remained almost exclusively analog at the transmission level [2,3].

In this chapter, we investigate CATV/HFC networks as an application for CWDM. Initially, the fundamentals of the CATV architecture are reviewed in Section 9.2, followed by the discussion of optical components for the forward and reverse path in the network. Here, we also address the challenges of analog modulation for directly modulated transmitters. In Section 9.3, the application of CWDM to the forward path is presented and in Section 9.4 to the return path.

9.2 CATV ARCHITECTURE

The evolution of CATV networks has taken them from a single office serving a few hundred subscribers to today's regional service providers having millions of customers. Although the basic topology has not changed over the years, the implementation of the network has been forced to accommodate the increase in the shear expanse of the systems as well as the tremendous increase in the bandwidth or number of channels. As illustrated in Figure 9.1, throughout the evolution an area is served by a small number of offices or head-ends, where most of the programming is placed on the network, and a distribution network that is largely in a hub and spoke topology.

At the start of CATV, purely RF-based coax networks were employed, as shown in Figure 9.1a. The bandwidths were in the 350-MHz range and the number of homes served by one feed from an amplifier was limited. Increased demand was met by increasing the number of amplifiers in a line, thus allowing for more taps. Soon, the quality of the signal at the end of a given line would not meet

CWDM in CATV/HFC Networks

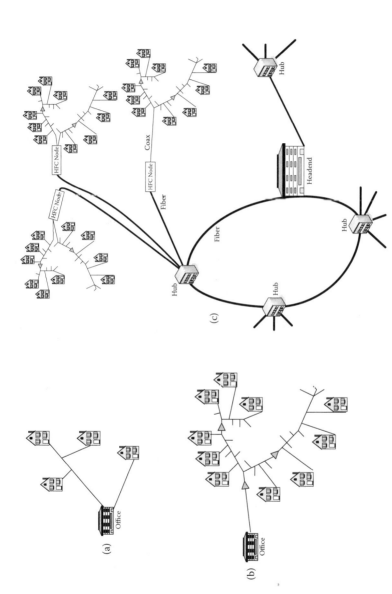

FIGURE 9.1 Three-step evolution to today's HFC networks with (a) a small local office serving a few homes via coax lines, (b) to larger numbers of subscribers by use of more and more RF amplifiers, and (c) very large areas with hundreds of thousands of subscribers served by a combination of fiber and coax distribution.

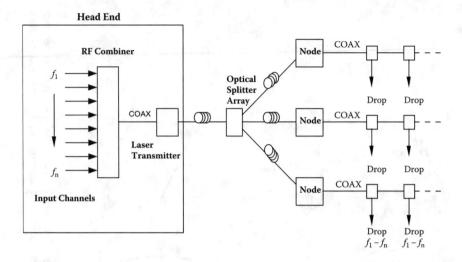

FIGURE 9.2 Primary distribution for CATV networks based on an optical star topology distributing the RF signals $f_1...f_n$ via fiber to remote nodes.

the standards that have been set for CATV drops into a home due to cascaded amplifiers and increased power signal splitting so that a new spoke would have to emerge from the central office or head end.

As coverage and bandwidth demands increased and fiber optic technologies emerged, CATV networks changed significantly into what is now referred to as HFC network. In this approach, the hub and spoke topology was first augmented by trunk signals to the hub where it was then converted to RF and distributed to subscribers via coax spokes with RF amplifiers and drops to the home. When subscription rates increased and the demand for even greater bandwidth emerged, it became necessary to extend the transmission of the CATV signals via fiber optical links out to a moderate-sized area around an HFC node, where the fiber optic signal was terminated and converted to the RF signals dropped in the home as illustrated in Figure 9.2. This network implementation exists today as the primary network configuration for almost all CATV systems. System operators embraced this approach enthusiastically because the bandwidths and transmission distances achievable using the HFC approach allowed them to vastly expand their coverage and services in a cost-effective manner.

As HFC networks evolved and the services expanded into more interactive products, they became bidirectional. This is made up of the forward path comprised of the video and other services delivered to the home and the reverse path allowing for the transmission of signals from the home and nodes back to the head end. Although the forward path signal quality requirements were stringent, the return path was required only to transport low-bandwidth digital format [e.g., frequency shift keying (FSK)] and performance could be greatly relaxed. However, because of the cost requirements the RF return path remained a challenge,

CWDM in CATV/HFC Networks 273

as the transmission equipment has to be installed for every user. Return path transmitters, for example, in the set-top box, were of low quality because of cost considerations and the return path signal was degraded significantly because of ingress caused by man-made RF signals and noise generated by household devices as simple as a hair drier. Optical transport of the return path traffic is not subject to ingress and was considered an excellent approach for transport from the home and inline RF components such as amplifiers from the HFC node to the hub and subsequently to the head end.

9.2.1 HFC Forward Path Optics

Performance requirements imposed on the forward path transmission system are the most demanding with regard to carrier to noise (CNR) and distortion. Hence, fiber optic transmission systems for the forward path have been designed to have low noise figures and predistorted for compensation of signal distortion to achieve the high performance demands for analog video. These links typically operate at 54 dB CNR, 65 dB CSO (composite second order), and 65 dB CTB (composite triple beat). Although it seems that there is an undue burden placed on the integrity of the optical transmission up to the HFC node, one must remember that the coax portion of the network contributes that vast majority of signal impairment. The fiber portion of the HFC forward path can also be divided into two regions: the trunk from the head end to the hub and the distribution from the hub to the node. In the trunk section, the optical signals are distributed to the hubs typically via a ring topology. Because of the distances traveled (40 to 100 km), external modulation is typically utilized because of the low chirp when compared to direct modulation. For the hub to node distribution, the distances are in the 5 to 40 km range and are therefore addressable using direct modulation.

9.2.2 HFC Reverse Path Optics

The reverse path payload usually consists of signals using the FSK, QPSK, or low order m-QAM modulation format. Typical channel bandwidths are only 1.5 MHz each while the total bandwidth traditionally allocated is 5 to 45 MHz or the region just below the video channels. There are many systems in use that employ larger reverse bandwidths from the node to the hub to accommodate the increase in cable modem and other digital interactive services. Reverse path optics are subject to much lower performance requirements except for two very important factors. Although cost is always an issue, there are many nodes in a system and they serve fewer homes than any other optical component. Hence, the cost of any optics dedicated to the node is allocated to fewer subscribers than any other optical system in the network. In addition, the environment in which the node exists is by far the harshest of any other optical element in the network. While the optics in hubs and head ends typically reside in climate-controlled buildings, HFC nodes are hung on the strand between telephone poles or are mounted in pedestals outdoors. Therefore, any lasers used in the node must be

9.2.3 ANALOG DIRECT MODULATION

As mentioned, direct modulation is the simplest approach for optical signal transmission. However, with any laser, CWDM or otherwise, there are mechanisms that will result in distortion of the modulated signal [4]. We can write the laser drive current as,

$$i_{drive} = -i_{threshold} + i_{bias} + i_{signal}(t), \qquad (9.1)$$

where $i_{threshold}$, i_{bias}, and i_{signal} are the threshold, bias, and time varying signal currents, respectively. If the signal current makes excursions below the laser threshold, this will shut off the laser emission, thus distorting the signal I_{ave} as shown in Figure 9.3. This will generally result in what appears to be second-order distortion and can be a severe transmission impairment.

Additionally, any deviation from a perfectly linear P-I curve of the laser will also add distortion, as shown in Figure 9.4. While this mechanism is usually

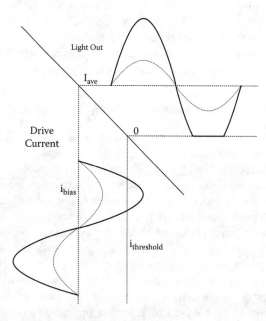

FIGURE 9.3 Impact of laser threshold on signal output of directly modulated laser. The large driving signal causes clipping at the output.

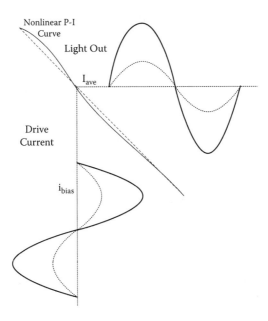

FIGURE 9.4 Drive current and laser output for a nonlinear P-I characteristic.

dominated by second order, there is also some third-order distortion present. This distortion is much harder to avoid and will limit the optical modulation index to a point generally too low for adequate CNR in most applications.

Designers of directly modulated laser transmitters for CATV applications can mitigate this distortion to a large degree by incorporating predistortion circuitry. A predistortion network will provide the same amplitude but opposite phase distortion products, which when added to the input signal will cancel out the distortion products caused by the laser. While this works quite well, the reduction of distortion products by 20 dB requires excellent phase and amplitude matching with the nonlinear characteristics over the whole RF bandwidth. The effectiveness of predistortion hinges on the stability of the nonlinear characteristics of the laser. This is due to the fact the predistortion network is producing the same distortion as the laser but opposite in phase so that in the end the distortion from the laser and predistortion network cancel out. Hence, if the predistortion circuit is tuned at a particular performance level, it is critical that the laser nonlinear characteristics remain relatively constant. Complicating this issue is the tendency of the nonlinear characteristics of distributed feedback (DFB) lasers to vary with temperature. Since the nonlinear characteristic of DFBs generally depends on the position of the longitudinal modes relative to the DFB resonance, the temperature of the laser must be kept constant. Thus, the cost benefit of uncooled CWDM lasers is not applicable and provides no advantage over other DFB lasers.

9.3 APPLICATIONS FOR CWDM

9.3.1 Forward Path

As wavelength division multiplexing (WDM) technologies evolved in the digital world, the availability, cost, and power of DFB lasers improved to the point where these were the optical source of choice for all directly and externally modulated transmitters. For the externally modulated transmitters, the choice was simple because of the need to replace the solid-state lasers and thus greatly reduce the cost and vastly increase the reliability of optical transport. The availability of amplifiers such as erbium-doped fiber amplifiers (EDFAs) also played an important role in the deployments of this topology. These types of transmitters were and are extensively utilized in the head end to hub trunk, as well as many applications where the optical path was extended to include the HFC nodes. Figure 9.5 shows an illustration of a typical network deployed using the optical trunk with extensions out to the nodes.

In this figure, EDFAs are used to overcome the losses due to transmission distance and taps at each hub. It is important to note that in the beginning of the use of DFBs in HFC networks the primary considerations were linewidth, power, RIN, and cost, not the need to multiplex many wavelengths on a single fiber. As things evolved, CWDM has never really captured much of the market as a source for externally modulated CATV transmission systems. Direct modulation has and does continue to play a crucial role in HFC transport. Relatively high performance transmission using DFB lasers became available mainly due to the development of linearization technologies that overcame the intrinsic second-order distortion imposed on the signal in directly modulated transmitter. As discussed earlier, predistortion combined with DFB lasers is the technology used exclusively in direct modulation HFC applications. Thus, the use of CWDM lasers in this application has little or no benefit because of the addition of thermo-electrical coolers and control circuitry and the electrical power required to operate these.

9.3.2 Forward Path Narrowcast Overlay

In an effort to provide more targeted programming on a demographic basis, narrowcasting or programming and advertising targeted to a particular audience has been deployed. This approach consisted of the transmission of common programming over the usual HFC network with the addition of the narrowcast overlay for local or targeted programming. This was accomplished using one wavelength for the common transport throughout the network and another for narrowcast that only covers a part of the whole network. Since in a typical CATV system there could be many different demographics, there needed to be many different narrowcast wavelengths. Figure 9.6 illustrates an example of such an application. In this arrangement, the narrowcast signals are generated via external modulation and simply require reasonable linewidth (~1 MHz). The wavelengths are dropped in pairs at each node, providing both common as well as narrowcast programming.

CWDM in CATV/HFC Networks

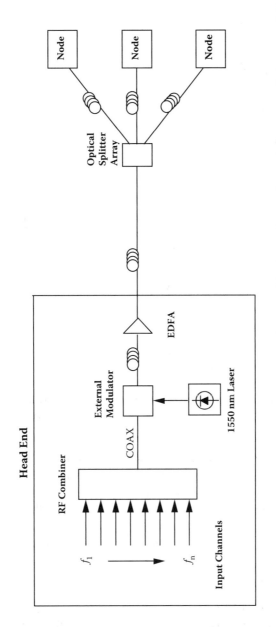

FIGURE 9.5 Trunk network with externally modulated 1550-nm transmitter, amplifier for reach extension, and subsequent splitting network distributing the RF signals to the nodes.

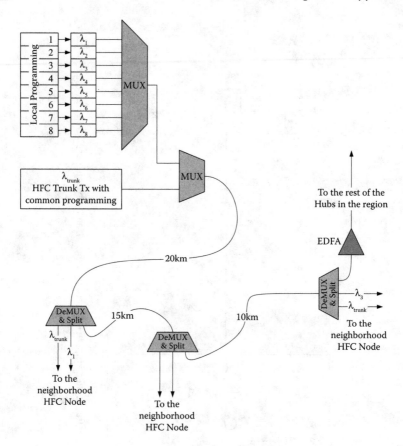

FIGURE 9.6 A narrowcast implementation with only eight CWDM narrowcast optical channels λ_1 to λ_8. At each hub, a portion of the λ_{trunk} is split off and λ_n is dropped completely. These are combined and sent to the HFC node in the neighborhood.

In this method of providing targeted programming, there could be requirements for many optical wavelengths. The potential number of wavelengths rules out the use of CWDM because of the limited channel count unless only a small number of local programs is fed into the network. In addition, there are concerns regarding stimulated Raman scattering (SRS) in these applications as well. This effect will be discussed further in combination with return path applications.

9.3.3 Return Path

Return path in modern HFC networks presents significant limitations to the ability to provide interactive and data services. In data services, for example, the need for greater upstream data transfer rates requires larger and larger bandwidth in the return path. Keep in mind that the RF spectrum available for return path is only from 5 to 50 MHz approximately, and that peak upload bandwidths are in

CWDM in CATV/HFC Networks

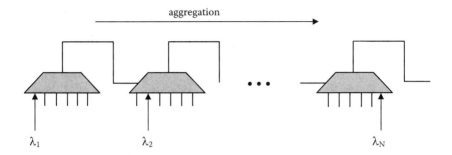

FIGURE 9.7 Return path aggregation using different CWDM wavelengths (1...N) for the upstream.

the Mbps region. One solution to this bottleneck is to stack the return path from different neighborhoods. This can be accomplished in three ways. The first is to provide a separate fiber for each neighborhood or aggregate group. This has obvious drawbacks due to the cost and availability of dark fiber. One RF approach to further aggregate the return path traffic is to shift the frequency of each return path block and stack them in the RF domain. This works well but is limited to a few blocks so that the total is in the 200-MHz range. Optical stacking or using separate wavelengths for each neighborhood has the advantage of providing a transmission path for several aggregate groups from one node as well as allowing for RF stacking to achieve further aggregation on each wavelength, as illustrated in Figure 9.7. Here, the emergence of CWDM technology allows for a low-cost solution to expand the capacity of the reverse path by allowing multiple return path transmitters that all share the same return fiber for aggregation of 16 or more neighborhoods or 50 MHz return path blocks.

9.3.4 Limitations Due to SRS

CWDM is particularly well-suited for the return path application because of the ability to operate uncooled in the HFC node and the relaxed requirements for the return path distortion performance.

There are some important considerations when designing such systems due to combining and splitting the wavelengths at the endpoints of the system as well as the presence of inter-channel interactions due to optical nonlinearities, particularly due to SRS. The degradation due to SRS arises due to the presence of Raman gain in the fiber [5]. This mechanism results in crosstalk between optical channels in the return path, which manifests itself as added noise limiting the system performance of CATV/HCF networks [6]:

- The CWDM channels at shorter wavelengths induce noise on low frequency RF channels.
- The signal at longer wavelengths (e.g., 1550 nm) can pull optical power from the shorter wavelengths, known as pump depletion.

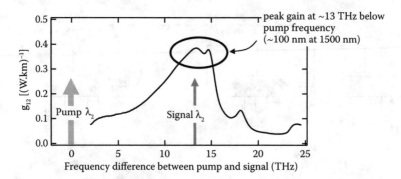

FIGURE 9.8 Interaction of two channels spaced 100 nm apart, for example, 1470 and 1570-nm CWDM channels by SRS. The short wavelength channel acts as a pump for the longer wavelength, resulting in nonlinear crosstalk; g_{12} is the Raman gain coefficient and reaches its maximum at wavelengths that are 100 nm longer than the pump.

To understand the SRS mechanism in more detail, the Raman gain characteristics must first be understood. Figure 9.8 illustrates the effect in a simplified situation in which there are two optical channels, λ_1 and λ_2, present in the fiber.

This figure illustrates the gain in the region of λ_2 due to the presence of λ_1. The absolute gain present at λ_2 clearly depends on the separation of the wavelengths and the channel power in λ_1. The wavelength dependence shows that there is gain for wavelengths greater than λ_1 and increases nearly linearly as the wavelength difference between the two sources increases up to about 100 nm. Hence, up to a point, the larger the wavelength or frequency difference, the greater the gain. This is a particular concern for CWDM since the wavelength difference between channels is 20 nm or 1/5th of the width of the Raman gain curve. Furthermore, for multiple optical carriers, each wavelength generates Raman gain that affects channels having longer wavelengths.

Further complicating the situation is the fact that each optical carrier is being modulated and because the gain characteristic generated by each channel depends on the amount of optical power in the channel, the gain is modulated, thus resulting in crosstalk. The process has been studied by Phillips and Ott [7] for CATV systems and an approximate expression for the amount of crosstalk is given by:

$$\text{Crosstalk (SRS)} \approx 10 \log \left[\left(\frac{g_{12} \rho_{SRS} P_2}{A_{eff}} \right)^2 \frac{1 + e^{-2\alpha z} - 2e^{-\alpha z} \cos(\omega d_{12} z)}{\left(\alpha^2 + (\omega d_{12})^2 \right)} \right] \quad (9.2)$$

where
λ_1 = wavelength of the channel being interfered with
λ_2 = wavelength of the interfering channel
g_{12} = Raman gain coefficient at λ_1 due to λ_2

A_{eff} = effective area of fiber
ρ_{SRS} = effective polarization overlap factor
P_2 = optical power of the signal at λ_2
α = lin. attenuation coefficient of the fiber
d_{12} = group velocity mismatch ($\approx D(\lambda_1 - \lambda_2)$ where D = dispersion coefficient)
z = fiber length
ω = RF angular frequency of the modulation of λ_2

A special concern to return path applications is the dependence of crosstalk on the RF modulation frequency, ω. The presence of the ω^2 term in the denominator of Equation 9.2 yields higher crosstalk at lower RF carrier frequencies. In Figure 9.9, the measured crosstalk is shown between 5 and 200 MHz introduced by SRS between two co-propagating CWDM channels spaced 100 nm apart at 1470 and 1570 nm, 3 dBm per channel launch power, and 14 km SSMF. The worst crosstalk of −45 dB was measured at 5 MHz with minima at higher frequencies due to the cosine term in Equation 9.2. The actual variation of the crosstalk with modulation frequency also depends on the fiber length and channel walk-off, that is, the product of wavelength spacing and fiber dispersion. Crosstalk levels not exceeding −60 dBc are found acceptable in practical systems [6].

Figure 9.9 illustrates the difficulties of multi-channel CWDM signals with modulation frequencies below 50 MHz. This plot shows an almost 20 dB increase from the average crosstalk in the 50 to 200 MHz region. Obviously, the reliance

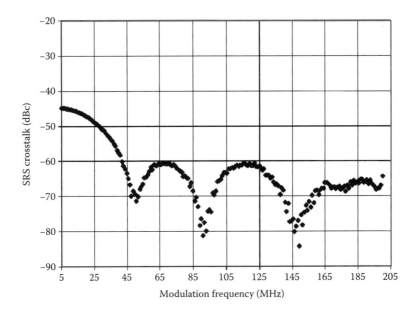

FIGURE 9.9 Measured SRS crosstalk as a function of modulation frequency for the 1570-nm channel in the presence of 1470-nm channel, 14 km SSMF, 3 dBm/channel, according to results in [8].

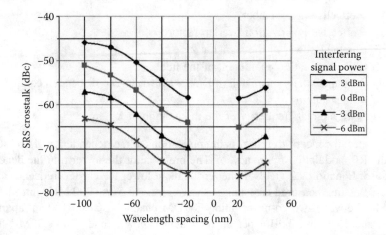

FIGURE 9.10 Measured SRS crosstalk as a function of CWDM channel spacing and launch power, high powers and wide channel spacing degrade the transmission performance, according to [8].

on the 5 to 50 MHz region for services sensitive to the noise presented by crosstalk is risky. Fortunately, the RF stacking approach as described earlier in Section 9.3.3 is well suited for this WDM approach.

Equation 9.2 further describes the expected crosstalk due to the wavelength spacing of the optical channels on the CWDM grid, through the Raman gain coefficient g_{12}. In addition, the amount of crosstalk also depends on the square of the optical power yielding a 2-dB increase in crosstalk for a 1-dB increase in optical power, P_2 making this effect more pronounced at higher launch powers. Figure 9.10 shows the crosstalk onto the 1570-nm channel as a function typical of CWDM channel power levels and wavelength spacing with a fixed modulation frequency of 5 MHz for worst-case estimation. The interfering channel is tuned between 1470 and 1610 nm and the crosstalk contribution is measured.

The result of this analysis does not rule out the use of CWDM lasers as sources but does however force some design considerations that must be acknowledged when prescribing the CWDM optical channel plan, launch powers, and the placement of critical services within the RF spectrum. The advantage of uncooled low-cost CWDM sources in WDM designs of this type outweighs the difficulties presented by SRS when the following design rules are met:

- Restrict launch powers to minimum levels for required CNR.
- When possible, do not utilize channels 100 nm apart, prefer narrower or wider channel spacing.
- Services with a high CNR requirement should be placed at the high end of the modulation spectrum.

9.4 CWDM IN FTTX AND FUTURE NETWORK DESIGNS

As CATV or broadband services expand with higher delivered bandwidths and interactive services such as IP television [9], the transport topologies will evolve toward different architectures. A variety of passive optical network (PON) implementations, including λ-PON or WDM-PON, may utilize CWDM in delivering digital services in a base band digital format and pure QAM. In this instance, all of the rigid specifications of analog modulation of the early fiber to the home builds are gone, and more forgiving distortion and noise specs will replace them. For these applications, the need for ultra stable performance will be relaxed and CWDM lasers will have applications simply because of the cost advantage of uncooled DFBs. The role of CWDM for PONs will be discussed in Chapter 10 in detail. With the lack of a cooling requirement and the electrical power requirements that accompany it, the installation of DFBs-based devices is made possible for outdoor pedestal and strand mount as well as homes.

REFERENCES

1. ITU-T Recommendation G.694.2 (2003), *Spectral Grids for WDM Applications: CWDM Wavelength Grid*.
2. Atlas, D.A. and Lenny, J.J., Multiwavelength Analog Video Transport Network, International Topical Meeting on Microwave Photonics 1999, Proceedings MWP'99, Vol. 1, pp. 189–192.
3. Feldman, R.D., Wood, T.H., Meester, J.P., and Austin, R.F., Broadband upgrade of an operating narrowband single-fiber passive optical network using coarse wavelength division multiplexing and subcarrier multiple access, *J. Lightwave Technol.*, 16(1), 1–8, 1998.
4. Petermann, K., *Laser Diode Modulation and Noise*, Kluwer Academic Publishers, Dordrecht 1991.
5. Agrawal, G.P., *Nonlinear Fiber Optics*, 2nd ed., Academic Press, 1995.
6. Piehler, D., Meyrueix, P., Rishi, G., Chen, L., Coppinger, F., and Thomas, R., Nonlinear Raman Crosstalk in a 125-Mb/s CWDM Overlay on a 1310 nm Video Access Network, Proceedings OFC 2004, paper FE8, February 2004.
7. Phillips, M.R. and Ott, D.M., Crosstalk due to optical fiber nonlinearities in WDM CATV lightwave systems, *J. Lightwave Technol.*, 17(10), 1782–1792, 1999.
8. Best, B., *Management of Stimulated Raman Scattering in CATV WDM Reverse Path Systems*, NCTA 2006, Atlanta.
9. Kawata, H., Ogawa, T., Yoshimoto, N., and Sugie, T., Multichannel video and IP signal multiplexing system using CWDM technology, *J. Lightwave Technol.*, 22(6), 1454–1462, 2004.

10 CWDM for Fiber Access Solutions

Carlos Bock and Josep Prat

CONTENTS

10.1 Introduction to Fiber-to-the-Home 285
    10.1.1 FTTH Basics and Standards..286
    10.1.2 Point-to-Point and Point-to-Multipoint 288
    10.1.3 Standards for FTTH-PON 290
    10.1.4 Advantages of CWDM in FTTx 291
    10.1.5 Is There Enough Bandwidth Using CWDM?........................292
    10.1.6 Limitations of CWDM in FTTx 293
10.2 Network Overlay 294
    10.2.1 Combining CWDM and TDM to Create High Density PONs 294
    10.2.2 Network Design...295
    10.2.3 Protocol Compatibility with Existing Standards 297
    10.2.4 Upstream Transmission 299
10.3 Service Overlay 300
    10.3.1 Offering Different Services over the Same Network Infrastructure 300
    10.3.2 Multi-Carrier Infrastructure...301
    10.3.3 Service Filtering 302
10.4 Advanced CWDM Network Implementations...................................305
    10.4.1 CWDM Overlay on EPON Systems.......................................305
    10.4.2 Ring-Based CWDM Metro/Access Network.........................305
    10.4.3 Wireless and CWDM Networks..306
    10.4.4 CWDM and D/UD-WDM Access Networks.........................306
10.5 Future Trends...308
References 309

10.1 INTRODUCTION TO FIBER-TO-THE-HOME

New requirements in terms of bandwidth are encouraging the evolution of access networks to provide more transmission capabilities to satisfy new application demands. This evolution inherently requires the deployment of an access optical infrastructure, as copper-based solutions cannot provide enough bandwidth. This evolution, however, needs to be gradual in order to have a viable economic model because deployment costs are very high. In this chapter, we will introduce different scenarios of fiber deployments in access and the possible applications of coarse wavelength division multiplexing (CWDM) to enhance fiber access solutions.

10.1.1 FTTH Basics and Standards

The fiber-to-the-home (FTTH) concept consists of the deployment of an all-optical infrastructure in the last mile/s segment of the network. It is a category inside the fiber-to-the-x (FTTx) approach, which gradually defines the deepness of the fiber reach toward the end user. Fiber-to-the-cabinet (FTTCab), fiber-to-the-curb (FTTC), fiber-to-the-building (FTTB), FTTH, and fiber-to-the-user (FTTU) are the main different options, sorted by the proximity level to the user as illustrated in Figure 10.1.

The FTTx concept is thus also an upgrade path from a copper-based infrastructure to an all-optical one, reaching the final step in the FTTH/U scenario, where there is an all-optical network path up to the end user. Applications, services, and transmission capabilities are different from one case to the other. The presence of copper generally limits the available bandwidth because of crosstalk and frequency-dependent effects. Furthermore, the achievable distance of the copper link is also limited by high attenuation factors and by a much worse bandwidth-by-distance factor than in the optical case. In other words, presence of optical fiber to the customer's premises means a huge boost in the available bandwidth without the distance limitation incurred by a copper-based link.

Although FTTH is the ideal solution to provide high transmission data rates, its deployment nowadays is mostly restricted to Greenfield scenarios due to the high deployment costs in terms of civil infrastructure such as trenches for the

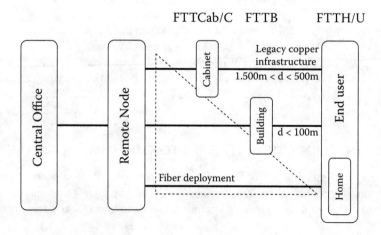

FIGURE 10.1 FTTx approaches — from FTTCab/C to FTTH/U.

TABLE 10.1
Distances and Available Bandwidths for Different Access Solutions

	Max Copper Length Distance	Commercial Transmission Technologies	Bandwidth per User
All copper	5 km	ADSL	128…1500 Kb/s (asymmetrical)
FTTCab	1.5 km	ADSL/VDSL/CATV	1…13 Mb/s (asymmetrical)
FTTC	300 m	VDSL/CATV	36…52 Mb/s (shared in CATV)
FTTB/H	0…100 m (FTTB)	EPON/GPON	1…2.5 Gb/s (shared)

Note: ADSL = asymmetric digital subscriber line. VDSL = very high data rate digital subscriber line.

fiber ducts. A possible solution to mitigate the high level of initial investment is to gradually implement the FTTH concept by using intermediate steps summarized as FTTx where the fiber–copper interface resides at different locations along the path between central office (CO) and end user.

The fiber can be moved closer to the end user in several steps, which are summarized in Table 10.1. The FTTCab architecture runs an optical fiber from the exchange center to a neighborhood cabinet, where the signal is converted to the electrical domain to feed the subscriber over a twisted copper pair. Typical distances of the copper segment are lower than 1.5 km. Transmission technologies over the twisted pair are normally from the ADSL family. Total loop distances vary from country to country. In Europe, typically 90% of the users are within 5 km of the CO, whereas in the U.S., this percentage is reduced to just 30% of the users.

The FTTC architecture expands the optical segment to a small curb-located cabinet, which is typically 300 m away from the subscriber. Signal is then converted to the electrical domain and distributed by twisted copper pair or coaxial cable (hybrid fiber coax approach) to the subscribers. In this scenario, VDSL technologies can deliver 50 Mb/s up to 300 m. The distance of the copper link is reduced, thus the transmission penalties induced by the copper media are minor. However, the maximum data rate needs to be checked individually for each twisted pair as it can vary depending on crosstalk, impairments, and channel transfer function.

FTTB/H are the only all-optical solutions and also the only ones that do not deploy active equipment in the outside plant (e.g., outdoor DSLAMs). This is a great advantage as there is no need to accommodate electronics, which normally require a controlled-environment location, outside the exchange center or the end-user premises. xDSL technologies use advanced digital signal processing techniques to maximize the available bandwidth of copper, which turns into high power consumption requirements and heat dissipation issues. In the fiber-to-the-building approach, the optical reception equipment is located at the building basement and then an internal copper network in the building/business reaches the final connection points. Standard CAT-5 wiring for data transmission can be deployed to give access to the end users. In the FTTH approach, each home has its own and dedicated optical receiver. There is a further evolution of this concept, which is the FTTU, also called fiber-to-the-desktop (FTTD), where fiber enters inside the home via indoor wiring with an optical termination at the individual network socket.

FTTH offers many benefits, which are listed subsequently:

- High bandwidth (GPON Standard with 2.5 Gb/s, >100 Mb/s planned for residential)
- Extended range (up to 20 km)
- Transparency and future upgradability
- Secure and robust transmission media

With optical fiber as the transmission medium, gigabit data rates can be easily transmitted over distances of tenths of kilometers where the requirements on the optical performance of the components is rather low compared to other optical transport systems. Transparency and future upgradeability are referred to the fact that the optical infrastructure of fibers and splitters, if totally passive, is totally bit-rate independent, which means that a future increase in bit-rate is perfectly supported by the existing optical infrastructure. Another important characteristic of optical fiber is its immunity against electromagnetic interference. In contrast to copper-based systems, this simplifies the transmitter and receiver electronics and makes the network more robust.

10.1.2 Point-to-Point and Point-to-Multipoint

Another fundamental differentiation in optical access networks is whether the optical fiber is shared or dedicated. The first case is the point-to-multipoint (P2MP) solution and the second case is the point-to-point (P2P) approach (see Figure 10.2). Both architectures have advantages and drawbacks. P2MP solutions have the advantage that less fiber and optical interfaces are required. On the other hand, there is a need to establish a multiplexing strategy to segment and distribute the available bandwidth among the different end users. Well known techniques use either time division multiplexing (TDM) or wavelength division multiplexing (WDM). This last approach, however, creates extra complexity for the multi-point solution. On the other side, single P2P architectures offer a dedicated fiber to each end subscriber, totally independent of the other network traffic. The advantage is also that the solution is very simple from the technological point of view and offer virtually unlimited bandwidth to each end with protocol independence and enhanced security. The main drawback of this approach is the amount of fiber required in the outside pant and of optical interfaces required at the exchange

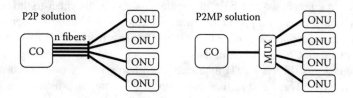

FIGURE 10.2 P2P and P2MP concept for FTTx.

CWDM for Fiber Access Solutions

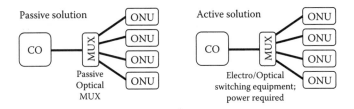

FIGURE 10.3 Active and passive access networks.

center for each user, as each user requires a dedicated transmitter at the CO [1]. Next, we will focus on P2MP solutions. Existing local area network (LAN) standards are usually applicable to P2P solutions, whereas P2MP present technological challenges that are intrinsic from the access domain and require the development of specific standards for this application.

Finally, another characterization of all optical access networks is the presence or absence of active equipment in the field (see Figure 10.3). Active optical networks (AONs) have active equipment in the field, which can be a drawback due to the required electrical power for those sites. AONs have an active switch, which is located infield, to route optical signals in an approach that is similar to a P2P solution but with the network switch outside of the exchange center. On the other hand, passive optical networks (PONs) have a completely passive outside plant which interconnects the exchange center with the end users by means of optical splitters or passive multiplexers. There is no power required infield. PONs offer the great advantage of a more robust and future-proof infrastructure and, therefore, are the focus of our study.

The basic model of a passive P2MP access network is presented in Figure 10.4. The main parts are the exchange center or head-end, also known as CO, the remote node (RN), and the end-user premises equipment or customer premises equipment (CPE). The access network optical transmission link is terminated by the equipment located at the CO and the optical network unit (ONU), which depending on the FTTx scenario is located in an outside cabinet (FTTCab and FTTC) or at the CPE (FTTB/H).

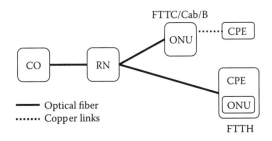

FIGURE 10.4 A PON model.

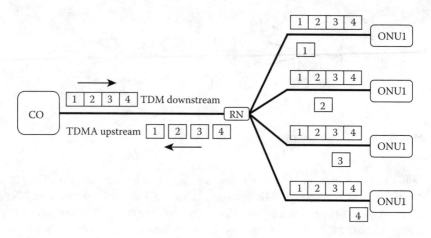

FIGURE 10.5 TDM downstream and TDMA upstream in TDM-PON.

10.1.3 STANDARDS FOR FTTH-PON

Present standards for FTTH-PON are all based on TDM to transmit downstream data and TDMA (TDM access) for upstream transmission (see Figure 10.5). When using a single fiber outside plant, data transmission is multiplexed using CWDM, transmitting downstream data on the 1490/1550-nm channel and upstream on the 1310-nm channel, as will be more deeply analyzed in the present chapter.

The history of PONs started parallel to the developments by the Full Service Access Network (FSAN) working group in the 1990s, formed by major telecommunications service providers and system vendors. The International Telecommunications Union (ITU) did further work that has led to two generation of standards. The older ITU-T G.983 standard is based on asynchronous transfer mode (ATM), and has therefore been referred to as APON (ATM PON). An evolution of the APON standard completed the ITU-T G.983 standard. This evolution was known as broadband PON, or BPON. Typical data rates of APON/BPON are 622 Mb/s of downstream bandwidth and 155 Mb/s of upstream traffic, although the standard defines and proposes higher rates [2].

ITU-T G.984 (GPON), the evolution of APON/BPON, represents a significant boost in both the total bandwidth and bandwidth efficiency. GPON delivers up to 2.488 Gb/s of downstream bandwidth and 1.244 Gb/s of upstream bandwidth. GPON encapsulation method (GEM) allows efficient packaging of user traffic, with frame segmentation to allow Quality of Service (QoS), which prepares the network to deliver multimedia content, such as voice and video streaming [3].

Further to the developments of the ITU, Institute of Electronic and Electrical Engineering (IEEE) also has developed a PON standard, based on Ethernet technology. Ethernet PON (EPON) standard, named IEEE 802.3ah, delivers 1 Gb/s of symmetrically transmitting Ethernet frames. It is not as efficient as GPON in terms of bandwidth management but provides a more cost-effective solution [4].

10.1.4 Advantages of CWDM in FTTx

TDM-PONs have simplicity as one of their features, especially in terms of optical interfaces. Those are very simple and inexpensive. However, TDM-PONs do not exploit the huge available bandwidth of optical fiber. Wavelength division multiplexing (WDM), on the other hand, permits the transmission of independent data channels on different wavelengths. Thus, WDM creates virtual P2P connections on a P2MP PON infrastructure. However, this requires optical interfaces for each of those channels and much more precise transmission equipment, which implies more expensive devices.

The use of CWDM (coarse WDM) technology in access, and more specifically on PONs, is potentially wide and comprises multiplexing of services and network overlay. CWDM is an interesting technology to be used in access because of the benefits from the economic point of view in comparison to the traditional concept of WDM. CWDM interfaces are more cost effective than their DWDM counterparts and cost effectiveness in access is critical to achieve a successful economic model because the cost of the equipment is directly assumed by the customer. Therefore, to develop inexpensive transmission systems is a very interesting approach to make PONs more attractive.

The use of several wavelengths intrinsically increments transmission capabilities and offers more flexibility to the network. Thus, CWDM offers substantial advantages over single-wavelength transmission in terms of capacity and applications that can be deployed over the network [5]. Typical approaches that can be deployed with CWDM and in general with any wavelength multiplexing technique are service or network overlay. This is to extend the number of users that can be connected to the same fiber trunk or to offer several services over the same network infrastructure. Both approaches will be discussed in this chapter.

The advantages of CWDM in access over dense WDM (DWDM) are mainly economic. CWDM uses channel spacing of typically 20 nm [6], whereas DWDM uses channel spacing of 50 GHz (0.4 nm) and multiples [7], generally 100 GHz (0.8 nm). This very narrow-channel spacing requires much more precise equipment. Transmission equipment for DWDM requires thermal stabilization and precision in the range of tenths of nanometers to achieve correct transmission. For CWDM, there is no need to use temperature-controlled lasers and multiplexers and design criteria for the optical devices are not so strict, which turns into a decrease in production costs.

Also, some advantages of DWDM are not so relevant in access. One of them, which is the possibility to amplify several channels simultaneously by means of optical amplification with erbium-doped fiber amplifiers (EDFAs) is normally not necessary, as the distance of the fiber segments is rarely larger than 20 km. Another advantage that is difficult to justify in access is the number of channels that can be multiplexed on the same fiber. Unless we are using WDM for routing purposes to assign a wavelength to each ONU, which is a very costly approach, there is generally no need to multiplex more than eight wavelengths on the same fiber. Even in the case we are using a single fiber and assign different wavelengths

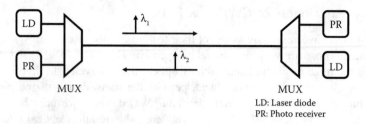

FIGURE 10.6 Single fiber bidirectional transmission using one shared fiber.

for upstream and downstream transmission (see Figure 10.6), the CWDM standard offers 18 wavelengths (16 practically).

In access, the use of CWDM for bidirectional transmission is a great advantage over the use of two different fibers for upstream and downstream transmission. PON standard products define standards to multiplex upstream and downstream links on the same fiber relying on CWDM. EPON, the Ethernet PON standard developed by the IEEE (IEEE 802.3ah), transmits the upstream channel on the 1310-nm wavelength and the downstream channel at 1490 nm. GPON, the PON standard developed by the ITU (ITU-T G.984), uses similar wavelengths and defines the upstream wavelength in the range of 1260 to 1360 nm and downstream wavelength in the range of 1480 to 1500 nm. Both standards leave the C-band free to allow cable television (CATV) overlay (see Chapter 9). In any case, those commercial approaches do not exploit all the advantages of CWDM and are just basic approaches of what the technology can offer.

In conclusion, advantages of CWDM over transmission using a single wavelength are the possibility to transmit several wavelengths on the same fiber. Compared to DWDM, CWDM offers less number of channels than DWDM systems but at a small fraction of the cost of DWDM. As economics are a primary factor in access, this reduction in costs of CWDM systems compared to DWDM offers a great advantage when implementing an optical access infrastructure.

10.1.5 Is There Enough Bandwidth Using CWDM?

The use of CWDM instead of DWDM limits the amount of wavelengths that can be simultaneously transmitted over a single fiber. As mentioned earlier, the ITU-T defined CWDM grid specifies 18 usable wavelengths that, in a bidirectional single-fiber system, turn into 9 independent data links (although typically just 16 and 8 wavelengths are practically used, respectively).

Another parameter that needs to be taken into account is the link data rate. Typically, CWDM laser sources operate at 2.5 Gb/s, whereas DWDM interfaces can operate at 10 Gb/s and more. The fabrication requirements of CWDM equipment are targeted to achieve the lowest possible cost, and electronics required to operate at 2.5 Gb/s are considerably more inexpensive than at 10 Gb/s. At 2.5 Gb/s, lasers can be directly modulated so there is no need to use an external modulator. Furthermore, at this data rate and distance range, the chromatic dispersion of the

CWDM for Fiber Access Solutions 293

optical fiber is still negligible so that there is no need to use any dispersion-compensation technique.

In any case, there is no technological limitation that would prevent the selection of 10-Gb/s transceivers using CWDM technology. At present, this is just a commercial limitation. Furthermore, at present, there is no commercial justification to implement 10-Gb/s interfaces in access at all CWDM wavelengths because there are no applications that can fully exploit this available bandwidth. Approaches to operate CWDM at bit-rates of 10 Gb/s have been described in Chapter 7.

The limited number of available wavelengths can be a restriction if we use the WDM-PON approach, which is to offer a dedicated wavelength to every single user. In this case, just nine users (typically eight) could be served by a single fiber. This can be a good solution for a small PON infrastructure for premium/business users. However, wavelength-routed PONs are inefficient in terms of wavelength utilization. In case one ONU is not transmitting, its assigned wavelength is unused and the optical interface at the CO is not used either. Furthermore, DWDM systems are extremely expensive to deploy because of the high cost of optical interfaces. Therefore, the WDM-PON model is very difficult to justify at present. To overcome the high cost of WDM interfaces, combined WDM/TDM techniques to share the WDM transmission equipment and reflective ONUs to reduce the cost of transmission equipment at the customer's premises are under study at present [8–10], but those are long-term solutions that are not deployable at present but are in research stage.

10.1.6 LIMITATIONS OF CWDM IN FTTx

As previously stated, the two inherent limiting factors of CWDM are the number of wavelengths that a single fiber can accommodate and the unavailability to optically amplify all the channels at the same time. The first limitation is a consequence of the 20-nm wide channel spacing of CWDM systems while the second one is due to the inability of EDFAs to amplify wavelengths outside the C- and L-bands. EDFAs can amplify C-band (1525 to 1565 nm) or L-band (1570 to 1610 nm) wavelengths, depending on how they are optimized. For other wavelength ranges, thulium-doped fiber amplifiers amplify in the S-band (1450 to 1490 nm) and praseodymium-doped amplifiers in the 1300-nm region. In any case, those regions have not seen any significant commercial use so far and, therefore, those amplifiers have not been the subject of as much development as EDFAs. Thus, when deploying a CWDM network, just a few wavelengths will be able to be amplified so wavelength assignment should be done carefully by reserving the wavelengths that can be amplified for those sections of the network with higher losses.

Earlier, we discussed the number of wavelengths that can be simultaneously transmitted and justified that for many applications this number is enough. In any case, there is a problem when trying to use all the 16 wavelengths that are defined in the CWDM standard over standard single mode fiber (ITU-T G.952). This is because of the attenuation peak that is due to the absorption by OH^- molecules in the 1400-nm region, severely limiting the transmission distance for the 1390- and 1410-nm CWDM channels as discussed in Chapter 2.

Those aspects are limitations in terms of transmission issues. However, the use of CWDM also presents management and logistic limitations. A critical problem of WDM systems, both coarse and dense, is that each ONU uses a specific wavelength to transmit upstream data. This means that colored transmitters are required and the stock of ONUs becomes very complex. N different ONUs are needed, where N is the number of different upstream wavelengths used in the network. Also, when planning the installation, a concise control is required to use the right ONU in each case. A possible solution to this stock issue would be to use tunable lasers at the ONUs but then the cost would increase dramatically. An alternative solution is to use modular ONUs, as will be presented in Section 10.2.3.

10.2 NETWORK OVERLAY

10.2.1 COMBINING CWDM AND TDM TO CREATE HIGH DENSITY PONs

A key application of CWDM in FTTx is the concatenation of CWDM and TDM to create a denser access infrastructure. By combining CWDM and TDM, several TDM-PON data signals are transmitted on the same trunk fiber. By means of a wavelength de/multiplexer those wavelengths are routed and separated and then distributed to different user groups in a classical TDM-PON approach. This solution is presented in Figure 10.7 and is called network overlay.

It is based on a feeder section and two distribution stages. The transport section is the single optical fiber span that interconnects the CO with the distribution stage and covers normally 70 to 90% of the total access network distance. The distribution stage is formed by a wavelength router and several power splitters. The wavelength router distributes the N wavelengths of the transport fiber to every output port, separating the different transmission channels. From this point, a classical TDM-PON access topology with a power splitter is deployed. We denote this as a TDM-PON sub-network. Each TDM-PON sub-network has an assigned downstream

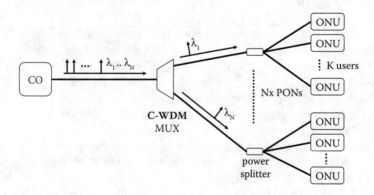

FIGURE 10.7 Network overlay combining CWDM and the TDM-PON.

CWDM for Fiber Access Solutions

wavelength and is completely independent from the other TDM-PONs. Downstream transmission is thus very simple, as it is a broadcast model with all the transmitters located at the CO. Upstream transmission is less straightforward because each sub-network requires a specific wavelength to transmit. Different ONUs are required, depending on the network segment they are attached to. This complicates stock control and increases maintenance costs. The upstream transmission issue will be discussed later.

By multiplexing several TDM-PONs on the same trunk fiber, the network overlay model allows increasing the number of users that are connected to the same feed fiber by a factor of N, where N is the number of CWDM wavelengths that are used in each direction. This approach allows a reduction in the number of COs as each CO can cover a wider distribution area. There is a tendency toward merging metropolitan and access networks in order to reduce operational costs and long-term investments. If we use different fibers for upstream and downstream transmission, so that there is no need to reserve CWDM channels for both directions, typically a maximum of 16 wavelengths could be used, multiplying by 16 the number of users that a single trunk fiber serves. The maximum power-splitting ratio of commercial TDM-PON for EPON and GPON technologies are 32 and 64, respectively, so in a combined CWDM / TDM-PON system, up to 512 or 1024 users can be served, depending on the technology that is implemented. This increase in transmission capabilities has its counterpart in a degradation of power budget of 5 dB, which are the typical losses of the CWDM multiplexing stages. To compensate these losses, the power-splitting ratio can be reduced or optical equipment with better sensitivity can be used. Also, different splitting ratios can be deployed on different channels, permitting the highest ones on those where the link budget can be enhanced by optical amplification, and the lowest splitting ratio and thus smallest splitting losses where the other link losses are higher.

10.2.2 Network Design

The design of the optical equipment located at the CO to deploy a combined CWDM/TDM PON is depicted in Figure 10.8. If the PON transmits on a single fiber upstream and downstream, a 2-port CWDM multiplexer or a power splitter separates both transmission directions. In this case, the number of sub-networks that can be overlaid is reduced to eight. In the double fiber approach, no additional CWDM multiplexer or power splitter is needed but then the outside plant infrastructure is duplicated for both directions. In any case, both models have the same design: an array of CWDM lasers, each of them transmitting at a specific CWDM wavelength, which are multiplexed to be transmitted on the same feeder fiber. The reception stage is the counterpart of the stack of lasers with a demultiplexer to separate the incoming upstream signals. An array of photodetectors then decodes each subsegment data separately.

The distribution outside plant is, as previously mentioned, divided into two different stages: first, a CWDM demultiplexing stage and then a conventional power-splitting stage. The demultiplexing stage is implemented with a CWDM

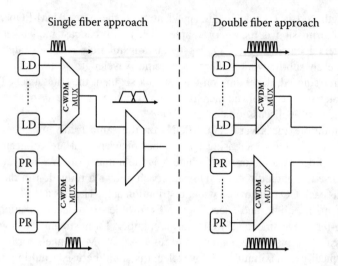

FIGURE 10.8 Central office design using the single-fiber and double-fiber approach.

demultiplexer, which routes the N incoming wavelengths to the appropriate output port, depending on the incoming wavelength.

After this wavelength-routed distribution stage, each secondary fiber carries a single wavelength in a model totally equivalent to a TDM-PON, as shown in Figure 10.9. The main element of the second stage is a power splitter, which

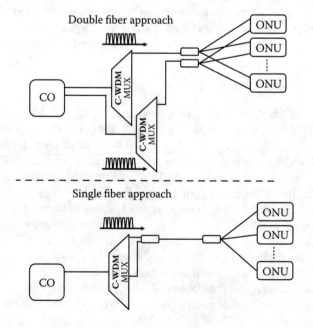

FIGURE 10.9 Outside plant for distribution using demultiplexer and power splitter.

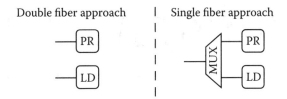

FIGURE 10.10 Two different ONU designs.

distributes the incoming signal equally to all the output ports. Losses in this case are much higher and are the more relevant limiting factor in terms of power budget of the entire link. Intrinsic insertion losses of a 1-to-N power splitter are $L = 3 \cdot \log_2(N)$ in dB. For a 32-port power splitter, this is 15 dB and 18 dB for a 64-port device. This is just a theoretical value, in practical, excess losses must also be computed and add $0.2 \cdot \log_2(N)$ dB of losses so the 15 and 18 dB turn into 16 and 19.2 dB, respectively.

The design of the ONU is very flexible and depends on whether the outside plant is single fiber or double fiber. In a single-fiber approach, different wavelengths would be required for upstream and downstream transmission, and a CWDM diplexer or a power splitter would be needed to separate incoming and outgoing signals accordingly. Then, the ONU would contain a simple photoreceiver (PR) and a CWDM transmitter (LD) at the specific wavelength channel to send upstream data, as illustrated in Figure 10.10. Upstream transmission starting from the ONU will be further discussed in Section 10.2.3.

10.2.3 Protocol Compatibility with Existing Standards

Compatibility with existing standards can be analyzed as how CWDM can coexist with legacy infrastructure or also how the available commercial standards can be adapted to work in a CWDM infrastructure.

Coexistence with legacy infrastructure requires the study of the optical spectrum that is already in use. Also, information about whether legacy fiber has the water absorption peak is critical to determine whether the 1390- and 1410-nm channels can be used. The normal upgrade path consists of the transition from a single-wavelength transmission system to a CWDM infrastructure. The only restriction of the upgrade is that a part of the spectrum that is already in use should not be interfered with, as shown in Figure 10.11. The rest can be used to multiplex several CWDM channels. The key in this upgrade is the requirement to be totally transparent to the existing infrastructure and not modify the already deployed equipment so that no service interruption occurs. Depending on the application, the upgrade path for the infrastructure will vary.

When the number of customers is to be increased but there are limitations to drop new trunk fibers, CWDM can be used to increase the number of available ports. By adding a CWDM multiplexer before the existing power splitter and

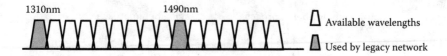

FIGURE 10.11 Spectrum of an upgrade for a 2-channel legacy network using a CWDM multi-wavelength extension.

selecting the appropriate multiplexer outside port to be used with the legacy infrastructure network, the upgrade can be done transparently and without affecting the existing users connected to the network. The same happens symmetrically with the upstream transmission. As an example, if a bidirectional EPON transmitting downstream at 1490 and upstream at 1310 is already deployed, the upgrade would consist of adding two multiplexers as per Figure 10.12, taking into account that 1490- and 1310-nm channels should be reserved for the legacy EPON system. With this improvement, we could install $(N-1)/2$ additional single-fiber PONs at that RN, where N is the number of ports of the CWDM multiplexer. This is an interesting upgrade as RNs in PONs are normally located close to the users. Drop fibers cover typically 5 to 15% of the total link distance, which means that the cost savings in terms of fiber cable deployment are substantial.

Compatibility of CWDM with existing standards can also be analyzed form the point of view of adapting existing PON standards to CWDM. The CWDM part is totally transparent to the TDM of commercial PON standards because CWDM just creates virtual P2P connections to multiplex several independent transmission channels on a single fiber. Thus, no specific adaptation is required to use CWDM with the currently available standards but selecting the wavelength of the optical transmission equipment. However, the power budget needs to be recalculated because of the extra losses introduced by the added CWDM multiplexing equipment. Depending on the technology, insertion losses of the CWDM multiplexer can be as low as 2 dB but in any case, a complete study of the power budget needs to be carried out to certify that the received power at the CO and ONUs are above the sensitivity levels.

The flexibility of CWDM offers a lot of possibilities to deploy customized solutions. Each wavelength of the CWDM is independent from the rest so a

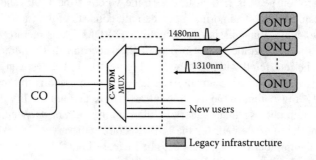

FIGURE 10.12 Possible upgrade designed over a legacy infrastructure.

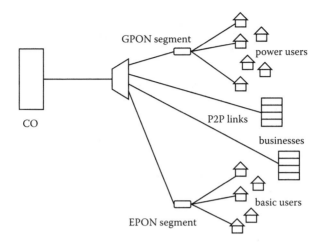

FIGURE 10.13 Deployment of a CWDM solution for G/EPON and P2P links.

combination of different solutions on the same trunk fiber is also possible. An example would be to combine P2P links for business and PON networks for end users over the same infrastructure or to mix GPON with EPON to serve customers with different quality requirements (see Figure 10.13).

10.2.4 Upstream Transmission

The design of the upstream transmission channel is particularly critical in PONs. In CWDM, as in standard PON systems, upstream transmission needs to be synchronized to avoid data collisions. However, in CWDM systems, the complexity of the upstream channel is greater because each network subsegment has a dedicated and specific upstream transmission wavelength, which may require the use of different laser sources, depending on the networks subsegment where the ONU is located. This may create stock problems and also increase maintenance costs.

To overcome this issue, one solution would be to use tunable CWDM lasers [11–13]. This solution would be costly and, at present, tunable CWDM lasers are still in an early development stage because there are not many present access applications that require this kind of device, although its implementation would be technologically feasible. One simple technology to implement, however, for emulating tunable lasers for CWDM is to use an array of discrete lasers and an array of switches to select the desired CWDM transmission channel. However, the cost of this discrete solution makes it unviable for access.

Other more advanced solutions are the use of reflective and wavelength-agnostic ONUs featuring remote modulation and remote transmission of upstream optical carriers. Those designs are based on the use of electro-absorption modulators (EAMs), reflective semiconductor optical amplifiers, and $LiNbO_3$ modulators [14]. Those designs are currently in research state and have as major drawback

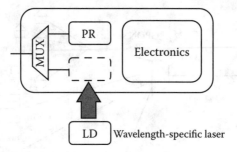

FIGURE 10.14 Modular ONU design.

that its wavelength response is restricted to the C- and L-band so that further research needs to be carried out to develop reflective ONUs, which can cover the entire CWDM wavelength range.

A simple solution, which solves the logistic complexity of having N different ONUs (where N is the number of upstream wavelengths), is to use modular ONUs. The concept is to have a common ONU with all the electronics and then different transmitters that are attached to the ONU during the installation process (see Figure 10.14). This is the same approach that at present is widely used in metro networks and switches by GBIC interfaces. This does not present any technological challenge and it does not solve to have different optical equipment, depending on the emitting wavelength. It just simplifies the management of stocks of ONUs and optical interfaces.

10.3 SERVICE OVERLAY

10.3.1 Offering Different Services over the Same Network Infrastructure

CWDM offers the possibility to create independent connections on different wavelengths, which are selected by detecting the appropriate optical channel. This feature of CWDM can be used to share a PON infrastructure and offer different services on the same network, which are selected at the CPE location by means of optical filters.

The underlying concept is very simple: on a classical TDM-PON, several CWDM channels are transmitted and then, at the ONU side, those channels are separated and detected by a CWDM optical filter and a photo receiver, as shown in Figure 10.15. As there are several interfaces working in parallel, network transmission capabilities are increased by a factor N, where N is the number of CWDM interfaces used.

Also, as upstream transmission is independent from downstream, the number of upstream interfaces may be different from the downstream ones, offering symmetric or asymmetric transmission data rates and channels.

CWDM for Fiber Access Solutions

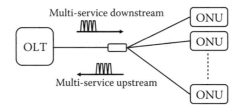

FIGURE 10.15 Proposed service overlay infrastructure of the TDM-PON.

Typically, in a multi-service ONU (see Figure 10.16) for home users, there would be several downstream interfaces and a single upstream transmitter. For this kind of user, present traffic patterns are asymmetric and downstream transmission requirements are normally higher than upstream ones. On the other hand, businesses and corporate users would require symmetric multi-service links having the same number of interfaces for downstream than upstream.

A typical approach for home users would be to offer different services like data, video-on-demand, and HDTV on different downstream wavelengths and offer different options to the customers depending on what they want to subscribe to [15]. As video transmission services are bandwidth demanding and, mostly, totally unidirectional, just one upstream wavelength is normally needed. Video conferencing is the only video service that is bidirectional. However, the video quality in this application case, and thus required bandwidth, is not as high as for entertainment or streaming video services. Also, each wavelength can transmit at different data rates, depending on the service. For corporate users, a typical approach would be to have different interfaces for intranet communications (optical VPNs) and for internet access. This separates and enhances intranet communications and offers an extra degree of security. Table 10.2 and Figure 10.17 present bandwidth requirements for different applications and also transmission requirements and traffic patterns.

10.3.2 Multi-Carrier Infrastructure

The service overlay model allows the transmission of different services independently on a common PON infrastructure, using different wavelengths.

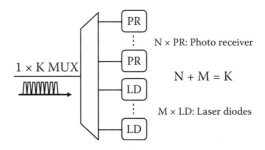

FIGURE 10.16 Details of a multi-service ONU design.

TABLE 10.2
Traffic Patterns and Bandwidth Requirements of Broadband Applications

Application	BW Requirements	Traffic Pattern	Latency
HDTV (High-definition TV)	20 Mb/s	Asymmetric, constant	Constant
DTV (Digital TV)	4–6 Mb/s	Asymmetric, constant	Constant
High-quality audio streaming	500 Kb/s	Asymmetric, constant	Constant
High-quality video conferencing	1 Mb/s	Symmetric, constant	<200 ms, constant
Virtual services (e-learning, virtual doctor...)	2–5 Mb/s	Symmetric	<200 ms, constant
Online gaming	1–10 Mb/s	Symmetric, variable	Low
Appliances remote control and sensoring	<500 Kb/s	Symmetric, variable	Low
Internet surfing	2–5 Mb/s	Asymmetric, variable	Not relevant
File sharing and P2P	5–10 Mb/s	Symmetric, variable	Not relevant

However, a more interesting approach is to share the PON infrastructure among several carriers. One of the barriers that FTTH is having at present is that deployment costs are extremely high, so to share the infrastructure among several operators is a good option to decrease capital expenses. This concept is also known as open-access networks. There is also another economic model in which this multi-carrier infrastructure solution has sense. In some geographical regions, municipalities are encouraging the use of broadband applications and internet access by deploying their own fiber infrastructure and then subcontracting the use of it to service providers. In this model, the CWDM multi-carrier model is a perfect solution. Each service provider is assigned a specific upstream and downstream wavelength and then the final customer has the choice to select the operator which best fulfills his requirements. Data streams of all the operators are transmitted on the same fiber. The selection of the individual carrier is very simple and is done at the CPE by means of the installation of the appropriate optical filter for downstream reception and the appropriate laser for the upstream transmission (see Figure 10.18). Another advantage of this model is that services are totally compatible so the selection of one does not limit the reception of other services at the same CPE. This economic model is known as network unbundling.

10.3.3 Service Filtering

At the ONU site, there are several solutions to simultaneously receive several channels. The straightforward solution is to install a CWDM multiplexer and then connect the photoreceiver to the port that provides the service that we want to subscribe to. However, this solution is costly and inefficient because, typically, just a few ports of the MUX will be used. This solution is only justified for

CWDM for Fiber Access Solutions

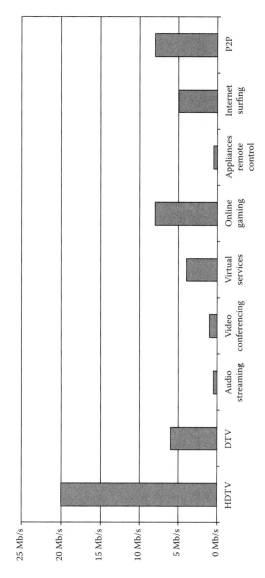

FIGURE 10.17 Typical bandwidth required for broadband applications.

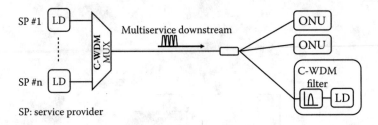

FIGURE 10.18 Multi-carrier model and service provider selection.

buildings and businesses in case we are installing the infrastructure in a building to give access to a large user base. On the other hand, this is the solution that offers the best power budget. Another solution would be to cascade Bragg grating filters and drop the required wavelength on each stage. This is a more cost-effective approach that also offers a good power budget. Finally, a third solution would be to use a power splitter and CWDM filters. This last approach is simpler but also offers more losses, limiting power budget. All these solutions are presented in Figure 10.19.

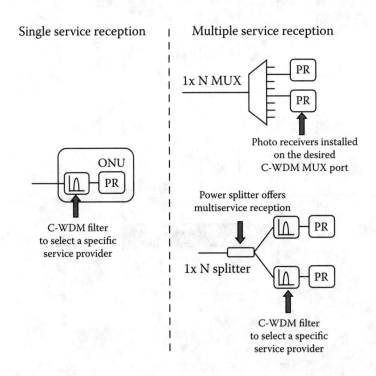

FIGURE 10.19 ONU for single and multiple service reception.

CWDM for Fiber Access Solutions

10.4 ADVANCED CWDM NETWORK IMPLEMENTATIONS

CWDM has a lot of potential in future access networks, as mentioned earlier. To demonstrate this potential, several network implementations and test-beds have been lately implemented. The objective of this section is to list and describe the most relevant work that has been carried out using CWDM in access.

10.4.1 CWDM Overlay on EPON Systems

In reference [16], a laboratory demonstration of an upgrade from a PON to a CWDM PON is implemented using zero water peak fiber to extend the number of CWDM channels. The concept that is developed is to expand a single fiber EPON/GPON transmitting at 1490 (downstream) and 1310 nm (upstream) by adding wavelength routers, thus allowing the multiplexing of additional channels. As this experiment highlights the use of zero peak fiber, it is focused on the wavelengths of the E-band (1370 to 1450 nm) where the water peak attenuation takes place. Figure 10.20 presents the proposed upgrade path. By using two E-band add and drop filters, the overlaid E-band wavelengths are combined and separated with/from the PON signals. This solution allows sharing the truck fiber of the access network, which is typically 90% of the total length of the access network, and thus reduces upgrade costs.

10.4.2 Ring-Based CWDM Metro/Access Network

A different approach is presented in reference [17]. Here, the network topology is based on a primary single-fiber distribution ring from which access nodes add and drop wavelengths to secondary trees, which interconnect the end users to the network. CWDM provides multiplexing capabilities to transmit several channels on the primary ring in a cost-effective way. Figure 10.21 presents the network

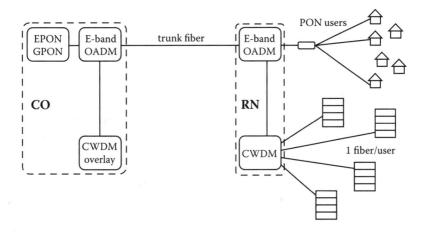

FIGURE 10.20 CWDM overlay on a legacy EPON/GPON.

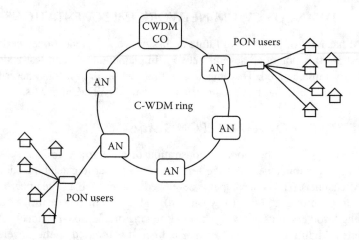

FIGURE 10.21 CWDM ring for network overlay.

topology. As an additional feature, the primary ring provides resilient capabilities in the event of a fiber cut.

10.4.3 Wireless and CWDM Networks

Wireless applications and optical networks can be combined to offer the advantages of both solutions and deploy a greatly optimized solution. The Radio-over-Fiber (RoF) concept consists of transmitting RF signals over a fiber link to simplify the remote antenna and thus concentrating all the network intelligence and signal processing at the CO while having very simple remote site antennas. CWDM has also a potential in RoF systems to enhance transmission capabilities. In references [18,19], CWDM is used to multiplex several RF channels on the same fiber and distribute them to different remote base stations sharing the optical infrastructure while keeping independent P2P CWDM channels between each remote station and the CO, as is depicted in Figure 10.22.

10.4.4 CWDM and D/UD-WDM Access Networks

Typical scenarios combine CWDM and TDM to expand the number of users that are covered by a network subsegment. However, there is another strategy to achieve high granularity which is to combine CWDM and dense WDM (DWDM) or even ultra dense WDM (UDWDM). The approach consists of subdividing the optical spectrum in different bands corresponding to the CWDM grid and then transmitting on every CWDM several D/UDWDM channels (see Figure 10.23). This allows P2P connections between the CO and each ONU offering much higher transmission capabilities. This approach uses CWDM multiplexer in the first routing stage and D/UDWDM multiplexer in the second stage. Its practical implementation nowadays

CWDM for Fiber Access Solutions

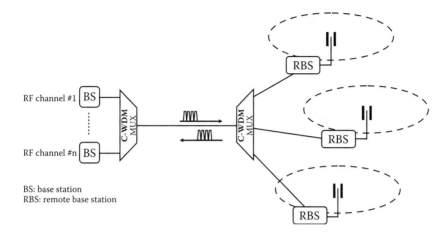

FIGURE 10.22 CWDM to enhance connectivity in RoF networks.

is restricted by the high costs of WDM equipment and the fact that each ONU uses a specific wavelength to transmit the upstream channel, which means that each ONU requires a wavelength-specific laser or a tunable source.

Another approach combining CWDM and DWDM is presented in reference [20] and consists of having first a DWDM stage and then CWDM at the secondary RNs. This approach exploits the free spectral range (FSR) periodicity of AWGs [21] and has as its main advantage the number of DWDM equipment is lower, thus it is a more cost-effective solution. By matching the FSR with the channel spacing of the CWDM stage (typically 20 nm), DWDM wavelengths are distributed to the CWDM stage and there they are separated to each destination port, as can be seen in Figure 10.24.

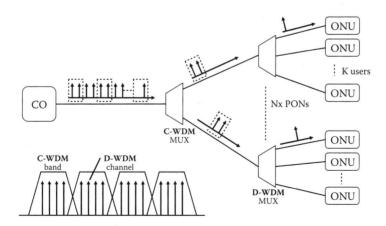

FIGURE 10.23 CWDM combined with D/UDWDM.

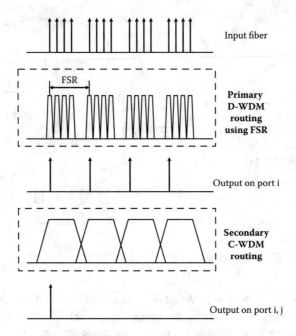

FIGURE 10.24 Routing profile of DWDM + CWDM using the FSR of the DWDM AWG router.

10.5 FUTURE TRENDS

Evolution paths in the research field on CWDM applied to access/FTTH are mainly focused on the following points:

- Delivery of more bandwidth per channel
- Extending the reach of access networks and increasing splitting ratio
- Development of colorless ONUs

The first point is linked to improving network throughput, the second to improving sharing factors, and the third to improving stock management. All three are related to economics because those improvements reduce deployment, operational and management costs, all of them crucial to making an access network economically viable.

The extension of CWDM to transmit at 10 Gb/s line rate has already been covered in Chapter 7 and is not exclusive of access but general to any transmission network [22,23]. There is no technical limitation to transmit using CWDM technology at 10 Gb/s because the optical principles are the same as in the DWDM case, where 10 Gb/s is a very common data rate. The contradiction is that CWDM is targeted for low-cost implementations of WDM to achieve cost-effective deployments and simply scaling to 10 Gb/s requires advanced electronics, external optical modulators, and generally dispersion-compensation techniques, which are costly equipment. Therefore, research on CWDM at 10 Gb/s is targeted to achieve 10-Gb/s transmission using just simple and low-cost electronics and optical equipment.

Another trend that deserves attention is to work on the extension of access networks toward the metropolitan area. This approach aims to suppress the intermediate COs or at least make them very simple in order to reduce the allocation and maintenance costs. By having a central access node, sharing factors are higher and thus the central equipment is shared by a larger number of users, so utilization is enhanced and cost per user is generally reduced [24]. On the other hand, the fact that the infrastructure is shared by a larger number of users means that in case of a network failure the number of users that are affected is much higher. Therefore, resilience techniques are essential to maintain network operation even in case of a network failure. Lately, optical resilience techniques that offer transparent protection to the end users are becoming popular in access [5,25]. CWDM is a key technology in the expansion of access networks toward MAN. At present, CWDM is used in metropolitan area networks (MANs) to offer more transmission capabilities per fiber so its extension to access is straightforward with the only limitation that just a few channels can be optically amplified, as discussed in Section 10.1.4.

After discussing the future trends in transmission and the network plant, the last point to be covered is the improvement of the optical unit at the end-user premises. In CWDM, there is a critical issue, which is that different wavelengths can be used to transmit, thus different lasers are required in each case. This creates stock problems, which can be solved by developing modular ONUs with a common mainframe and then a CWDM upstream laser module. However, in the research field, there is a lot of work involved in the development of colorless ONUs, which are mainly targeted for WDM applications. Most of them are based on the remodulation of an incoming signal coming from the CO. DWDM has several candidates to become future reflective ONUs [10], but for CWDM further research needs to be carried out to find optical devices or advanced techniques that are able to cover the full CWDM spectrum.

REFERENCES

1. Ramaswami, R. and Sivarajan, K.N., *Optical Networks, a Practical Perspective*, Morgan Kaufmann, San Francisco, 1998.
2. ITU-T recommendations G.983.x [x =1..9]: Broadband optical access systems based on Passive Optical Networks (PON).
3. ITU-T recommendations G.984.x [x = 1..4]: Gigabit-capable passive optical networks (GPON).
4. IEEE 802.3ah Ethernet in the First Mile Task Force, *Part 3: Carrier Sense Multiple Access with Collision Detection (CSMA/CD) Access Method and Physical Layer Specifications Amendment: Media Access Control Parameters, Physical Layers, and Management Parameters for Subscriber Access Networks*, IEEE STD 802.3ah-2004.
5. Langer, K.-D., Grubor, J., and Habel, K., Promising Evolution Paths for Passive Optical Access Networks, Proceedings International Conference on Transparent Optical Networks, vol. 1, pp. 202–207, 4–8 July 2004.

6. ITU-T recommendation G.694.1: Spectral grids for WDM applications: DWDM frequency grid.
7. ITU-T recommendation G.694.2: Spectral grids for WDM applications: CWDM wavelength grid.
8. Bock, C., Prat, J., and Walker, S.D., Hybrid WDM/TDM PON using the AWG FSR and featuring centralized light generation and dynamic bandwidth allocation, *J. Lightwave Technol.*, 23(12), 3981–3988, December 2005.
9. Bock, C. and Prat, J., WDM/TDM PON experiments using the AWG free spectral range periodicity to transmit unicast and multicast data, *OSA Optics Express*, 13(8), 2887–2891, April 2005.
10. Prat, J., Arellano, C., Polo, V., and Bock, C., Optical network unit based on a bidirectional reflective semiconductor optical amplifier for fiber-to-the-home networks, *Photon. Technol. Lett.*, 17(1), 250–252, January 2005.
11. Full-Band Tunable Laser TSL-210F, http://www.santec.com/products/lasers TSL 2 10F.htm.
12. Bookham 2.5Gb/s CWDM Buried Het. Laser 80km reach, www.bookham.com/datasheets/transmitters/LC25WC.cfm.
13. Agilent 81600B Tunable Laser Source Family, www.home.agilent.com/USeng/nav/-536891695.536883260/pd.html.
14. Prat, J., Arellano, C., Polo, V., and Bock, C., Optical network unit based on a bidirectional reflective semiconductor optical amplifier for fiber-to-the-home networks, *Photon. Technol. Lett.*, 17(1), 250–252, January 2005.
15. Kawata, H., Ogawa, T., Yoshimoto, N., and Sugie, T., Multichannel video and IP signal multiplexing system using CWDM technology, *J. Lightwave Technol.*, 22(6), 1454–1462, June 2004.
16. Eichenbaum, B.R., Thiele, H.J., Nelson, L.E., Matthews, M.J., and Van Veen, D.T., CWDM Overlay on an EPON System — A Demonstration and Economic Ramifications for FTTH, Proceedings NFOEC 2003, pp. 574–580.
17. Wang, Z., Lin, C., and Chan, C.-K., Demonstration of a single-fiber self-healing CWDM metro access ring network with unidirectional OADM, *Photon. Technol. Lett.*, 18(1), 163–165, January 2006.
18. Martinez, A., Munoz, P., Capmany, J., Sales, S., Ortega, B., and Pastor, D., Multiservice hybrid radio over fiber and baseband AWG-PON using CWDM and spectral periodicity of arrayed waveguide gratings, *Photon. Technol. Lett.*, 16(2), 599–601, February 2004.
19. Ismail, T., Liu, C.P., Mitchell, J.E., Seeds, A.J., Qian, X., Wonfor, A., Penty, R.V., and White, I.H., Full-Duplex Wireless-over-Fiber Transmission Incorporating a CWDM Ring Architecture with Remote Millimetre-Wave LO Delivery Using a Bi-directional SOA, Proceedings OFC/NFOEC 2005, vol. 4, March 2005.
20. Parker, M.C., Farjady, F., and Walker, S.D., Wavelength-tolerant optical access architectures featuring N-dimensional addressing and cascaded arrayed waveguide gratings, *J. Lightwave Technol.*, 16(12), 2296–2302, December 1998.
21. Smit, M.K. and Van Dam, C., PHASAR-based WDM-devices: principles, design and applications, *J. Selected Topics in Quantum Electronics*, 2(2), 236–250, June 1996.
22. Thiele, H.J., Nelson, L.E., and Das, S.K., Capacity-enhanced coarse WDM transmission using 10 Gbit/s sources and DWDM overlay, *Electron. Lett.*, 39(17), 1264–1266, August 2003.

23. Winzer, P.J., Fidler, F., Matthews, M.J., Nelson, L.E., Thiele, H.J., Sinsky, J.H., Chandrasekhar, S., Winter, M., Castagnozzi, D., Stulz, L.W., and Buhl, L.L., 10-Gb/s upgrade of bidirectional CWDM systems using electronic equalization and FEC, *J. Lightwave Technol.*, 23(1), 203–210, January 2005.
24. Davey, R., The Future of Optical Transmission in Access and Metro Networks — an Operator's view, Proceedings ECOC 2005, Glasgow, September 2005, We2.1.3.
25. Bock, C., Arellano, C., Lázaro, J.A., and Prat, J., Resilient single-fiber ring access network using coupler-based OADMs and RSOA-based ONUs, Proceedings OFC 2006, JThB78, March 2006.

List of Abbreviations

µm	micrometer
2R	re-amplification, re-shaping
3R	re-amplification, re-shaping, re-timing
ADSL	Asymmetric digital subscriber line
ADM	Add/drop multiplexer
Alpha	loss
AM	amplitude modulation
AOI	angle of incidence
AON	active optical network
APC	automatic power control
APD	avalance photodiode
APON	ATM over PON
AR	anti-reflection
ASE	amplified stimulated emission
ATM	asynchronous transfer mode
AWG	arrayed waveguide
BCH	Bose-Chandhuri-Hocquenghem
BER	bit error rate
BPON	broadband PON
CAT5	cable standard
CATV	cable television
C-band	transmission band
CD	chromatic dispersion
CDR	clock data recovery
CMOS	complementary metal oxide semiconductor
CNR	carrier to noise ratio
CO	central office
CPE	customer premises equipment
CSMF	conventional single mode fiber
CSO	composite second order
CTB	composite triple beat
CW	continuous wave
CWDM	coarse wavelength division multiplexing
dB	power ratio
dBm	power level

DBR	distributed Bragg reflector
DCF	dispersion-compensating fiber
DCM	dispersion-compensating module
Deg	degree
DEMUX	demultiplexer
DFB	distributed feedback
DFE	differential feed equalization
DGD	differential group delay
DML	directly modulated laser
DSF	dispersion-shifted fiber
DSLAM	digital subscriber line access multiplexer
DTV	digital television
DWDM	dense wavelength division multiplexing
EA-EML	electro-absorption externally modulated laser
E-band	transmission band
EDFA	erbium-doped fiber amplifier
EDTFA	erbium-doped tellurite fiber amplifier
EML	externally modulated laser
EPON	Ethernet PON
ER	extinction ratio
ESCON	enterprise systems connection
ETDM	electrical time division multiplexing
FBG	fiber Bragg grating
FBGL	fiber Bragg grating laser
FBT	fused biconic taper
FC	flat connector type
FEC	forward error correction
FFE	forward feed equalization
FGL	fiber-grating laser
FM	frequency modulation
FOM	figure of merit
FP	Fabry-Perot
FSAN	full service access network
FS-CWDM	full-spectrum CWDM
FSK	frequency shift keying
FTTB	fiber-to-the-building
FTTC	fiber-to-the-curb
FTTCab	fiber-to-the-cabinet
FTTH	fiber-to-the-home
FTTU	fiber-to-the-user
FTTX	fiber-to-anywhere
FWHM	full width half maximum
FWM	four-wave mixing

List of Abbreviations

GaAs	gallium arsenide
Gb/s	gigabit per second
GbE	gigabit Ethernet
GBIC	gigabit interface converter
GEM	GPON encapsulated method
GFF	gain flattening filter
GHz	gigahertz
GPON	gigabit PON
GRIN	gradient index
HDTV	high definition television
HFC	hybrid fiber coax
IEC	International Electrotechnical Commission
IEEE	Institute of Electronic and Electrical Engineering
IP	internet protocol
IPTV	internet protocol television
IR	infrared
ISI	intersymbol interference
ISL	insertion loss
ITU-T	International Telecommunications Union
Lambda	wavelength
LAN	local area network
L-band	transmission band
LC	Lucent connector type
LD	laser diode
LDF	low dispersion fiber
LED	light emitting diode
LH	transceiver standard
LMC	low mode coupled
LOA	linear optical amplifier
LR	long range
LWPF	low water peak fiber
LX	transceiver standard
LX4	4-channel transceiver standard
MAN	metropolitan area network
MBE	molecular beam epitaxy
Mbit/s	megabit per second
MCVD	modified chemical vapor deposition
MDF	medium dispersion finer
MLSE	maximum likelihood sequence estimation
MMF	multimode fiber
m-QAM	m-fold quadrature amplitude modulation

MQW	multi-quantum well
MSA	multisource agreement
MUX	multiplexer
MZI	Mach Zehnder interferometer
MZM	Mach Zehnder modulator
NA	numerical aperture
NF	noise figure
NIC	network interface card
Nm	nanometer
NRZ	nonreturn to zero
NZDSF	nonzero dispersion-shifted fiber
OADM	optical add-drop multiplexer
O-band	transmission band
OC-192	optical carrier level, 10 Gb/s
OC-3	optical carrier level, 155 Mb/s
OC-48	optical carrier level, 2.5 Gb/s
OC-768	optical carrier level, 40 Gb/s
OD	optical demultiplexer
OD	outside diameter
ODD	overclad during draw
OEO	optical-electrical-optical
OH	hydrogen
OIF	optical interface
OLT	optical line terminal
OM	optical multiplexer
ONU	optical network unit
OOK	on-off keying
OTDM	optical time division multiplexing
OVD	outside vapor deposition
OXC	optical cross connect
P2MP	point-to-multi-point
P2P	point-to-point
PBC	polarization beam splitter
PCVD	plasma chemical vapor deposition
PDG	polarization-dependent gain
PDL	polarization-dependent loss
PIN	P-intrinsic-N diode
PMD	polarization-mode dispersion
PON	passive optical network
PR	photo receiver
PRBS	pseudo-random bit sequence

List of Abbreviations

QPSK	quadrature phase shift keying
QSFP	quad small form factor pluggable
RBS	remote broadcasting station
RDF	reverse dispersion fiber
RF	radio frequency
RIT	rod in tube
RMS	root mean square
RN	remote node
ROADM	reconfigurable optical add-drop multiplexer
RoHS	restriction of hazardous substances
RS	Reed Salomon
RSOA	reflective semiconductor optical amplifier
RX	receiver
RZ	return to zero
SAN	storage area network
S-band	transmission band
SBS	stimulated Brillouin scattering
SDH	synchronous digital hierarchy
SDM	space division multiplexing
SFF	small form factor
SFP	small form factor pluggable
SIR	signal to interference
SMF	single mode fiber
SMF-28	standard single mode fiber
SMSR	side mode suppression ratio
SNR	signal-to-noise ratio
SOA	semiconductor optical amplifier
SONET	synchronous optical network
SPM	self-phase modulation
SR	short range
SRS	stimulated Raman scattering
SSMF	standard single mode fiber
ST	straight tip connector type
SX	transceiver standard
T1	telecommunications connection type
TDFA	thulium-doped fiber amplifier
TDL	temperature-dependent loss
TDM	time division multiplexing
TDMA	time division multiple access
TDM-PON	time division multiplexing PON
TE	transversal electrical
TEC	thermoelectrical cooler

TF	tunable filter
TFF	thin-film filter
TIA	transimpedance amplifier
TM	transversal magnetic
TX	transmitter
VAD	vapor axial deposition
VCSEL	vertical cavity surface emitting laser
VDSL	very high bitrate digital subscriber line
VoD	video on demand
VSB	vestigial sideband
WAN	wide area network
WDM	wavelength division multiplexing
WWDM	wide wavelength division multiplexing
XAUI	10 Gb/s attachment interface
xDSL	any DSL type
XFP	10 Gb/s transceiver standard
XPM	cross-phase modulation
ZWP	zero water peak

Index

A

Abbreviations, list of, 313–317
Absolute gain, 280
Access network(s), 289
   CWDM, ring-based, 305
   extension toward metropolitan area, 309
Access solutions, available bandwidths, 287
Accumulated dispersion, 226
Active access networks, 289
Active optical networks (AONs), 289
Add/drop MUX, 93
Adiabatic chirp, 29, 214, 239
Adjacent channel isolation, ITU-T G.671, 17
ADSL transmission technologies, 287
Alkali contamination, 49
Allowable input power, ITU-T G.671, 15
AllWave fiber
   deuterium treatment process for, 48, 49
   DWDM overlay and, 223
   FEC-enabled transmission over, 236
   performance, RIC-ODD process and, 44
   spectral attenuation, 128
AM, *see* Amplitude modulation
Amplification, *see* Optical amplification
Amplified spontaneous emission (ASE), 175, 178
Amplifier(s), 171–197, *see also* Semiconductor optical amplifier
   amplified CWDM transmission line, 193–194
   challenges of amplifying CWDM, 177–180
      influence of broadband noise, 178–180
      wideband amplification, 177–178
   configuration, EDFA–TDFA, 181, 182
   distributed, 174
   doped fiber-based CWDM amplifiers, 180–182
   erbium-doped fiber, 39, 107, 171, 276, 291
   gain, 174, 186
   hybrid, 190
   limiting, 123
   lumped element, 174
   principles of optical amplification, 172–177
      gain saturation and gain dynamics, 176–177
      noise, 175–176
      optical amplifier types, 172–174

   properties, 177
   radio frequency, 270
   Raman, 172, 174, 178, 190–192
   rare earth doped fiber, 173
   semiconductor-based CWDM amplifiers, 182–190
      device specifics, 182–183
      dynamics, gain clamping, 184–186
      fiber transmission, 187–190
   thulium-doped fiber, 178
   transimpedance, 80, 123
   transmission distance, 171
Amplitude modulation (AM), 270
Angle of incidence (AOI), 106–107, 113, 122
Antireflection (AR) coating, 106, 113
AOI, *see* Angle of incidence
AONs, *see* Active optical networks
APC circuitry, *see* Automatic power control circuitry
APD receiver, *see* Avalanche photodiode receiver
Apodized grating, 94, 97
APON, *see* ATM PON
Application code(s)
   achievable transmission distance for, 11
   bidirectional applications, 8
   ITU-T optical interface, 3
   summary, ITU-T G.695, 15, 16
AR coating, *see* Antireflection coating
Arrayed waveguide(s)
   demultiplexer, structural elements of, 100
   most commonly used process for, 99
   response function of, 100
Arrayed waveguide gratings (AWGs), 92
   design, ASIC design and, 99
   free spectral range periodicity of, 307
   narrow passband of, 101
   router, DWDM, 308
ASE, *see* Amplified spontaneous emission
ATM PON (APON), 290
Attenuation
   coefficient, SSMF, 201
   dependent loss path, 130
   losses, nonfiber, 136, 137
   slope compensating wavelength assignment, 147, 152

319

Automatic power control (APC) circuitry, 72
Avalanche effect, APD diodes and, 70
Avalanche photodiode (APD) receiver, 202, 234, 262
  circuits, 77
  input power, 11
  sensitivity of, 68
AWGs, see Arrayed waveguide gratings

B

Bandpass filters, 107, 109, 110
Bandwidth
  addition of, 260
  demand
    access networks, 253
    growth of, 251
    video transmission services and, 301
  requirement(s)
    broadband applications, 302, 303
    capacity-demanding applications and, 267
    SANs, 263
BCH code, see Bose-Chandhuri-Hocquenghem code
BER, see Bit error ratio
Bidirectional crosstalk attenuation, ITU-T G.671, 17
Bidirectional CWDM link configuration, transceivers and, 80, 81–82
Bidirectional isolation, ITU-T G.671, 17
Bidirectional transmission, 71
Bit error ratio (BER), 13, 202, 210, 234
Bit rate
  distance product, 62
  operation, increased, 218
  transmission, mixed, 227
Black box
  applications, optical multiplexer and, 11
  approach, 3, 5
  internal optical signals, 6
  model, 5
  transmitter and receiver, 6
Black link(s)
  application(s)
    codes, 7–8
    optical multiplexer insertion losses, 11
  approach, 3, 5, 7
  concept, 7
  model, optical characteristics, 7
  optical power penalty and, 8
  systems, plug-and-play nature of, 6
  virtual dark fiber services and, 7
Booster amplification, 193

Bose-Chandhuri-Hocquenghem (BCH) code, 237
BPON, see Broadband PON
Broadband
  applications
    bandwidth requirements, 302, 303
    traffic patterns, 302
  noise, 178–180
  PON (BPON), 290

C

Cable effective cut-off wavelength
Cable television (CATV), 253, see also CATV/HFC networks
  overlay, C-band and, 292
  systems, primary network configuration, 272
Capacitance, bit rate and, 70
Capacity-demanding applications, 267
CAPEX intense high precision machinery, 99
Carrier to noise (CNR), 273
CATV, see Cable television
CATV/HFC networks, 269–283
  applications for CWDM, 276–282
    forward path narrowcast overlay, 276–278
    limitations due to SRS, 279–282
    return path, 278–279
  CATV architecture, 270–275
    analog direct modulation, 274–275
    HFC forward path optics, 273
    HFC reverse path optics, 273–274
  future network designs, 283
  noise limiting system performance of, 279
  optical star topology, 272
  trunk network, 277
C-band, 20, 65, 173
  CATV overlay and, 292
  EDFA, 172
  split-band amplifier for, 180
CD, see Chromatic dispersion
Center wavelength deviation parameter, 13
Central office (CO), 287, 289, 295
Channel
  density, design strategy for increasing, 119
  insertion loss deviation, ITU-T G.671, 15
  launch power, 281
  overlay, CWDM, 206
  power, receiver sensitivity and, 133
  wavelength range, ITU-T G.671, 17
Chirp, see also Laser chirp
  adiabatic, 29, 214, 239
  components, 214
  dispersion and, 30
  DML, 67

Index

dynamic, 29, 30, 214
free lasers, signal bandwidth and, 28
low laser, 211
negative, 214
parameters, 29
positive, 214
values, pulse broadening and, 30
Chirped grating, 94
Chromatic dispersion (CD), 27–31, 200
  chirp and, 30
  effects, 28
  laser chirp and, 216
  SSMF, 206
  tolerance, 28
  wavelength versus, 28
  when transmitter has chirp, 29
Cladding
  index, 23, 24
  mode propagation constants, 25
  ovality, 35
CNR, *see* Carrier to noise
CO, *see* Central office
Coarse wavelength division multiplexing (CWDM), 1, 91, 125, 200
  amplification, 172, 177–180
    influence of broadband noise, 178–180
    Raman effect and, 35
    wideband amplification, 177–178
  amplifiers
    doped fiber-based, 180–182
    semiconductor-based, 182
  black link ring
    hubbed, 9
    nonhubbed, 10
  CATV/HFC networks, 276–282
    forward path narrowcast overlay, 276–278
    limitations due to SRS, 279–282
    return path, 278–279
  channel(s)
    laser wavelength drift for, 3, 4
    optical crosstalk between, 13
    overlay, 206
    plan, ITU-based, 126, 127
    spacing, 269
    transmission distance, 241
  coexistence with legacy infrastructure, 297
  demultiplexer(s), 93
    insertion losses, 11
    ITU-T G.671 parameters, 15
    metropolitan networks and, 254, 255
  devices
    center wavelength thermal stability, 102
    directivity, 102
    insertion loss, 102

    isolation, 102
    passband ripple, 102
    passband width, 102
    polarization-dependent loss, 102
    return loss, 102
  dispersion-compensating fiber for, 51
  DWDM over, 206
  filter(s)
    dielectric coating structure, 106
    example, 109
    reflectance spectrum, 114
  FTTx
    advantages, 291
    limitations, 293
  full-spectrum, 234
  hubbed ring network, 126, 129, 135, 136
  important features for using in network, 91
  international standards, 2
  link(s)
    linear add-drop, 9
    point-to-point, 8
  meshed network, N-node, 156
  metropolitan networks, 254–258, 262
    add/drop multiplexer, 256–257
    cross-connect, 257
    disaster recovery, 263–264
    multiplexer and demultiplexer, 254–256
    primary transceiver, 259
    rings, 258, 259
    scalability with DWDM over CWDM, 265–266
    single wavelength networks, 264–265
    transceiver, 257–258
  modules, size of, 119
  multi-carrier model, 302, 304
  multiplexer(s), 93
    filter, 74
    insertion loss, 11
    ITU-T G.671 parameters, 15
    metropolitan networks and, 254
    parameters, 15
    passband of, 59
    service filtering and, 302
    upgraded channel, 228
  narrowcast optical channels, 278
  passband center, uncooled lasers, 75
  pay as you grow philosophy, 203, 253, 260
  rings using, 258
  ring topology, metropolitan networks and, 254
  scaling to multiple wavelengths, 260
  semiconductor-based amplifiers and, 179
  set transceivers for use in, 79
  solutions, price erosion of, 267

spectrum, SSMF macrobending loss, 25
standards, 1
system(s)
  full-spectrum, 201
  limiting factor in, 27
  link budget for, 3
  multiple bounce concept and, 121
  photodetectors commonly used in, 70
  reach limitation, 169
TF filters for, 104
transceiver(s)
  data loss from, 261
  integrated, 89
transmission
  bidirectional, 242, 244, 245
  equalization, 209
  forward error correction and, 209
  line, amplified, 193
  link, 78
uncooled transmitters for, 73, 74
wavelength(s)
  channels, filter losses for, 147
  chromatic dispersion, 13
  grid, ITU-T G.694.2, 5
  range, permissible, 60
  return path aggregation using different, 279
Coarse wavelength division multiplexing laser(s)
  advances in, 211–223
    increased bit-rate operation, 218–222
    laser chirp, extinction ratio, and chromatic dispersion, 213–218
    laser output power, 211–213
    other laser types for 10-Gb/s CWDM systems, 222–223
  power penalties of, 224
  temperature variation, 2
  uncooled, 219
  wavelength, narrow passband of, 101
Coarse wavelength division multiplexing network(s), 128–135
  advanced implementations, 305–307
    CWDM and D/UD-WDM access networks, 306–307
    CWDM overlay on EPON systems, 305
    ring-based CWDM metro/access network, 305–306
    wireless and CWDM networks, 306
  attenuation slope compensating wavelength assignment, 128–131
  channel power, 133
  impact of loss, 133–135
  ring perimeter, 131–133
  spectral range of, 98

Composite triple beat (CTB), 273
Continuous wave (CW) laser output, 211, 212, 270
Copper link, achievable distance with, 286
Core-cladding concentricity, 36
Core networks, 252
Corning
  LEAF® NZDSF, 40
  MetroCor® NZDSF, 40
Coupling coefficient, microbending loss, 24
CPE, see Customer premises equipment
Cross-phase modulation (XPM), 35, 232
CTB, see Composite triple beat
Customer premises equipment (CPE), 289
Cut-off wavelength
  change in, 34
  effective, 32, 33, 34
  theoretical, 32
CWDM, see Coarse wavelength division multiplexing
CW laser output, see Continuous wave laser output

D

Damping factor, 214
Data
  recovery, 237
  storage, CWDM wavelengths for, 266
  streams, transmission of, 302
DBRs, see Distributed Bragg reflectors
DCF, see Dispersion-compensating fiber
Decision feedback equalization (DFE), 237
Demultiplexer (DEMUX), 58, 93
  arrayed waveguide, structural elements of, 100
  4-channel, 123
  insertion loss sequence, 119, 120
  wavelength-routed distribution and, 296
DEMUX, see Demultiplexer
Dense wavelength division multiplexing (DWDM), 2, 20, 58, 125, 171, 251, 306
  amplification, 177
  AWG router, 308
  channel overlay, 207, 208
  CWDM and, 200
  laser(s)
    transmitters, 72
    uncooled, 220
  multi-channel, 269
  multiplexer, 207
  over CWDM, 206
    metropolitan networks and, 265
    network upgrade using, 266

Index

overlay, 203, 223
sub-band, transmission penalties, 227
systems, filters used in, 126
transmission equipment for, 291
upgrade channels, bit-rate of, 208
Depressed cladding fiber, 38
Detectors, comparison of, 69
Deuterium treatment process, 48, 49
DFB laser, *see* Distributed feedback laser
DFE, *see* Decision feedback equalization
DGD, *see* Differential group delay
Dielectric coatings, 104
Differential gain, 214
Differential group delay (DGD), 15
Directly modulated laser (DML), 21, 29, 60, 202
   chirp, 67, 215, 225
   frequency chirp, 29
   laser chirp and dispersion for, 67
   modulation behavior of, 214
   modulation characteristics, 221
   modulation response, 220
   overlay channels, optical spectrum of, 233
   positive chirp of, 214
   static characteristics of, 212
   transmission penalty, 229
   uncooled, 205, 211, 218, 234
Disaster recovery, metropolitan networks, 263
Dispersion
   accumulated, 226
   anomalous, 67
   DML, 67
   fiber, negative, 216
   shifted fibers (DSFs), 27
   slope, 52
   tolerance, 66, 68
      fiber gating laser, 64
      SSMF, 63
Dispersion-compensating fiber (DCF), 50
   figure of merit for, 51
   loss, CWDM muxes/demuxes, 226
   module, performance of, 52
   Raman-pumped, 53
Distributed amplifier, 174
Distributed Bragg reflectors (DBRs), 186
Distributed feedback
   direct modulation of, 67
   emission wavelength, 215
   resonance, laser temperature and, 275
Distributed feedback (DFB) laser, 60, 63, 62, 201
   common use for, 64
   cost of, 62
   device data, 64
   performance of, 62

restricted use of, 64
use in WDM systems, 63
wavelength detuning in, 216
DML, *see* Directly modulated laser
Doped fiber-based CWDM amplifiers, 180–182
Dry fiber, 12
DSFs, *see* Dispersion-shifted fibers
Dual fiber collimator, 115
DWDM, *see* Dense wavelength division multiplexing
Dynamic chirp, 29, 30, 214

E

EA effect modulator, *see* Electro absorption effect modulator
EA-EMLs, *see* Electro-absorption-based externally modulated lasers
EAMs, *see* Electro-absorption modulators
E-band, 13, 22, 305
Economic model, 302
EDFA, *see* Erbium-doped fiber amplifier
Edge filters, 107, 108
EDTFA, *see* Erbium-doped tellurite fiber amplifier
EFFA, *see* Erbium-doped fiber amplifier
Effective cut-off wavelength, 32, 33, 34
Electro-absorption-based externally modulated lasers (EA-EMLs), 61
Electro absorption (EA) effect modulator, 65
Electro-absorption modulated lasers (EMLs), 228
Electro-absorption modulators (EAMs), 229, 299
Electronic equalization, CWDM transmission and, 210
Electronics, integration of optics with, 122
EMLs, *see* Electro-absorption modulated lasers
EPON, *see* Ethernet PON
ER, *see* Extinction ratio
Erbium-doped fiber amplifier (EDFA), 39, 107, 171, 276, 291
   constant-power signal, 176
   excited state of erbium atom in, 176
   standard amplification band of, 173
   TDFA amplifier configuration, 181, 182
Erbium-doped tellurite fiber amplifier (EDTFA), 180
Ethernet, 3, 253
   Gigabit, 203
   ITU-T G.695 and, 3
   OADM connection, 261, 262
   PON (EPON), 290

GPON, CWDM overlay on, 305
standard, 292
systems, CWDM overlay on, 305
router, 265
transceivers, types of, 84
European Community for Restriction of
Hazardous Substances, 119
Extinction ratio (ER), 30, 67, 212
laser chirp and, 218
power penalty and, 219
Eye diagrams, back-to-back, 221, 222, 243

F

Fabry Perot (FP) laser, 62–63, 222
cost of, 62
replacement options for, 63
FBG, *see* Fiber Bragg grating
FBGLs, *see* Fiber Bragg grating lasers
FEC, *see* Forward error correction
Feed forward equalization (FFE), 237
FFE, *see* Feed forward equalization
FGLs, *see* Fiber Bragg grating lasers
Fiber(s), *see also* Optical fiber
AllWave, 223
deuterium treatment process for, 48, 49
spectral attenuation, 128
bidirectional transmission, 71
coupling loss, 184
depressed cladding, 38
design(s)
low bend loss, 38
problem, for wideband, high bit rate
systems, 20
development, shift of focus, 20
dispersion
compensating, 50, 51
equation, 66
shifted, 27
dry, 12
effective cut-off wavelength, 34
exhaust, 267
gating laser, dispersion tolerance, 64
geometry requirements, ITU-T G.652, 37
germanium-doped, 47, 49
grating, index profile of, 98
hydrogen aging loss problem, 42
loss distribution, histogram of, 45
loss requirements, ITU-T G.652, 38
low water peak, 48, 37–38, 200
matched-cladding, 37
negative dispersion, 216
non-low-water-peak, assumed attenuation
coefficients for, 12

nonzero dispersion-shifted, 206
PMD, 46
properties, bending-induced loss, 22
macrobending loss, 23–24
microbending loss, 24–27
PureAccess, 40
RIC-ODD, 45
single mode
cut-off wavelength of, 31
dispersion in, 27
material dispersion, 27
profile dispersion, 27
waveguide dispersion, 27
zero-OH$^-$, 41–50
standard single mode, 22
step index, theoretical cut-off wavelength for,
32
strength, interface quality and, 46
water-peak, assumed attenuation coefficients,
12
wet, 12
zero water peak, 37–38, 200
EPON systems and, 305
hydrogen aging loss and, 48
Fiber access solutions, 285–311
advanced CWDM network implementations,
305–307
CWDM and D/UD-WDM access
networks, 306–307
CWDM overlay on EPON systems, 305
ring-based CWDM metro/access network,
305–306
wireless and CWDM networks, 306
fiber-to-the-home, 285–294
advantages of CWDM in FTTx, 291–292
bandwidth using CWDM, 292–293
basics and standards, 286–288
limitations of CWDM in FTTx, 293–294
point-to-point and point-to-multipoint,
288–289
standards for FTTH-PON, 290
future trends, 308–309
network overlay, 294–300
combining CWDM and TDM to create
high density PONs, 294–295
network design, 295–297
protocol compatibility with existing
standards, 297–299
upstream transmission, 299–300
service overlay, 300–304
multi-carrier infrastructure, 301–302
offering different services over same
network infrastructure, 300–301
service filtering, 302–304

Index

Fiber-to-anywhere (FTTx), 253, 286, 288, 291
Fiber attenuation, 21
Fiber Bragg grating (FBG), 92, 94
  athermalized package design, 96
  methods to obtain, 94, 95
  multichannel WDM signal reaching, 95
  refractive index modulation, 97
  results of athermalization, 96
  spectral loss, 98
Fiber Bragg grating lasers (FBGLs), 61, 64, 204, 222
  CWDM upgrade using, 223
  mode hopping, 65
Fiber-to-the-building (FTTB), 42, 286, 287
Fiber-to-the-cabinet (FTTCab), 286, 287
Fiber Channel, 3, 253
Fiber-to-the-curb (FTTC), 286
Fiber-to-the-desktop (FTTD), 287
Fiber-to-the-home (FTTH), 39, 286
  PON, standards, 290
  traffic, 41–42
Fiber-to-the-user (FTTU), 286
Figure of merit (FOM), 51, 226
Filter(s)
  bandpass, 107, 109, 110
  comparison of, 102
  CWDM
    dielectric coating structure, 106
    example, 109
    reflectance spectrum, 114
  edge, 107, 108
  gain flattening, 98, 175
  loss(es)
    calculation, 143, 148, 150
    CWDM wavelength channels, 147
    derivations, 135
  thin-film, 92, 133
    bandpass characteristics, 106
    coating layer structure of, 105
    implementation of OADM, 145–146
    market domination by, 101
    performance data for, 104
    price per channel, 101
    properties of, 104–113
    reflection losses for, 143
    scalability provided by, 101
  wide channel spacing, 125
FM, *see* Frequency modulation
FOM, *see* Figure of merit
Forward error correction (FEC), 172, 205
  board, CMOS-based, 238
  correction threshold, 210
  CWDM transmission and, 209

Fourier Transform Infrared (FTIR) spectroscopy, 44
4-skip-0, 107, 110
Four-wave mixing (FWM), 35, 186, 232
FP laser, *see* Fabry-Perot laser
Free spectral range (FSR) periodicity, 307
Frequency
  chirp, DML, 29
  modulation (FM), 214
  shift keying (FSK), 272
FSAN, *see* Full Service Access Network
FSK, *see* Frequency shift keying
FSR periodicity, *see* Free spectral range periodicity
FTIR spectroscopy, *see* Fourier Transform Infrared spectroscopy
FTTB, *see* Fiber-to-the-building
FTTC, *see* Fiber-to-the-curb
FTTCab, *see* Fiber-to-the-cabinet
FTTD, *see* Fiber-to-the-desktop
FTTH, *see* Fiber-to-the-home
FTTU, *see* Fiber-to-the-user
FTTx, *see* Fiber-to-anywhere
Full Service Access Network (FSAN), 290
Full-spectrum CWDM system, 201
Full width half maximum (FWHM) pulse width, 30
FWHM pulse width, *see* Full width half maximum pulse width
FWM, *see* Four-wave mixing

G

GaAs/InP photodetectors, 70
Gain
  absolute, 280
  amplifier, 174
  clamping, 186
  curve, Raman, 280
  differential, 214
  dynamics, 176
  flattening filters (GFFs), 98
  medium, SOA, 174
  Raman, 192
  ripple, 175
  saturation, 176
  signal wavelength and, 184
  spectrum, SOA, 175
GbE, *see* Gigabit Ethernet
GBIC, *see* Gigabit interface converter
GEM, *see* GPON encapsulation method
Germanium-doped fibers, hydrogen aging losses in, 47
GFFs, *see* Gain-flattening filters

Gigabit Ethernet (GbE), 84, 203
Gigabit interface converter (GBIC), 7, 84
Glass
  defect sites, 22
  packages, design approach, 119
  processing techniques, OH⁻ contamination and, 22
GPON encapsulation method (GEM), 290
Gradient index lens (GRIN), 115
Grating fiber, index profile of, 98
Greenfield scenarios, 286
GRIN, *see* Gradient index lens

H

HFC networks, *see* CATV/HFC networks; Hybrid fiber coax networks
Hubbed ring network, CWDM, 129, 135, 136
Hub loss model, 144
Hybrid amplifiers, 190
Hybrid fiber coax (HFC) networks, 270, 271, 272
Hydrogen aging loss(es)
  long wavelength loss and, 49
  LWP fibers, 42
  RIC-ODD process and, 47
  types of, 47
  ZWP fibers, 42

I

IEEE, *see* Institute of Electronic and Electrical Engineering
In-band crosstalk, 243
Infrared (IR) absorption loss, 21
Institute of Electronic and Electrical Engineering (IEEE), 85, 290
  PON standard, 290, 292
  transceiver performance standards, 85
International Telecommunications Union (ITU), 59
  CWDM channel plan standardized by, 126, 127
  FTTH-PON standards, 290
  grid, 59
  optical interface, application codes, 3, 6
  optical transmission recommendations, path penalty, 13
  proposal for CWDM systems, 22
  Recommendation G.652
    depressed cladding fiber, 38
    fiber geometry requirements, 37
    fiber loss requirements, 38

low bend loss fiber designs, 38–39
low water peak and zero water peak fiber, 37–38
single mode fiber cut-off wavelength, 32
Recommendation G.655, 39
Recommendation G.656, NZDSF and, 41
Recommendation G.671 (Transmission Characteristics of Optical Components and Subsystems), 2, 15
  adjacent channel isolation, 17
  allowable input power, 15
  bidirectional crosstalk attenuation, 17
  bidirectional isolation, 17
  channel insertion loss, 15
  channel insertion loss deviation, 15
  channel wavelength range, 17
  nonadjacent channel isolation, 17
  polarization dependent loss, 15
  polarization dependent reflectance, 15
  polarization mode dispersion, 15
  reflectance, 15
  ripple, 17
  unidirectional crosstalk attenuation, 17
Recommendation G.694.2 (Spectral Grids for WDM Applications), 2–3
Recommendation G.695 (Optical Interfaces for Coarse Wavelength Division Multiplexing Applications), 2, 3–15
  application code example, 6
  application code nomenclature, 3–5
  application code summary, 15, 16
  black box and black link approaches, 5–8
  black box interface, 6
  center wavelength deviation, 13–15
  center wavelength deviation parameter, 13
  channel plans, 13, 14
  distance, 11–13
  path penalty, 13
  power budget, 10–11
  topologies, 8–10
  unidirectional applications, 8
  unidirectional and bidirectional transmission, 8
Recommendation G.952, 293
Internet protocol television (IPTV), 253
Inter-symbol interference (ISI), 176
Intranet communications, 301
IPTV, *see* Internet protocol television
IR absorption loss, *see* Infrared absorption loss
ISI, *see* Inter-symbol interference
ITU, *see* International Telecommunications Union

Index

K

Kerr effect, 34, 35

L

LAN, *see* Local area network
Laser(s)
  cavity, refractive index of, 214
  chirp(s)
    chromatic dispersion and, 216
    directly modulated, 213
    DML, 67
    extinction ratio and, 218
    free, signal bandwidth and, 28
    intrinsic property of, 221
  CWDM
    advances in, 211–223
    uncooled, 219
  differential gain, 212
  diode, frequency modulation of, 214
  directly modulated, 21, 29, 60, 202
    chirp, 215, 225
    modulation behavior of, 214
    positive chirp of, 214
    static characteristics of, 212
    transmission penalty, 229
    uncooled, 205, 211, 218, 234
  distributed feedback, 60, 63, 62, 201
    common use for, 64
    cost of, 62
    device data, 64
    performance of, 62
    restricted use of, 64
    use in WDM systems, 63
    wavelength detuning in, 216
  DWDM, uncooled, 220
  electro-absorption modulated, 61, 228
  externally modulated, 65
  Fabry Perot, 62–63, 222
  fiber Bragg grating, 61, 204, 222
  L-band, 65, 173, 180
  operating wavelengths, variation of, 59
  oscillation frequency, 219
  output power, 11
    continuous wave, 211
    variation of, 76
  P-I curve, 274
  pre-bias condition, 214
  Raman pump, 41
  slope efficiency, 211, 212
  transmitters, 72
  uncooled, 61, 75, 125
  vertical cavity surface emitting, 63, 186

wavelength(s)
  drift, 3, 4, 76
  narrow passband of, 101
  statistical distribution of, 75
  variation, tolerance to, 59
  worst-case reach for different, 68
LDF, *see* Low dispersion NZDSF
LEAF® NZDSF, Corning, 40
LEDs, *see* Light emitting diodes
Legacy infrastructure
  CWDM coexistence with, 297
  upgrade designed over, 298
LH systems, *see* Long haul systems
LIAs, *see* Limiting amplifiers
Light emitting diodes (LEDs), 58
Light source(s)
  overview of, 61
  technique for modulating, 269–270
Limiting amplifiers (LIAs), 123
Linearly polarized (LP) modes, 24
Linear refractive index, 34–35
Linewidth enhancement factor, 214
Local area network (LAN), 289
Logical mesh network, configuration of, 156
Logical star network, 138, 139
Long haul (LH) systems, 60, 252
Long wavelength loss, 49
Low dispersion NZDSF (LDF), 40
Low water peak (LWP) fiber, 37–38, 41, 200
  chromatic dispersion, 206
  hydrogen aging loss, 42, 48
  properties of, 66
LP modes, *see* Linearly polarized modes
Lucent/OFS, TrueWave® RS, 40
Lumped element amplifiers, 174
LWP fiber, *see* Low water peak fiber

M

Mach-Zehnder interferometer (MZI), 186
Mach Zehnder modulator, 65
Macrobending loss, 21, 23–24
MANs, *see* Metropolitan area networks
Marcuse microbend model, 26
Matched-cladding fiber, 37
Material dispersion, single mode fibers, 27
Maximum-likelihood sequence detectors (MLSEs), 210
MCVD, *see* Modified chemical vapor deposition
MDFs, *see* Medium dispersion fibers
Medium dispersion fibers (MDFs), 20, 41

Meshed network
  optimal wavelength assignment in, 166
  topology of, 160
  wavelength assignment in, 162
MetroCor® NZDSF, 40
Metropolitan area networks (MANs), 309
Metropolitan networks, 251–267
  CWDM, ring-based, 305
  CWDM in metro, 253–254
  CWDM network building blocks, 254–258
  CDWM systems deployed in, 40
  definitions, 252–253
  key issues for optical network engineers, 259–263
  metro CWDM applications, 263–267
    adding CWDM to single wavelength networks, 264–265
    disaster recovery, 263–264
    future applications, 266–267
    scalability with DWDM over CWDM, 265–266
  properties, 262
  protocol transparency, 261
  rings using CWDM, 258–259
  transport connectivity, 262
Metro ring, poor man's way of building, 252
Microbending, 22
  fiber core model, 26
  loss, 24–27
Mixed bit-rate transmission, example, 227
MLSEs, *see* Maximum-likelihood sequence detectors
Model
  black box, 5
  black link, optical characteristics, 7
  CWDM multi-carrier, 302, 304
  economic, 302
  hub loss, 144
  Marcuse microbend, 26
  node loss, 147
  PON, 289
  WDM-PON, 293
Modified chemical vapor deposition (MCVD), 37
Modulation
  amplitude, 270
  cross-phase, 35, 232
  direct
    analog, 274
    DFB, 67
    HFC transport and, 276
  DML, 220, 221
  frequency, 214, 281
  nonreturn to zero, 77

  refractive index, 97
  self-phase, 35
MQW structures, *see* Multi quantum well structures
MSA, *see* Multi-source agreement
Multiple bounce concept, 120, 121
Multiplexer (MUX), 58, 93
  add/drop, 93
  DWDM, 207
  filter, CWDM, 74
  insertion loss sequence, 119
  losses, 202
  technology, CWDM channel spacing, 201
Multiplexing
  basic concepts of, 92
  CWDM spectrum after, 235
  lossless, 193
Multi-ports, 117
Multi quantum well (MQW) structures, 63, 184, 212
Multi-source agreement (MSA), 77, 83–84
MUX, *see* Multiplexer
MUX/DEMUX
  CWDM passive, 260
  East, 259
  functionality, 101
  ring construction with, 258
  West, 259
MZI, *see* Mach-Zehnder interferometer

N

Negative chirp, 214
Negative penalty, 217
Network(s), *see also* Metropolitan networks; Nonamplified networks
  access, 289, 305, 309
  active optical, 289
  configuration, CATV systems, 272
  core, 252
  design, power budgets and, 10
  hybrid fiber coax, 270, 271, 272
  infrastructure, different services over same, 300
  local area, 289
  logical star, 138, 139
  long-haul, 252
  meshed
    optimal wavelength assignment in, 166
    topology of, 160
    wavelength assignment in, 162
  metro access, 253
  N-node CWDM meshed, 156
  open-access, 302

Index

overlay, 294, 306
protection, 261
regional, 252
trunk, 277
unbundling, 302
VoD, 260
wireless, 306
Networking functionality, demand for devices supporting, 91
NF, *see* Noise figure
Node
  assignment, 157, 158
  loss model, 147
  set of wavelengths in, 162
Noise
  broadband, amplifiers and, 178–180
  CATV/HFC network performance and, 279
  figure (NF), 174, 190
  optical amplification and, 175–176
Nonadjacent channel isolation, ITU-T G.671, 17
Nonamplified networks, 125–169
  analysis of results for 4-node ring, 152–156
    attenuation slope compensating wavelength assignment, 152–154
    extension to 8-node ring, 154–156
    impact of ring perimeter, 154
    nonattenuation slope compensating wavelength assignment, 152
  application, 138–151
    calculation of filter losses, 143–151
    wavelength assignment, 138–143
  future work, 168–169
  generalized case of CWDM design, 135–138
  N-node CWDM meshed network, 156–168
    optimization techniques, 162–168
    wavelength assignment, 156–161
  optimized design of CWDM networks, 128–135
    attenuation slope compensating wavelength assignment, 128–131
    channel power, 133
    impact of loss, 133–135
    ring perimeter, 131–133
Nonattenuation slope compensating wavelength assignment, 152, 161
Nonfiber attenuation losses, 136, 137
Nonlinear refractive index, 34–35
Non-low-water-peak fibers, assumed attenuation coefficients for, 12
Nonreturn to zero (NRZ), 77, 216, 218
Nonzero dispersion-shifted fiber (NZDSF), 39, 206
  index of refraction profile, 40
  LEAF®, 40
  link, extinction for, 227
  low dispersion, 40, 52
  MetroCor®, 40
  spectral attenuation, 225
  transmission over, 234
  TrueWave® RS, 40, 230
  wideband optical transport, 41
  zero dispersion wavelength, 40, 230
NRZ, *see* Nonreturn to zero
NZDSF, *see* Nonzero dispersion-shifted fiber

O

OADM, *see* Optical ring topologies with add/drop multiplexing
O-band
  filter losses in, 168
  nonfiber losses, 155
  ring perimeter, 152
  wavelengths, 126
ODD process, *see* Overclad-during-Draw process
ONU, *see* Optical network unit
Open-access networks, 302
Optical amplification, 172–177
  gain saturation and gain dynamics, 176–177
  noise, 175–176
  optical amplifier types, 172–174
Optical confinement factor, 214
Optical fiber(s), 19–55, *see also* Fiber
  alignment, basic concepts, 117
  common transmission fibers used for CWDM, 37–41
    nonzero dispersion-shifted fiber, 39–41
    NZDSF for wideband optical transport, 41
    standard single mode fiber, 37–39
  core alignment of, 36
  dispersion-compensating fibers, 50–53
    basic principles, 50–51
    DCF for CWDM, 51–52
    Raman-pumped DCF, 53
  fiber properties and effects on CWDM system performance, 21–37
    bending-induced loss, 22–27
    chromatic dispersion, 27–31
    cut-off wavelength, 31–34
    fiber attenuation, 21–22
    geometric properties, 35–37
    nonlinear effects, 34–35
    polarization mode dispersion, 31
  geometric properties of, 35
  sensitivity to UV light, 94
  single mode, cut-off wavelength of, 31, 32

step index, 23
zero-OH⁻ single mode fibers for CWDM applications, 41–50
  manufacturing process for zero-OH⁻ AllWave® fiber, 43
  RIC-ODD process and AllWave fiber performance, 44–50
Optical filter, multiple bounce concept, 120
Optical Kerr effect, 34
Optical links, retrofitting of, 122
Optical multiplexers, insertion losses, 11
Optical network unit (ONU), 289
  colorless, 308
  design(s)
    different, 297
    modular, 300
    multi-service, 301
  multiple service reception, 304
  single service reception, 304
  wavelength-agnostic, 299
Optical power penalty, 8
Optical receivers, 76
Optical ring topologies with add/drop multiplexing (OADM), 8, 95
  add/drop wavelength band, 144
  add-function of, 119
  applications, grating performance, 95
  configuration, pass-channels in, 98
  connection to Ethernet, 261, 262
  insertion losses, 263
  metropolitan networks and, 256
  TFF implementation of, 145
  thin film filter implementation of, 134
Optical signal-to-noise ratio (OSNR), 210
Optical stacking, HFC networks, 279
Optical transceiver, evolution of, 70
Optical VPNs, 301
OSNR, see Optical signal-to-noise ratio
Outdoor DSLAMs, 287
Outside plant environments, equipment installed in, 122
Outside vapor deposition (OVD), 37
OVD, see Outside vapor deposition
Overclad-during-Draw (ODD) process, 43

P

Passive access networks, 289
Passive optical network (PON), 253, 283, 289
  creation of high density, 294
  IEEE standard, 290
  infrastructure, sharing of, 302
  model, 289
  upstream transmission channel in, 299
Passive product, reliability of, 101
PCVD, see Plasma chemical vapor deposition
PDG, see Polarization-dependent gain
PDL, see Polarization-dependent loss
Per-channel bit-rates, increasing, 204
Photoreceiver (PR), 297
PIN, see P-intrinsic-n photodiode
P-intrinsic-n photodiode (PIN), 68
  capacitance, 70
  diodes, 69
  receiver(s), 202
    input power, 11
    sensitivity, 68, 257
Plasma chemical vapor deposition (PCVD), 37
Plate-mapping process, 112
Plug and play solution, metropolitan networks and, 251
PMD, see Polarization mode dispersion
P2MP solution, see Point-to-multipoint solution
Point-to-multipoint (P2MP) solution, 288
Point-to-point (P2P) approach, 255, 288, 299
Polarization
  controller, 244
  dependent gain (PDG), 174
  dependent loss (PDL), 15, 115, 174
    CWDM, 102
    ITU-T G.671, 15
  dependent reflectance, ITU-T G.671, 15
  mode dispersion (PMD), 13, 31
    ITU-T G.671, 15
    measurement, low-mode-coupled method for, 46
    RIC-ODD process, 46
  overlap factor, 280–281
PON, see Passive optical network
Positive chirp, 214
Postcompensation, 50
Power
  budgets, optical link and, 10
  penalty(ies)
    CWDM laser, 224
    extinction ratio and, 219
P2P approach, see Point-to-point approach
PR, see Photoreceiver
PRBS, see Pseudo-random bit sequence
Preamplification, 193
Precompensation, 50
Profile dispersion, single mode fibers, 27
Pseudo-random bit sequence (PRBS), 215

Index

Pulse broadening
  dynamic chirp, 30
  source of, 31
Pump
  depletion, 279
  pump FWM, 41
  wavelength, 174–175

Q

QoS, *see* Quality of Service
QSFP, *see* Quad small form pluggable
Quad small form pluggable (QSFP), 86
Quality of Service (QoS), 290
Quartz, natural, alkali contamination and, 49

R

Radio frequency (RF) amplifiers, 270
Radio-over-Fiber (RoF), 306, 307
Raman amplifier, 172, 174, 178, 190
Raman effect, 35
Raman gain coefficient, 282
Raman pump lasers, 41
Rare earth doped fiber amplifier, 173
Rayleigh scattering, 21, 200, 224–225
Reactive silica defects, 48
Receiver(s)
  APD, 68, 234, 262
  avalanche photodiode, 202
  broad noise band into, 194
  early, 71
  multi-channel, 88
  optical, 76
  PIN, sensitivity of, 68
  sensitivity, 238, 239, 240
    channel power and, 133
    impact of accumulated dispersion on, 31
    penalties, 244
    transceiver, 257
    transmission distance and, 217
Reed-Solomon (RS) code, 237
Reflectance
  insertion loss, 115
  ITU-T G.671, 15
Reflect port, 115
Refractive index
  modulation, FBG, 97
  optical Kerr effect and, 34
  profile, step index fiber, 23
Regional networks, 252
Remote node (RN), 289
Reverse path optics, HFC, 273

RF amplifiers, *see* Radio frequency amplifiers
RIC-ODD process
  advantages, 44
  AllWave fiber performance and, 44
  cost advantages, 43
  fiber geometry, 46
  fiber loss, 44
  fiber PMD, 46
  hydrogen aging losses, 47
  interface quality and fiber strength, 46
RIC process, *see* Rod-in-Cylinder process
Ring
  losses, calculation of, 154
  perimeter(s)
    attainable, 164
    calculated, 153, 159
    CWDM network, 131–133
    impact of, 154
    O-band, 152
    unidirectional hubbed ring, 149, 151
    wavelength assignment and, 141
Ripple, ITU-T G.671, 17
RIT process, *see* Rod-in-Tube process
RMS amplitude, *see* Root mean square amplitude, 25
RN, *see* Remote node
Rod-in-Cylinder (RIC) process, 43
Rod-in-Tube (RIT) process, 43, 44
RoF, *see* Radio-over-Fiber
Root mean square (RMS) amplitude, 25
Router
  AWG, 308
  Ethernet, 265
RS code, *see* Reed-Solomon code

S

SANs, *see* Storage Area Networks
SBS, *see* Stimulated Brillouin scattering
Scattering processes, inelastic, 35
SDM, *see* Space division multiplexing
Self-phase modulation (SPM), 35
Semiconductor optical amplifier (SOA), 172
  access architecture, 189
  amplified spontaneous emission spectrum of, 183
  bandwidth, 178, 183
  designs, high power, 184
  fiber transmission using, 187
  gain-clamped, 186
  gain medium, 174
  gain spectrum, 175
  gain versus output power curve of, 185
  photon emission in, 173

transmission of WDM channels, 188
ultrafast gain compression of, 185
Service filtering, 302
SFF transceiver, *see* Small form factor transceiver
SFP transceiver, *see* Small form pluggable transceiver
Signal
  power, transmission distance and, 217
  wavelength, dependence of gain on, 184
Signal-to-interference ratio (SIR), 242, 243
Signal-to-noise ratio (SNR), 51, 175
  maximizing of, 177
  optical, 210
Silica defects, reactive, 48
Silicondioxide, 104
Single Channel Interface Specifications, 8
Single mode fiber(s)
  collimator, 115
  cut-off wavelength of, 31, 32
  dispersion in, 27
  zero-OH⁻, 41–50
SIR, *see* Signal-to-interference ratio
6-port device, conceptual drawing of, 118
Small form factor (SFF) transceiver, 83
Small form pluggable (SFP) transceiver, 7, 85
SNR, *see* Signal-to-noise ratio
SOA, *see* Semiconductor optical amplifier
SONET, 253
  ITU-T G.695 and, 3
  link, single wavelength, 265, 266
  oriented data rate, 203
Space division multiplexing (SDM), 92, 126
SPM, *see* Self-phase modulation
Spooling, macrobending loss and, 21
SRS, *see* Stimulated Raman scattering
SSMF, *see* Standard single mode fiber
Standards, 1–17
  ITU-T recommendation G.671, 15–17
  ITU-T recommendation G.694.2, 2–3
  ITU-T recommendation G.695, 3–15
    application code nomenclature, 3–5
    application code summary, 15
    black box and black link approaches, 5–8
    center wavelength deviation, 13–15
    channel plans, 13
    distance, 11–13
    path penalty, 13
    power budget, 10–11
    topologies, 8–10
    unidirectional and bidirectional transmission, 8
Standard single mode fiber (SSMF), 22, 32, 37, 40, 200

alternative to, 66
attenuation coefficient, 201
chromatic dispersion, 206
dispersion tolerance, 63
macrobending loss of, 24, 25, 39
step index core shapes, 37
transmission distances, 29
typical mode field radius for, 36
zero dispersion region of, 269
Step index fiber
  index profile, 23
  theoretical cut-off wavelength for, 32
Stimulated Brillouin scattering (SBS), 34
Stimulated emission, 172
Stimulated Raman scattering (SRS), 34, 278
  crosstalk, 280, 281, 282
  limitations, 279
Storage application, addition of wavelengths, 266
Storage Area Networks (SANs), 260
  disaster recovery and, 263
  extension, 264
Storage servers, metropolitan network, 264
Sumitomo Electric, PureAccess fiber, 40
System cost reductions, 1

T

Tantalpentoxide, 104
TDFA, *see* Thulium-doped fiber amplifier
TDL, *see* Temperature-dependent loss
TDM, *see* Time division multiplexing
TDM-PON sub-network, 294–295
TEC, *see* Thermoelectric cooling
Telcordia GR-1221, reliability testing per, 119
Temperature-dependent loss (TDL), 117
TFF, *see* Thin-film filter
Theoretical cut-off wavelength, 32
Thermoelectric cooling (TEC), 73, 76, 211
Thin-film coating(s)
  chamber, *in situ* monitoring capability of, 112
  process flowchart for, 111
Thin-film filter (TFF), 92, 133
  bandpass characteristics, 106
  coating layer structure of, 105
  implementation of OADM, 134, 145–146
  market domination by, 101
  miniature design concept and, 121
  optical elements of, 116
  packaging solutions, 113–122
    cascaded 3-port packages, multi-ports, and glass package, 113–120
    multiple bounce concept, 120–122

Index

performance data for, 104
price per channel, 101
properties, 104–113
    manufacturing process steps, 111–113
    thin-film multi-cavity structures, 104–110
reflection losses for, 143
scalability provided by, 101
3-port package
Thulium-doped fiber amplifier (TDFA), 178
TIA, *see* Transimpedance amplifier
Time division multiplexing (TDM), 92, 126, 288
Transceiver(s), 57–90
    bidirectional CWDM link configuration, 80, 81–82
    building blocks, 79–82
    CWDM, data loss from, 261
    detectors for CWDM, 68–70
        APD diodes, 70
        PIN diodes, 69–70
    DWCM wavelength-selective, 255
    elements of, 79
    Ethernet, types of, 84
    evolution of optical transceiver, 70–72
    GBIC, 84
    integrated, 88, 89
    laser transmitters, 72–73
    metropolitan networks and, 259
    optical receivers, 76–78
    receptacle design, 82
    SFF, 83
    SNAP 12 MSA, 88
    sources for CWDM, 61–68
        application to CWDM, 66–68
        laser types and their properties, 62–66
    standards, 82–85, 87
    transceivers for 10 Gb/s, 85–86
    trends for future CWDM transceivers, 86–89
    uncooled transmitters for CWDM, 73–76
    wavelength agnostic, 86
Transimpedance amplifier (TIA), 70, 77, 123
    distance, signal power and, 217
    link, laser wavelengths used in, 59
    mixed fiber-type, 230
    penalty(ies)
        DML, 229
        DWDM sub-band, 227
    technologies, ADSL, 287
Transmit port, 115
Transmitter(s)
    chirp, chromatic dispersion and, 29
    chirp-free, 28
    early, 71
    laser, 72
    multi-channel, 88
    uncooled, 73–76
TrueWave® RS NZDSF, Lucent/OFS, 40, 230
    dispersion slope, 40
    REACH, 20
    span, DWDM overlay with, 232
Trunk network, 277

U

UDWDM, *see* Ultra dense WDM
ULH applications, *see* Ultra long-haul applications
Ultra dense WDM (UDWDM), 306
Ultra long-haul (ULH) applications, 58
Uncooled lasers, 61, 75, 125
Uncooled transmitters, CWDM, 73–76
    power variation, 76
    wavelength drift, 74–76
Unidirectional crosstalk attenuation, ITU-T G.671, 17
Unidirectional hubbed ring
    filter loss calculation, 148, 150
    ring perimeter, 149, 151
    wavelength assignment, 142, 156
Upgrade paths and toward 10 Gb/s, 199–249
    advances in CWDM lasers, 211–223
        increased bit-rate operation, 218–222
        laser chirp, extinction ratio, and chromatic dispersion, 213–218
        laser output power, 211–213
        other laser types for 10-Gb/s CWDM systems, 222–223
    CWDM capacity upgrade options, 203–210
        channel overlay, 206–208
        equalization and FEC for CWDM transmission, 209–210
        increasing per-channel bit-rates, 204–206
    CWDM system upgrade demonstrations, 223–230
        example for DWDM overlay, 223–227
        example for mixed bit-rate transmission, 227–230
    full-spectrum CWDM at 10 Gb/s, 234–245
        FEC-enabled transmission over AllWave, 236–245
        transmission over NZDSF, 234–236
    mixed fiber-type transmission, 230–233

10-Gb/s DML over AllWave and TrueWave-RS, 230
10-Gb/s DWDM overlay with TrueWave-RS, 232
10-Gb/s EML with AllWave and multiple spans of TrueWave-RS, 232–233
UV light, sensitivity of optical fiber to, 94

V

VAD, *see* Vapor axial deposition
Vapor axial deposition (VAD), 37, 46
VCSEL, *see* Vertical cavity surface emitting laser
Vertical cavity surface emitting laser (VCSEL), 61, 63, 186
Vestigial sideband (VSB), 270
Video conferencing, 301
Video on Demand (VoD), 253
  bandwidth requirement, 267
  networks, 260
Video streaming, 290
Virtual dark fiber services, black link and, 7
Visual Basic, wavelength assignment in meshed network, 163
V-number, waveguide, 32
VoD, *see* Video on Demand
VSB, *see* Vestigial sideband

W

Water peak, 22, 38
  definition of, 200
  fibers, assumed attenuation coefficients, 12
  optical loss caused by, 41
Waveguide
  dispersion, single mode fibers, 27
  V-number, 32
Wavelength(s)
  agnostic transceivers, 86
  assignment
    algorithm, 152–154, 168, 169
    application of, 141
    attenuation slope compensating, 128, 147, 152
    4-node ring, 138
    meshed network, 162
    nonattenuation slope compensating, 152, 161
    optimal, 166
    ring perimeter and, 141
    steps in Visual Basic, 163
    unidirectional hubbed ring, 142, 156

C-band, 65
channel, loss incurred by, 130
cut-off, 32, 33
CWDM, return path aggregation using different, 279
detuning, DFB laser, 216
drift, 74
E-band, 305
laser, statistical distribution of, 75
O-band, 126
optical communications link, 59
pump, 174–175
red-shift of, 214
upgraded channel, 227
variation of, 59
zero dispersion, 40, 230
Wavelength division multiplexing (WDM), 20, 58, 92, 180, 288
  channel(s)
    crosstalk, 234
    transmission of, 188
  high-power pump LD unit, 191
  low-cost implementations of, 308
  PON, 283, 293
  system, upgrade paths for, 204
  ultra dense, 306
  virtual P2P connections, 291
  wide, 200
Wavelength division multiplexing filters, 91–123
  basic network concepts, 92–93
  future trends and requirements, 122–123
  properties of thin-film filters, 104–113
    manufacturing process steps, 111–113
    thin-film multi-cavity structures, 104–110
  requirements for CWDM filters, 93
  technical options for wavelength division multiplexing, 94–104
  thin-film filter packaging solutions, 113–122
    cascaded 3-port packages, multi-ports, and glass package, 113–120
    multiple bounce concept, 120–122
WDM, *see* Wavelength division multiplexing
West-East propagating detected signal, 244
Wet fiber, 12
Wide wavelength division multiplexing (WWDM), 58, 200
Wireless backbone networks, 260

Index

Wireless networks, 306
WWDM, *see* Wide wavelength division multiplexing

X

XPM, *see* Cross-phase modulation

Z

Zero dispersion
  region, SSMF, 269
  wavelength, NZDSF, 40
Zero-OH⁻ single mode fibers, 41–50
  manufacturing process, 43
  RIC-ODD process, 44–50
Zero water peak, 22
Zero water peak (ZWP) fiber, 37–38, 41, 126, 200
  attenuation, 126, 127, 128
  DWDM overlay and, 223
  EPON systems and, 305
  hydrogen aging loss, 42, 48
  logical star network, 138
  standard dispersion, 241
ZWP fiber, *see* Zero water peak fiber